About Access Archaeology

Access Archaeology offers a different publishing model for specialist academic material that might traditionally prove commercially unviable, perhaps due to its sheer extent or volume of colour content, or simply due to its relatively niche field of interest.

All *Access Archaeology* publications are available in open-access e-pdf format and in (on-demand) print format. The open-access model supports dissemination in areas of the world where budgets are more severely limited, and also allows individual academics from all over the world the chance to access the material privately, rather than relying solely on their university or public library. Print copies, nevertheless, remain available to individuals and institutions who need or prefer them.

The material is professionally refereed, but not peer reviewed. Copy-editing takes place prior to submission of the work for publication and is the responsibility of the author. Academics who are able to supply print-ready material are not charged any fee to publish (including making the material available in open-access). In some instances the material is type-set in-house and in these cases a small charge is passed on for layout work.

This model works for us as a publisher because we are able to publish specialist work with relatively little editorial investment. Our core effort goes into promoting the material, both in open-access and print, where *Access Archaeology* books get the same level of attention as our core peer-reviewed imprint by being included in marketing e-alerts, print catalogues, displays at academic conferences and more, supported by professional distribution worldwide.

Open-access allows for greater dissemination of the academic work than traditional print models, even lithographic printing, could ever hope to support. It is common for a new open-access e-pdf to be downloaded several hundred times in its first month since appearing on our website. Print sales of such specialist material would take years to match this figure, if indeed it ever would.

By printing 'on-demand', meanwhile, (or, as is generally the case, maintaining minimum stock quantities as small as two), we are able to ensure orders for print copies can be fulfilled without having to invest in great quantities of stock in advance. The quality of such printing has moved forward radically, even in the last few years, vastly increasing the fidelity of images (highly important in archaeology) and making colour printing more economical.

Access Archaeology is a vehicle that allows us to publish useful research, be it a PhD thesis, a catalogue of archaeological material or data, in a model that does not cost more than the income it generates.

This model may well evolve over time, but its ambition will always remain to publish archaeological material that would prove commercially unviable in traditional publishing models, without passing the expense on to the academic (author or reader).

Ceramic manufacturing techniques and cultural traditions in Nubia from the 8th to the 3rd millennium BC.

Examples from Sai Island

Giulia D'Ercole

ARCHAEOPRESS PUBLISHING LTD
Summertown Pavilion
18-24 Middle Way
Oxford OX2 7LG

www.archaeopress.com

ISBN 978 1 78491 671 8
ISBN 978 1 78491 672 5 (e-Pdf)

© Archaeopress and G D'Ercole 2017

Cambridge Monographs in African Archaeology 96
Series Editors: Laurence Smith, Brian Stewart, Stephanie Wynne-Jones

All rights reserved. No part of this book may be reproduced or transmitted,
in any form or by any means, electronic, mechanical, photocopying or otherwise,
without the prior written permission of the copyright owners.

*To my family and to my grandmother, Elena,
who taught me how to tell a story*

Contents

Foreword xv

Acknowledgements xvii

Introduction 1

The invention of pottery and its cultural significance 1

Sai Island in northern Upper Nubia (northern Sudan) 2

Methodological approach and research questions: style, traditions and change 3

Structure of the research 4

1. Nubia and its cultural sequences between the 8th and the 3rd millennium BC: Khartoum Variant, Abkan and Pre-Kerma 5

Introduction 5

Climate, environmental conditions and human occupation during the Early Holocene along the Nile Valley and in the Egyptian Western Desert 5

The Khartoum Variant culture: definition, chronology and settlements 9

    Khartoum Variant sites in the region of Abka – Wadi Halfa (Lower Nubia) 9

    Khartoum Variant sites at Sai Island (northern Upper Nubia) 10

Climate, environmental conditions and human occupation during the Middle Holocene along the Nile Valley and in the Egyptian Western Desert 13

The Abkan culture: definition, chronology and settlements 16

    Abkan sites in the region of Abka – Wadi Halfa (Lower Nubia) 16

    Abkan sites at Sai Island (northern Upper Nubia) 17

Climate, environmental conditions and human occupation during the Late Holocene along the Nile Valley and in the Egyptian Western Desert 20

The Pre-Kerma culture: definition, chronology and settlements 22

    Pre-Kerma sites at Kerma (Upper Nubia) 22

    Pre-Kerma sites at Sai Island (northern Upper Nubia) 23

2. Sai Island: archaeological research and cultural sequence 25

Introduction 25

Geological setting and paleo-climatic reconstruction 25

The archaeological sites on the island 26

Selected sites 29

    Site 8-B-10C 29

    Site 8-B-76 32

    Site 8-B-52a 32

    Site 8-B-10A 35

3. Ceramic productions on Sai Island: analysis of the macroscopic data 37
Materials and methods 37
Site 8-B-10C 38
State of preservation 38
Preparation: clay processing and addition of non-plastic inclusions 39
Production: shaping 42
Production: decoration techniques, implements, elements, motifs and structure of the decoration 42
Production: surface treatment 46
Spatial distribution 48
Site 8-B-76 50
State of preservation 50
Preparation: clay processing and addition of non-plastic inclusions 50
Production: shaping 51
Production: decoration techniques, implements, elements, motifs and structure of the decoration and surface treatment 52
Site 8-B-52A 53
State of preservation 53
Preparation: clay processing and addition of non-plastic inclusions 54
Production: shaping 54
Production: decoration techniques, implements, elements, motifs and structure of the decoration 57
Production: surface treatment 60
Site 8-B-10A 62
State of preservation 62
Preparation: clay processing and addition of non-plastic inclusions 63
Production: shaping 65
Production: decoration techniques, implements, elements, motifs and structure of decoration .. 66
Production: surface treatment 69

4. Archaeometric analysis 95
Introduction 95
Selection and preparation of the samples 95
The ceramic samples 95
The sediment samples 99
Laboratory analyses 101
Optical microscopy (OM) 101

X-ray powder diffraction analysis (XRPD) 101

X-ray fluorescence analysis (XRF) 102

Results: the ceramic samples 102

X-ray powder diffraction analysis 102

Optical microscopy 105

X-ray fluorescence analysis 109

Results: the sediment samples 119

X-ray powder diffraction analysis 119

Optical microscopy 121

X-ray fluorescence analysis 122

The ceramic and sediment samples: classification criteria and comparison of results 125

5. Comparing chaînes operatoires: continuity and discontinuity in the ceramic assemblages of Sai Island 139

Introduction 139

Discussion and comparison of the macroscopic data 139

Preparation: clay processing and addition of non-plastic inclusions 139

Production: shaping and surface treatment 141

Stylistic traditions 142

Discussion and comparison of the archaeometric data 146

Compositional variables 147

Technological variables 150

Preparation: addition of organic inclusions 150

Production: shaping (manufacturing techniques) 151

Firing 152

Final considerations 153

Change between the Khartoum Variant and the Abkan horizon 154

Continuity between the Abkan and the Pre-Kerma 156

6. The Sai Island sequence and the Nubian and Sudanese traditions 157

Introduction 157

Section 1: Comparison of decorative styles 157

Khartoum Variant horizon 157

Dotted Wavy Line 157

Packed zigzags of dots and dashes 158

Alternately Pivoting Stamp 159

Simple impression 159

    Decorations on the rim 159

  Abkan horizon 160

  Pre-Kerma horizon 160

Section 2: Comparison of technological traditions 162

  Khartoum Variant horizon 162

  Abkan horizon 164

  Pre-Kerma horizon 167

Final considerations 168

Bibliography 173

List of Figures

Figure 1.1 A synopsis of the chronological sequences during the Holocene in the regions of Egypt and Sudan mentioned in the text (figure by G. D'Ercole). 6

Figure 1.2 Map of Egypt and Sudan during the Early Holocene (c. 8000–5000 BC) with geographical features and sites cited in the text (figure by G. D'Ercole). 8

Figure 1.3 Map of Egypt and Sudan during the Middle Holocene (c. 5000–3500 BC) with geographical features and sites cited in the text (figure by G. D'Ercole). 15

Figure 1.4 Map of Egypt and Sudan during the Late Holocene (c. 3500–2500 BC) with geographical features and sites cited in the text (figure by G. D'Ercole). 21

Figure 2.1 Geographical location of Sai Island and schematic geological and morphological map of northern Sudan (modified after D'Ercole et al. 2015). 26

Figure 2.2 Geological map of Sai Island with the location of the four sites from which the pottery samples come (modified after D'Ercole et al. 2015). 27

Figure 2.3 View of the interior landscape of Sai Island with the profile of the Jebel Abri in the background (photo by G. D'Ercole). 28

Figure 2.4 View of the western riverbank of Sai Island with young alluvial sediments and typical riverine vegetation (photo by G. D'Ercole) 28

Figure 2.5 Excavation at site 8-B-10C (photo by R. Ceccacci). 30

Figure 2.6 Plan of levels 1 (on the left) and 2 (on the right) of site 8-B-10C showing hut floors, features and post holes (modified after Garcea 2011–2012). 31

Figure 2.7 Excavation at site 8-B-76 (photo by E. A. A. Garcea). 33

Figure 2.8 Estimated extension of site 8-B-52A (on the left) and schematic map (on the right) showing the three silos from which the ceramic sample analysed come (map modified after Hildebrand and Shilling 2016). 34

Figure 2.9 Site 8-B-52A: detail of a 'two level units' pit with the slab of schist used to seal the opening of the pit (photo by G. D'Ercole). 35

Figure 2.10 Excavation at site 8-B-10A (photo by R. Ceccacci). 36

Figure 3.1 Angularity of mineral inclusions from site 8-B-10C. 40

Figure 3.2 Distribution of angular and rounded inclusions in relation to the different types of textures from site 8-B-10C. 41

Figure 3.3 Distribution of mineral and organic inclusions in relation to the different types of textures from site 8-B-10C. 42

Figure 3.4 Sherd thicknesses from site 8-B-10C. 43

Figure 3.5 Ranges of the lengths of decorative motifs from site 8-B-10C (minimum, maximum and average). 46

Figure 3.6 Ranges of the teeth numbers of combs with evenly serrated edges from site 8-B-10C (minimum, maximum and average). 46

Figure 3.7 Percentages of burnishing from site 8-B-10C. 47

Figure 3.8 Spatial distribution of ceramic sherds from the surface (on the left) and from level 1 (on the right) at site 8-B-10C. Blue symbols (from 1 to 5 sherds/square); green symbols (from 6 to 10 sherds/square); orange symbols (from 11 to 20 sherds/square); red symbols (from 21 to 45 sherds/square) (figure by G. D'Ercole). ... 49

Figure 3.9 Spatial distribution of ceramic sherds from levels 2 (on the left) and 3 (on the right) at site 8-B-10C. Blue symbols (from 1 to 5 sherds/square); green symbols (from 6 to 10 sherds/square) (figure by G. D'Ercole)............ 49

Figure 3.10 Spatial distribution of ceramic sherds with fine (in red on the left), medium (in green in the middle) and coarse texture (in blue on the right) at site 8-B-10C (figure by G. D'Ercole). 50

Figure 3.11 Angularity of mineral inclusions from site 8-B-76. 51

Figure 3.12 Sherd thicknesses from site 8-B-76. 52

Figure 3.13 Percentages of burnishing from site 8-B-76. 53

Figure 3.14 Angularity of mineral inclusions from site 8-B-52A. 55

Figure 3.15 Sherd thicknesses from site 8-B-52A. 56

Figure 3.16 Ranges of the lengths of combs and double-pronged tools from site 8-B-52A (minimum, maximum and average). 59

Figure 3.17 Percentages of burnishing from site 8-B-52A. 61

Figure 3.18 Angularity of mineral inclusions from site 8-B-10A. 64

Figure 3.19 Sherd thicknesses from site 8-B-10A. 66

Figure 3.20 Ranges of the lengths of decorative motifs from site 8-B-10A (minimum, maximum and average). 68

Figure 3.21 Percentages of burnishing from site 8-B-10A. 70

Figure 4.1 Profile of site 8-B-76. Arrows indicate the depths from which the sediments samples were taken (figure by G. D'Ercole). 100

Figure 4.2 Diffractograms representative of the three identified XRD groups of ceramics: QPl (quartz-plagioclase); Q (quartz) and QKfs (quartz-K-feldspar). 105

Figure 4.3 Thin section and SEM images showing an Abkan sample from site 8-B-76 with ilmenite (Ilm), zircon (Zrn) and rutile (Rut) inclusions (Images a–a1) and a micrite calcite aggregate (Image b) (modified after D'Ercole et al. 2017). 108

Figure 4.4 EDS spectrum showing the presence of ilmenite ($FeTiO_3$) (on top) and zircon ($ZrSiO_4$) (below). 109

Figure 4.5 Thin section and SEM images showing a Pre-Kerma sample from site 8-B-10A with amphibole (Amp) and collophane (Col) inclusions (Images a–a1) and carbon inclusions (Image b) (modified after D'Ercole et al. 2017)... 110

Figure 4.6 EDS spectrum showing the presence of collophane (microcrystalline hydrated fluoroapatite) (on top), amphibole (in the middle) and carbon inclusions (below). 111

Figure 4.7 Thin sections and SEM images showing an Abkan sample (Image a) with a metamorphic rock fragment (Mrf) and a charcoal inclusion (Chr); and a Khartoum Variant sample from site 8-B-10C (Image b) with zircon (Zr) and ilmenite (Ilm) inclusions (modified after D'Ercole et al. 2017). 112

Figure 4.8 CaO vs. K_2O plot (wt %). The colour of symbols refers to the XRD groups: blue = QPl group; orange = QKfs group; green = Q group. The shape of the symbols refers to the sites: triangles are samples from site 8-B-10A; diamonds are samples from site 8-B-52-A; squares are samples from site 8-B-76; circles are samples from site 8-B-10C; crosses are the clay lumps from site 8-B-10A. 115

Figure 4.9 MgO vs. K_2O plot (wt %). The colour of symbols refers to the XRD groups: blue = QPl group; orange = QKfs group; green = Q group. The shape of the symbols refers to the sites: triangles are samples from site 8-B-10A; diamonds are samples from site 8-B-52-A; squares are samples from site 8-B-76; circles are samples from site 8-B-10C; crosses are the clay lumps from site 8-B-10A. 115

Figure 4.10 Fe_2O_3 vs. TiO_2 plot (wt %). The colour of symbols refers to the XRD groups: blue = QPl group; orange = QKfs group; green = Q group. The shape of the symbols refers to the sites: triangles are samples from site 8-B-10A; diamonds are samples from site 8-B-52-A; squares are samples from site 8-B-76; circles are samples from site 8-B-10C; crosses are the clay lumps from site 8-B-10A. 116

Figure 4.11 Rb vs. Y plot. The colour of symbols refers to the XRD groups: blue = QPl group; orange = QKfs group; green = Q group. The shape of the symbols refers to the sites: triangles are samples from site 8-B-10A; diamonds are samples from site 8-B-52-A; squares are samples from site 8-B-76; circles are samples from site 8-B-10C; crosses are the clay lumps from site 8-B-10A. 119

Figure 4.12 Rb vs. Zr plot. The colour of symbols refers to the XRD groups: blue = QPl group; orange = QKfs group; green = Q group. The shape of the symbols refers to the sites: triangles are samples from site 8-B-10A; diamonds are samples from site 8-B-52-A; squares are samples from site 8-B-76; circles are samples from site 8-B-10C; crosses are the clay lumps from site 8-B-10A. 120

Figure 4.13 Diffractograms representative of the two identified XRD groups of local sediments: QPl (quartz-plagioclase) and QCal (quartz-calcite). 121

Figure 4.14 CaO vs. Al_2O_3 plot (wt %). The colour of symbols refers to the XRD groups: blue = QPl group; grey = QCal group. The shape of the symbols refers to the sites: triangles are samples from site 8-B-10A; squares are samples from site 8-B-76; circles are samples from site 8-B-10C 123

Figure 4.15 MgO/K_2O vs. SiO_2/Al_2O_3 plot. The colour of symbols refers to the XRD groups: blue = QPl group; grey = QCal group. The shape of the symbols refers to the sites: triangles are samples from site 8-B-10A; squares are samples from site 8-B-76; circles are samples from site 8-B-10C 124

Figure 4.16 Fe_2O_3/TiO_2 vs. SiO_2/Al_2O_3 plot. The colour of symbols refers to the XRD groups: blue = QPl group; grey = QCal group. The shape of the symbols refers to the sites: triangles are samples from site 8-B-10A; squares are samples from site 8-B-76; circles are samples from site 8-B-10C 124

Figure 4.17 Rb vs. Y plot. The colour of symbols refers to the XRD groups: blue = QPl group; grey = QCal group. The shape of the symbols refers to the sites: triangles are samples from site 8-B-10A; squares are samples from site 8-B-76; circles are samples from site 8-B-10C 126

Figure 4.18 Y vs. Nb plot. The colour of symbols refers to the XRD groups: blue = QPl group; grey = QCal group. The shape of the symbols refers to the sites: triangles are samples from site 8-B-10A; squares are samples from site 8-B-76; circles are samples from site 8-B-10C 126

Figure 4.19 Rb vs. Zr plot. The colour of symbols refers to the XRD groups: blue = QPl group; grey = QCal group. The shape of the symbols refers to the sites: triangles are samples from site 8-B-10A; squares are samples from site 8-B-76; circles are samples from site 8-B-10C 127

Figure 4.20 CaO vs. K_2O plot (wt %). The colour of symbols refers to the XRD groups: blue = QPl group; orange = QKfs group; green = Q group. The shape of the symbols refers to the sites: triangles are samples from site 8-B-10A; diamonds are samples from site 8-B-52-A; squares are samples from site 8-B-76; circles are samples from site 8-B-10C; crosses are the clay lumps from site 8-B-10A. Empty triangles are sediments from site 8-B-10A; empty squares are sediments form site 8-B-76; empty circles are sediments from site 8-B-10C. 130

Figure 4.21 MgO vs. K_2O plot (wt %). The colour of symbols refers to the XRD groups: blue = QPl group; orange = QKfs group; green = Q group. The shape of the symbols refers to the sites: triangles are samples from site 8-B-10A; diamonds are samples from site 8-B-52-A; squares are samples from site 8-B-76; circles are samples from site 8-B-10C; crosses are the clay lumps from site 8-B-10A. Empty triangles are sediments from site 8-B-10A; empty squares are sediments form site 8-B-76; empty circles are sediments from site 8-B-10C. 130

Figure 4.22 MgO/K_2O vs. SiO_2/Al_2O_3. The colour of symbols refers to the XRD groups: blue = QPl group; orange = QKfs group; green = Q group. The shape of the symbols refers to the sites: triangles are samples from site 8-B-10A; diamonds are samples from site 8-B-52-A; squares are samples from site 8-B-76; circles are samples from site 8-B-10C; crosses are the clay lumps from site 8-B-10A. Empty triangles are sediments from site 8-B-10A; empty squares are sediments form site 8-B-76; empty circles are sediments from site 8-B-10C. 131

Figure 4.23 Biplot of the PCA analysis (PC1 versus PC2) for derived pottery and clayey sediments, showing the clusters identified. The colour of symbols refers to the XRD groups: blue = QPl group; = QKfs group; green = Q group. The shape of the symbols refers to the sites: triangles are samples from site 8-B-10A; diamonds are samples from site 8-B-52-A; squares are samples from site 8-B-76; circles are samples from site 8-B-10C; crosses are the clay lumps from site 8-B-10A. Empty triangles are sediments from site 8-B-10A; empty squares are sediments form site 8-B-76; empty circles are sediments from site 8-B-10C (modified after D'Ercole et al. 2015). 132

Figure 5.1 Percentages of types and dimensions of mineral inclusions from sites 8-B-10C, 8-B-76, 8-B-52A and 8-B-10A. 140

Figure 5.2 Percentages of organic inclusions from sites 8-B-10C, 8-B-76, 8-B-52A and 8-B-10A. 141

Figure 5.3 Sherd thicknesses from sites 8-B-10C, 8-B-76, 8-B-52A and 8-B-10A. 142

Figure 5.4 Percentages of burnishing from sites 8-B-10C, 8-B-76, 8-B-52A and 8-B-10A. 143

Figure 5.5 Percentages of polishing from sites 8-B-10C, 8-B-76, 8-B-52A and 8-B-10A. 143

Figure 5.6 Sherd decorated with dotted wavy line motifs from site 8-B-10C (photo by R. Ceccacci). 144

Figure 5.7 Sherd decorated with a roulette by using the rocker stamp technique from site 8-B-10C (photo by R. Ceccacci). 144

Figure 5.8 Sherd decorated with milled impressions on the rim from site 8-B-76 (photo by R. Ceccacci). 145

Figure 5.9 Black-topped sherd resembling A-Group pottery decorated by using the alternatively pivoting stamp technique from site 8-B-52A (photo by R. Ceccacci). 145

Figure 5.10 Sherd decorating by using the rouletting technique from site 8-B-10A (photo by R. Ceccacci). 146

Figure 5.11 Photos of two thin sections in which it is recognizable the manufacturing technique used by the potter can be seen. Image a: coiling technique; image b: drawing out a lump of clay (figure by G. D'Ercole). 152

Figure 5.12 Examples of the different effects of oxidation-reduction resulting from firing. Image a: completely oxidized fracture for a sample lacking organic content; b: fracture with a dark core and oxidized surfaces; c: completely oxidized fracture for a sample with organic content (figure by G. D'Ercole). 153

Figure 6.1 Evolution of the main techniques and decorative motifs of ceramics in the regions of Egypt and Sudan mentioned in the text. (Acronyms: DW-sw; DW-lw; APS; IWL stand for: Dotted Wavy Line-Short Waves; Dotted Wavy Line-Long Waves; Alternately Pivoting Stamp; Incised Wavy Line) (figure by G. D'Ercole). 161

Figure 6.2 Evolution of the main technological features of ceramics in the regions of Egypt and Sudan mentioned in the text. Symbols > stand for an increase of mineral/organic inclusions; < a decrease of mineral/organic inclusions; >> a significant increase of mineral/organic inclusions. (figure by G. D'Ercole) 169

List of Plates

Plate 3.1 Abkan pottery from site 8-B-76. (1-3) Globular and straight walled jars. Drawings by T. D'Este. 72

Plate 3.2 Pre-Kerma pottery from site 8-B-52A. (1-4) Bowls with straight walls; (5-7) slightly convex bowls; (8) saucer-shaped bowl; (9-10) hemispherical bowls. Drawings by T. D'Este. 73

Plate 3.3 Pre-Kerma pottery from site 8-B-52A. (1-5) Ovoid jars; (6-9) jars with straight walls. Drawings by T. D'Este... 74

Plate 3.4 Pre-Kerma pottery from site 8-B-10A. Sector East. (1-12) Slightly convex bowls and bowls with straight walls. Drawings by T. D'Este. 75

Plate 3.5 Pre-Kerma pottery from site 8-B-10A. Sector East. (1-4) Hemispherical and straight walled bowls. Drawings by T. D'Este. 76

Plate 3.6 Pre-Kerma pottery from site 8-B-10A. Sector East. (1-9) Hemispherical and straight walled bowls. Drawings by T. D'Este. 77

Plate 3.7 Pre-Kerma pottery from site 8-B-10A. Sector East. (1-3) Hemispherical and straight walled bowls. Drawings by T. D'Este. 78

Plate 3.8 Pre-Kerma pottery from site 8-B-10A. Sector East. (1-11) Jars with straight walls. Drawings by T. D'Este........ 79

Plate 3.9 Pre-Kerma pottery from site 8-B-10A. Sector East. (1-7) Closed forms. Drawings by T. D'Este. 80

Plate 3.10 Pre-Kerma pottery from site 8-B-10A. (1-5) Sector East. Closed forms; (6-12) Sector West. Globular or ovoid jars and hemispherical bowls. Drawings by T. D'Este.. 81

Plate 3.11 Pre-Kerma pottery from site 8-B-10A. Sector West. (1-2) Slightly convex bowls; (3) bowl with straight walls; (4-7) jars with straight walls. Drawings by T. D'Este.. 82

Plate 3.12 Khartoum Variant pottery from site 8-B-10C. (1) Rim decorated with milled impressions; (2) ceramic sherd possibily used as comb; (3-7) dotted wavy line decorations. Photos by R. Ceccacci. 83

Plate 3.13 (1-6) Khartoum Variant pottery from site 8-B-10C. Zigzag patterns produced using the rocker stamp technique with a roulette; (7-8) Abkan pottery from site 8-B-76. Photos by R. Ceccacci. 84

Plate 3.14 Pre-Kerma pottery from site 8-B-52A. Large sherd decorated using the rocker stamp technique with a plain edge tool. Photos by R. Ceccacci.. 85

Plate 3.15 (1) External and (2) internal surface of a ceramic sherd possibly re-used as 'spoon'. Photos by R. Ceccacci. 86

Plate 3.16 Pre-Kerma pottery from site 8-B-52A. Sherds decorated using the rocker stamp technique with (1-3) a plain edge tool; (4-5) a comb with an evenly serrated edge; (6) a comb with an unevenly serrated edge. Photos by R. Ceccacci. 87

Plate 3.17 Pre-Kerma pottery from site 8-B-52A. (1-7) Rimsherds decorated with simple incisions or impressions. Photos by R. Ceccacci. 88

Plate 3.18 Pre-Kerma pottery from site 8-B-52A. (1) Undecorated sherd with a little hole close to the rim; (2-6) sherds decorated using the alternately pivoting stamp technique (APS). Photos by R. Ceccacci. 89

Plate 3.19 Pre-Kerma pottery from site 8-B-10A. (1-13) Sherds decorated with incisions with (1-8) simple; (9) criss-cross; (10-13) herring-bone patterns. Photos by R. Ceccacci. 90

Plate 3.20 Pre-Kerma pottery from site 8-B-10A. (1-6) Black-topped rims; (7-8) sherds decorated using the rocker stamp technique; (9-11) sherds decorated with simple impressions. Photos by R. Ceccacci. 91

Plate 3.21 Pre-Kerma pottery from site 8-B-10A. (1-14) Sherds decorated using the rouletting technique. Photos by R. Ceccacci. 92

Plate 3.22 Pre-Kerma pottery from site 8-B-10A. (On top) Sherd decorated using the rouletting technique; (below) sherd showing marks of burnishing on the internal surface. Photos by R. Ceccacci. 93

Plate 3.23 Photos of fractures showing (1) angular inclusions; (2) rounded inclusions; (3) calcareous inclusions; (4) organic inclusions. Photos by R. Ceccacci. 94

Plate 4.1 Microphotographs of examples from Pre-Kerma samples from fabric QPl-Veg. (a, c, e) Cross Polarized Light; (b, d, f) Plane Polarized Light. Photographs by G. Eramo. 133

Plate 4.2 Microphotographs of examples from Pre-Kerma samples (a–d) from fabric QPl (a–b) and QPl-Col (c–d) and from Abkan samples (e–f) from fabric QPl-Chr. (a, c, e) Cross Polarized Light; (b, d, f) Plane Polarized Light. Photographs by G. Eramo. 134

Plate 4.3 Microphotographs of examples from Pre-Kerma samples (a–d) and Khartoum Variant samples (e–f) showing mineral phases and organic structures recognized under the microscope. (a, e, f) Cross Polarized Light; (b, c, d) Plane Polarized Light. Photographs by G. Eramo. 135

Plate 4.4 Microphotographs of examples from Khartoum Variant samples from fabric QKfs-Bt (a–b); QKfs-Ms (c–d) and Q (e–f). (a, c, e) Cross Polarized Light; (b, d, f) Plane Polarized Light. Photographs by G. Eramo. 136

Plate 4.5 Microphotographs of examples from local sediment samples from fabric QPl (a–d); and QPl-Cal (e–f). (a, c, e) Cross Polarized Light; (b, d, f) Plane Polarized Light. Photographs by G. Eramo. 137

List of Tables

Table 1.1 Radiocarbon dates from Khartoum Variant (KV) and Shamarkian (SHK) sites in the Abka-Wadi Halfa region (Lower Nubia). Calibrations were obtained by the author using OxCal v. 4.2.4 Bronk Ramsey (2013); IntCal13 atmospheric curve (Reimer et al. 2013)..................10

Table 1.2 Radiocarbon dates from Khartoum Variant (KV) sites on Sai Island (northern Upper Nubia). Calibrations in Garcea et al. (2016a) based on Riemer et al. (2013). 11

Table 1.3 Radiocarbon dates from Early Neolithic (El Nabta/Al Jerar phases) sites in the Nabta-Kiseiba region (Western Desert). For periods when several dates are available, only the earliest and the most recent are provided. Calibrations were obtained by the author using OxCal v. 4.2.4 Bronk Ramsey (2013); IntCal13 atmospheric curve (Reimer et al. 2013). 12

Table 1.4 Radiocarbon dates from Mesolithic sites in the Kerma region (Upper Nubia). For periods when several dates are available, only the earliest and the most recent are provided. Calibrations in Honegger and Williams (2015) based on Riemer et al. (2013). 13

Table 1.5 Radiocarbon dates from Abkan (ABK) and Post-Shamarkian (Post-SHK) sites in the Abka-Wadi Halfa region (Lower Nubia). For periods when several dates are available, only the earliest and the most recent are provided. Calibrations were obtained by the author using OxCal v. 4.2.4 Bronk Ramsey (2013); IntCal13 atmospheric curve (Reimer et al. 2013). 17

Table 1.6 Radiocarbon dates from Abkan (ABK) sites on Sai Island (northern Upper Nubia). Calibrations in Garcea et al. (2016a) based on Riemer et al. (2013)..................18

Table 1.7 Radiocarbon dates from Early Nubian sites in the Laqiya region (Northwest Sudan). When more dates were available, for each site is provided the oldest and the youngest date. Calibrations in Lange and Nordström (2006)..................18

Table 1.8 Radiocarbon dates from Middle and Late Neolithic (El Ghanam, Ru'at El Baqar phases) sites in the Nabta-Kiseiba region (Western Desert). FFor periods when several dates are available, only the earliest and the most recent are provided. Calibrations were obtained by the author using OxCal v. 4.2.4 Bronk Ramsey (2013); IntCal13 atmospheric curve (Reimer et al. 2013)..................19

Table 1.9 Radiocarbon dates from Neolithic sites in the Kerma region (Upper Nubia). For periods when several dates are available, only the earliest and the most recent are provided. Calibrations in Honegger and Williams (2015) based on Riemer et al. (2013)..................19

Table 1.10 Radiocarbon dates from Pre-Kerma (PK)/Kerma sites in the Kerma region (Upper Nubia). For periods when several dates are available, only the earliest and the most recent are provided. Calibrations in Honegger and Williams (2015) based on Riemer et al. (2013)..................23

Table 1.11 Radiocarbon dates from Pre-Kerma (PK)/Kerma sites on Sai Island (northern Upper Nubia). Calibrations in Hildebrand and Schilling (2016) by OxCal v. 4.2.4 Bronk Ramsey (2013); IntCal13 atmospheric curve (Reimer et al. 2013)..................24

Table 1.12 Radiocarbon dates from A-Groups sites in the Laqiya region (Northwest Sudan). Calibrations in Lange (2003)..................24

Table 3.1 Distribution of classifiable and unclassifiable, by site..................37

Table 3.2 Stratigraphic distribution of classifiable and unclassifiable sherds from site 8-B-10C...................39

Table 3.3 Stratigraphic distribution of types of inclusions from site 8-B-10C...................39

Table 3.4 Frequency, sphericity and angularity of inclusions from site 8-B-10C...................40

Table 3.5 Stratigraphic distribution of types of textures from site 8-B-10C...................41

Table 3.6 Stratigraphic distribution of body parts of the vessel from site 8-B-10C 42
Table 3.7 Stratigraphic distribution of decorated and undecorated sherds from site 8-B-10C 43
Table 3.8 Stratigraphic distribution of decorative techniques from site 8-B-10C 43
Table 3.9 Stratigraphic distribution of tools used for rocker stamping from site 8-B-10C 44
Table 3.10 Stratigraphic distribution of tools and decorative motifs from site 8-B-10C 45
Table 3.11 Stratigraphic distribution of decorative structures from site 8-B-10C 47
Table 3.12 Stratigraphic distribution of inside and outside burnishing from site 8-B-10C 47
Table 3.13 Stratigraphic distribution of polishing from site 8-B-10C 48
Table 3.14 Distribution of polishing in relation to decorated and undecorated sherds from site 8-B-10C 48
Table 3.15 Distribution of textures in relation to the presence /absence of polishing from site 8-B-10C 48
Table 3.16 Stratigraphic distribution of types of textures from site 8-B-76 50
Table 3.17 Frequency, sphericity and angularity of inclusions from site 8-B-76 51
Table 3.18 Stratigraphic distribution of types of inclusions from site 8-B-76 52
Table 3.19 Stratigraphic distribution of body parts of the vessel from site 8-B-76 52
Table 3.20 Stratigraphic distribution of decorated and undecorated sherds from site 8-B-76 53
Table 3.21 Stratigraphic distribution of inside and outside burnishing from site 8-B-76 53
Table 3.22 Classifiable and unclassifiable sherds from site 8-B-52A 54
Table 3.23 Types of inclusions from site 8-B-52A 54
Table 3.24 Frequency, sphericity and angularity of inclusions from site 8-B-52A 55
Table 3.25 Types of textures from site 8-B-52A 56
Table 3.26 Body parts of the vessel from site 8-B-52A 56
Table 3.27 Decorated and undecorated sherds from site 8-B-52A 57
Table 3.28 Types of decorative techniques from site 8-B-52A 57
Table 3.29 Tools used for rocker stamping from site 8-B-52A 58
Table 3.30 Tools and decorative motifs from site 8-B-52A 58
Table 3.31 Types of decorative structures from site 8-B-52A 59
Table 3.32 Inside and outside burnishing from site 8-B-52A 60
Table 3.33 Types of polishing from site 8-B-52A 61
Table 3.34 Distribution of polishing in relation to decorated and undecorated sherds from site 8-B-52A 62
Table 3.35 Classifiable and unclassifiable sherds from site 8-B-10A 62
Table 3.36 Stratigraphic distribution of classifiable and unclassifiable sherds from sector west of site 8-B-10A .. 62
Table 3.37 Stratigraphic distribution of classifiable and unclassifiable sherds from sector east of site 8-B-10A ... 63
Table 3.38 Types of inclusions from site 8-B-10A 63
Table 3.39 Frequency, sphericity and angularity of inclusions from site 8-B-10A 64
Table 3.40 Types of textures from site 8-B-10A 65
Table 3.41 Body parts of the vessel from site 8-B-10A 65
Table 3.42 Decorated and undecorated sherds from site 8-B-10A 66

Table 3.43 Types of decorative techniques from site 8-B-10A. 67

Table 3.44 Stratigraphic distribution of decorative techniques from sector east of site 8-B-10A. 67

Table 3.45 Types of decorative structures from site 8-B-10A. 68

Table 3.46 Inside and outside burnishing from site 8-B-10A. 69

Table 3.47 Types of polishing from site 8-B-10A. 70

Table 3.48 Distribution of textures in relation to the presence /absence of polishing from site 8-B-10A. 70

Table 3.49 Distribution of polishing in relation to decorated and undecorated sherds from site 8-B-10A. 71

Table 4.1 Total number of analysed samples (Acronyms ABK; KV; PK stand for: Abkan; Khartoum Variant; Pre-Kerma) .. 95

Table 4.2a Provenance and macroscopic features of the analysed samples (Acronyms ABK; APS; BT; DWL; KV; PK; S stand for: Abkan; Alternately Pivoting Stamp; Black Topped; Dotted Wavy Line; Khartoum Variant; Pre-Kerma; Surface). 96

Table 4.2b Provenance and macroscopic features of the analysed samples (Acronyms ABK; APS; BT; DWL; KV; PK; S stand for: Abkan; Alternately Pivoting Stamp; Black Topped; Dotted Wavy Line; Khartoum Variant; Pre-Kerma; Surface). 97

Table 4.2c Provenance and macroscopic features of the analysed samples (Acronyms ABK; APS; BT; DWL; KV; PK; S stand for: Abkan; Alternately Pivoting Stamp; Black Topped; Dotted Wavy Line; Khartoum Variant; Pre-Kerma; Surface). 98

Table 4.2d Provenance and macroscopic features of the analysed samples (Acronyms ABK; APS; BT; DWL; KV; PK; S stand for: Abkan; Alternately Pivoting Stamp; Black Topped; Dotted Wavy Line; Khartoum Variant; Pre-Kerma; Surface). 99

Table 4.3 Provenance of the analysed local sediment samples (Acronyms ABK; KV; PK stand for: Abkan; Khartoum Variant; Pre-Kerma). 100

Table 4.4a Mineral composition of the analysed ceramic samples and clay lumps (C.M.: Clay Minerals; Qtz: Quartz; Kfs: K-feldspar; Pl: Plagioclase; Cal: Calcite; Px: Pyroxene; Amp: Amphibole; Zrn: Zircon). Quantities: XXXXX – predominant; XXXX – abundant; XXX – good; XX – moderate; X – scarce; tr – traces 103

Table 4.4b Mineral composition of the analysed ceramic samples and clay lumps (C.M.: Clay Minerals; Qtz: Quartz; Kfs: K-feldspar; Pl: Plagioclase; Cal: Calcite; Px: Pyroxene; Amp: Amphibole; Zrn: Zircon). Quantities: XXXXX – predominant; XXXX – abundant; XXX – good; XX – moderate; X – scarce; tr – traces 104

Table 4.5a Major elements composition of the analysed ceramic samples and clay lumps in relation to the identified mineralogical/petrographic groups (x: average composition, σ: standard deviation). Major elements and LOI (Loss on ignition) in weight percent (w%). 113

Table 4.5b Major elements composition of the analysed ceramic samples and clay lumps in relation to the identified mineralogical/petrographic groups (x: average composition, σ: standard deviation). Major elements and LOI (Loss on ignition) in weight percent (w%). 114

Table 4.6a Trace elements composition of the analysed ceramic samples and clay lumps in relation to the identified mineralogical/petrographic groups (x: average composition, σ: standard deviation). Trace elements in parts per million. 117

Table 4.6b Trace elements composition of the analysed ceramic samples and clay lumps in relation to the identified mineralogical/petrographic groups (x: average composition, σ: standard deviation). Trace elements in parts per million. 118

Table 4.7 Mineral composition of the analysed local sediment samples (C.M.: Clay Minerals; Chl: Chlorite; Qtz: Quartz; Kfs: K-feldspar; Pl: Plagioclase; Cal: Calcite; Dol: Dolomite; Px: Pyroxene). Quantities: XXXXX – predominant; XXXX – abundant; XXX – good; XX – moderate; X – scarce; tr – traces. 120

Table 4.8 Major elements composition of the analysed local sediment samples in relation to the identified mineralogical/petrographic groups (x: average composition, σ: standard deviation). Major elements and LOI (Loss on ignition) in weight percent (w%). 122

Table 4.9 Trace elements composition of the analysed local sediment samples in relation to the identified mineralogical/petrographic groups (x: average composition, σ: standard deviation). Trace elements in parts per million. 125

Table 4.10 Main groups, fabrics and sub-groups of the analysed ceramic samples and clay lumps. Main groups: QPl: quartz-plagioclase; QKfs: quartz-K-feldspar; Q: quartz. Fabrics: QPl-Veg: quartz-plagioclase-vegetal; QPl-Col: quartz-plagioclase-collophane; QPl-Cha: quartz-plagioclase-charcoal; QKfs-Bt: quartz-K-feldspar-biotite; QKfs-Ms: quartz-K-feldspar-muscovite. Mineral symbols mentioned are as in Kretz (1983). M: metamorphic rock fragments; V: volcanic rock fragments. Symbols (+), (-) e (~) express the abundance of the mineralogical phases; tr, traces (after D'Ercole et al. 2015). 128

Table 4.11 Main groups and fabrics of the analysed local sediment samples. Main groups: QPl: quartz-plagioclase; QCal: quartz-calcite. Fabrics: QPl-Col: quartz-plagioclase-collophane; QPl-Cha: quartz-plagioclase-charcoal. Mineral symbols mentioned are as in Kretz (1983). M: metamorphic rock fragments; V: volcanic rock fragments. Symbols (+), (-) e (~) express the abundance of the mineralogical phases; tr: traces (after D'Ercole et al. 2015). 129

Table 5.1 Percentages of decorative techniques from sites 8-B-10C, 8-B-76, 8-B-52A and 8-B-10A 144

Table 5.2 Synoptic description of the macroscopic technological and stylistic features of the different ceramic assemblages (after D'Ercole 2015). 147

Table 5.3 Synoptic description of the compositional and technological features of the different ceramic assemblages (after D'Ercole 2015). 153

Table 6.1 Major elements composition of the Khartoum Variant (KV) ceramics analysed from Sai Island (group QKfs: 20 samples) in comparison with the Early Khartoum (EK) productions from el-Kadada (20 samples), Saggai (17 samples), Umm Marrahi (4 samples), Sorurab (5 samples) and el-Qoz (4 samples). Major elements in weight percent (w%). 165

Table 6.2 Major elements composition of the Abkan (ABK) ceramics analysed from Sai Island (group QPl: 16 samples) in comparison with the Neolithic productions from el-Kadada (90 samples) and el-Qoz (6 samples). Major elements in weight percent (w%). 167

Table 6.3 Major elements composition of the Pre-Kerma (PK) ceramics analysed from Sai Island (group QPl: 30 samples) in comparison with the Kerma productions from Kerma: cemetery (51 samples) and town (16 samples). Major elements in weight percent (w%). 168

Table 6.4 Major elements composition of the local sediment samples analysed from the Abkan and Pre-Kerma sites on Sai Island (11 samples) in comparison with the Nile alluvia from the regions of the VI Cataract (27 samples) and Kerma (7 samples). Major elements in weight percent (w%). 168

Foreword

'It often happens that I have to explain to young people why it is good to study. Telling them that it is for the passion of learning is useless, if they do not have the passion of learning' (U. Eco). These words by Umberto Eco are prophetic, but when students express their interest in one of our subjects, we cannot know their degree, if any, of passion of learning. Such an enthusiasm may not exist in the beginning of their studies as most students usually develop it with time and with the progress of their research; some other students, however, immediately discover their passion of learning. Giulia D'Ercole has been one of the latter and did not take long to convey her passion of learning to me. I do not know when she developed it, but I can certainly say that it was already there when I met her.

I accepted Giulia's invitation to write this Foreword to her book with great pleasure and honour. When this book sees the light, I will have known her for 10 years. I met her in 2007, after her MA supervisor at the University of Rome La Sapienza, Barbara Barich, asked me if I could take a prospective PhD student with a strong desire to study ceramic assemblages. At the time, I had begun to collaborate with Francis Geus, the then director of the Sai Island Archaeological Mission (SIAM) of the University of Lille III-Charles De Gaulle, who in 2004 had entrusted me with the research on the later prehistory of Sai Island, in northern Sudan. My first excavation on Sai was at the Khartoum Variant site, 8-B-10C, and I happened to have quite a considerable sample at hand that I could offer Giulia immediately for a thorough study.

Giulia explained to me that her major interests were in the technological modes of production according to their manufacturing sequences, by observing the macroscopic features (decorations, morphologies, surface treatments, pastes, and tempers), as well as the petrographic, mineralogical, and chemical components. For the former, I introduced her to a database I had specifically created for Saharan and Sudanese pottery and trained her on what I had in turn learnt from my former professor Isabella Caneva on the classification of incised and impressed pottery. However, in spite of my readiness to share my knowledge on the macroscopic systems of pottery classification, I was unable to entirely satisfy Giulia's interests. We then agreed to turn to Italo Muntoni, who at the time was still attached to the University of Rome La Sapienza, and I commended her to him and to his colleague Giacomo Eramo of the Department of Earth and Geoenvironmental Sciences at the University of Bari. All together, we assisted Giulia in crafting her research plan and identifying the most suitable materials and methods.

Once methodological procedures were set in place, it was time for me to prepare my next field season at Sai Island and I proposed to Giulia that she join my team in Sudan. With the enthusiasm and the unconsciousness of a graduate student, she immediately accepted my proposal, and followed me in my expedition. In the field, I had the opportunity to get to know her in depth and discovered a very motivated, determined, persevering, and collaborative person at work, and a sensitive, thoughtful and passionate team member. She also became quickly fond of the awesome beauties of Sudan and the friendly openness of Sudanese people. For Giulia, this was the first of five consecutive field seasons with my research team in Sudan, of which she had become a regular member. Over the years I could see the student Giulia gradually become a trustworthy and responsible colleague. She took part in the excavations of the lower levels at the Khartoum Variant site 8-B-10C, the Abkan deposit at 8-B-76, and the Pre-Kerma site 8-B-10A. These sites cover a long time period spanning from the 8th to the 3rd millennium BC. Beside the excavations, she painstakingly worked on the collected ceramic samples in the laboratory of the SIAM's guest house. In addition, Elisabeth Hildebrand kindly gave her the permission to study the pottery excavated at the Pre-Kerma site 8-B-52A, of which she was in charge, in order to let her have a reliable comparative sample of Pre-Kerma pottery from a well-dated site.

Back in Italy, Giulia spent several months in Bari preparing her samples for the petrographic, mineralogical and chemical analyses, processing the data, and learning to interpret the results under the valuable supervision of Italo Muntoni, Giacomo Eramo, and their collaborators and students in the Department of Earth and Geoenvironmental Sciences. She adopted an integrated approach, combining the macroscopic observations with petrographic analyses of thin section by an optical microscope (MO), mineralogical analyses with powder X-ray diffraction (PXRD), and chemical analyses with X-ray fluorescence (XRF) of major, minor and trace elements of a significant selection of samples from the entire Khartoum Variant, Abkan and Pre-Kerma ceramic assemblages, which she classified in the database.

Giulia defended her thesis in 2012 and it was an easy task for me to present and comment on her successful results, which were fully appreciated by the rest of the jury. The technological approach for the reconstruction of the diachronically varying manufacturing sequences of the ceramic productions from the 8th to the 3rd millennium BC in Sudan that Giulia carried out brought Sudanese pottery into the general debate on the understanding of *chaînes opératoires*. Although this has become a standard essential scientific approach with regard to pottery manufacturing techniques from other parts of the world, it still is in its infancy in Africa.

In the past, different ceramic productions associated with foragers, early pastoralists and agro-pastoralists had been clearly distinguished from a macroscopic point of view. However, this systematic archaeometric analysis cast new light on the modes of pottery production by the diverse human groups and offered a contribution that qualitatively improved our knowledge on pottery production, function, and its social roles during the different periods. For example, it was well-established even with the naked eye that Khartoum Variant and 'Mesolithic' pottery in general was different from later productions. However, these analyses explained why and how they differed in raw material acquisition, clay processing, paste preparation, and use of the vessels. On the other hand, later, Abkan and Pre-Kerma, productions, which show different decorations and surface treatments not only in comparison with earlier assemblages, but also between each other, turned out to be more similar than expected from a mineralogical, petrographic and chemical point of view. These results have allowed to make new and deeper reflections on the makers of these ceramic assemblages, and on their skills, traditions, and concept of pottery as storage, cooking, serving, or ritual vessels. They have also provided insightful suggestions on the ways initial practices of pottery manufacturing were replaced by long-lasting cultural traditions.

This book is not the publication of Giulia D'Ercole's PhD thesis. Four more years of experience with a post-Doc with the ERC Project Across Borders, directed by Julia Budka, at the Austrian Academy of Sciences in Vienna and at the Ludwig Maximilians University in Munich provided her with further experience and professionalism, which she was also able to extend to Egyptian Pharaonic ceramic assemblages. Over the years, Giulia took the time to entirely revise her thesis, cut certain parts, expand others, update all of it, and transform it into a volume for specialists in the international scientific community.

This volume can be appreciated by multiple readers. Colleagues working in Sudan are the most obviously concerned, but certainly not the only ones. Other Africanist archaeologists can have the opportunity to look at a model example of an effective technological approach to African ceramic assemblages; ceramologists working in other parts of the world may be finally relieved to find a systematic and critical discussion on the technological features of 6000 years of pottery production; finally, PhD students, including those who think they will never make it to the end, can be inspired, encouraged, and reassured that a serious work has greater chances to be rewarded.

Elena A. A. Garcea

Acknowledgements

The writing of a book is a creative process made up of inspired mornings and miserable days in which it can be a challenge even to put a few words down on the page. Actually, it was mainly in those 'off days' which I spent for the most part walking around with my dog in the park, that I got some of the best ideas and found out the key to get through a controversial section of my research. The good thing is that I have never been alone during this process but indeed could always count on many colleagues and friends who have supported me with generosity and competence in this work.

The first person I would like to thank is Elena Garcea, my PhD supervisor, and the person who first introduced me to the archaeology of Sudan, proposing I take part in her research on the later prehistory of Sai Island and offering me the opportunity to study the ceramic material from the sites where she directed the research. When I first met her in 2007 I had I strong passion for pottery and ceramic technology, but I had never held before in my hands a Khartoum Variant dotted wavy line sherd or studied by myself a ceramic sample. I am sincerely grateful to Elena Garcea for everything she taught me about ceramics and, in general, for sharing with me a speculative method of work and logic of thought, which is rare and precious, and which I hope I have absorbed. I was lucky to have found in her a mentor and a friend.

My gratitude goes also to Barbara Barich, my PhD supervisor at the University of Rome - La Sapienza. I thank her for the trust and the appreciation that she showed me during the years of my studies, supporting me in my choices and encouraging me to never lose sight of the anthropological and cultural aims which are at the foundation of every archaeological reconstruction.

Sincere thanks go to the Mission Archéologique de l'Ile de Saï and to its former and current directors, Didier Devauchelle and Vincent Francigny, and its field directors Florence Doyen, until 2012, and Julia Budka, from 2013 onwards. I am greatly indebted to Julia Budka for giving me the exceptional chance to continue to work on Sai Island even after my PhD, offering me a post doctoral position in her European Research Council Starting Grant no. 313668 'AcrossBorders' for the technological study of the New Kingdom pottery of the island. I would like to thank her for generously welcoming me into her research team, first in Vienna, at the Austrian Academy of Science, and later, at the Ludwing-Maximilians University in Munich. Thanks to her, and to my new Egyptologist colleagues, I could also enlarge my perspectives as a prehistorian and develop new skills and significant competencies.

I would also thank all the colleagues and friends in the field I have got to know and with whom I shared the good times and the difficulties of the research. I am particularly grateful to Elisabeth Hildebrand for giving me the opportunity to study the ceramic sample of the Pre-Kerma site 8-B-52A and to Jennifer Smith for generously providing me with the local sediment samples used in this study and for having exchanged with me valuable information about the geology of Sai.

I also thank the National Corporation for Antiquities and Museums (NCAM) in Khartoum, its directors, previously Hassan Hussein Idris and, currently, Abdelrahman Ali Mohamed, and all the inspectors of NCAM who have accompanied and assisted us in Sai. A very special thank you goes to the inspector Huda Magzoub and with her to all the exceptional Sudanese people and friends I met during my research on Sai Island.

My sincere gratitude goes also to the Department of Earth and Geoenvironmental studies at the University of Bari, Italy, where I carried out the archaeometric laboratory analyses presented in this book. I especially thank Italo M. Muntoni and Giacomo Eramo who have coordinated my work during this delicate task, passing on to me all their experience and precious knowledge about ceramic technology, petrography and geochemistry. It was a rare and lucky opportunity for me to have learnt this expertise from two experienced scholars with a deep knowledge both in archaeology and natural sciences.

The translation of most of this volume from Italian to English is the result of the painstaking work of William Ayres and Maria Serena Concilio. I am truly grateful to both of them for the time and the care they dedicated helping me in this work. I would also like to thank Roberto Ceccacci and Tiziana D'Este, respectively, for the photographs and the drawings of the ceramic material, and my friend, Paola Canzonetta, for helping me in realizing the geographical maps presented in this book.

Finally, this volume is the result of ideas and thoughts shared over the years with many colleagues and friends during conferences, in the field or simply at home or at the office. I am grateful to all of them and to the the people we have possibly never met but whose ideas have indirectly inspired me from their papers and books.

Lastly, I want to thank my husband, Lorenzo Verginelli, who is so modest and generous as not even to want to be thanked, but who is actually the person who supported me most in this work, helping me tirelessly during so many weekends to prepare the illustrations and tables for publication. He is also the one who never stopped encouraging me to believe in this project and not to lose my passions and ambitions.

This research project was funded thanks to a three-year full tuition scholarship I obtained from the Graduate School of the University of Rome - La Sapienza. This work has benefited, in addition, for contributions to the realization of archaeometric analyses and other external services, of the PRIN 2008 funds (Research projects of relevant national interest) assigned to Elena Garcea of the University of Cassino and Southern Latium and Barbara Barich of the University of Rome - La Sapienza. Fieldwork at the prehistoric sites at Sai Island was conducted thanks to an agreement between the Sudan's National Corporation for Antiquities and Museums and Elena Garcea, who has been entrusted with the Research Unit for the Later Prehistory within the Sai Island Archaeological Mission of the University of Charles de Gaulle-Lille 3 since 2004. In 2011 and 2012 the researches on the recent prehistory of Sai Island were financed by the University of Cassino and Southern Latium and the National Geographic Society - Committee for Research and Exploration (Grants # 8715-09 and 9201-12 awarded to Elena Garcea).

Giulia D'Ercole

Introduction

'Material objects and technologies are concrete expressions and embodiments of human thought and ideas.' (V.G. Childe 1956: 1)

The invention of pottery and its cultural significance

The invention of pottery is the first case in human history of the manufacture of an artefact by transforming a raw material with particular physical and chemical characteristics – clay – into a product – ceramic – with new and different characteristics (Childe 1936). Because of its own unique properties and for the exceptional versatility of its uses, ceramic has always exercised an undeniable attraction and its study has attracted scholars of all periods. The act of kneading moist clay with water and attentively selecting, initially by trial and error, the most suitable tempering materials; and then again the creative gesture of decoration and, often unrewarded, effort of firing the pot reveal human labour in its whole fragility, strength and greatness. Pottery is also a class of artefacts has changed considerably both in technology and in style over time and space, so that the many stylistic and technological choices adopted in the manufacturing sequence of a vessel do not simply own an aesthetic or functional value, but are primarily of cultural significance (Gosselain 1998, 2000; Sillar and Tite 2000). We could paraphrase the words of Gordon Childe, cited at the beginning, and say that 'the act itself of manufacturing a pot and the final product which results from that act are concrete expressions and embodiments of human thoughts and ideas'.

Another fascinating aspect of pottery is that, in contrast to stone tools which, in a sense, existed since the appearance of the human species, the manufacture of clay objects and the invention of ceramics only occurred hundreds of millennia later, at dates and in ways that varied depending on the different geographic and cultural contexts.

In the Near East, the earliest ceramic containers appeared in approximately 7000 BC, at almost the same time in both Central Anatolia and Upper Mesopotamia (Thissen 2007). From north to south, the functions of this initial pottery could vary (Nieuwenhuyse 2010), but these regions have in common the fact that ceramic was first introduced there when the 'Neolithic process' was well under way and the local populations had already passed through the so-called agricultural 'revolution' (Childe 1936). In fact, although sedentism and agriculture had been precociously adopted in this area at the very beginning of the Holocene and Pre-Pottery Neolithic societies in the Near East already possessed the tools, raw materials and technological knowledge to produce pottery (Hodder 2006; Nieuwenhuyse 2010; Thissen 2007), the invention of ceramic was for them a relatively late phenomenon, apparently independent of the adoption of the new sedentary style of life and of the first control on food production.

Totally different is the case of the African continent and of Sudan. Here the first ceramic containers appeared at the end of the 10th/beginning of the 9th millennium BC, with the earliest dates at c. 8700 BC from Sorourab 2, in Central Sudan (Hakem and Khabir 1989; Mohamed and Khabir 2003), and at c. 8600 BC from the district of Amara West, in Northern Sudan (northern Upper Nubia), few kilometers north of Sai Island (Garcea et al. 2016a), where the present research was carried out. In Sudan, as in other contemporary contexts in north-western Africa (i.e., Huysecom et al. 2009; Roset 1982, 1987) – and differently from what can be observed in the Near East – the earliest ceramic productions appeared before and independently of the adoption of a fully productive economy, that is within hunter-fisher-gatherer communities living in permanent or semi-permanent settlements with a 'delayed-return foraging' economy (Garcea 2016b).

In the Near East, sedentism was therefore a prerequisite for agriculture and many scholars consider it the spur that would have triggered such a crucial economic change. However, a direct link between sedentism and pottery does not exist and the adoption of pottery seems rather to respond to the need felt by the new agricultural communities for improved storage facilities and for more efficient ways of processing food (Nieuwenhuyse 2010: 72; cf., also Moore1995; Redman 1978). In the African continent, this paradigm appears to be overturned: sedentism constituted a prerequisite for pottery technology but not necessary for agriculture which was instead introduced only at an advanced stage and only in a few specific areas of the country.

These arguments lead us to review some of the theoretical assumptions accepted until now on the invention of pottery and to reflect in a more critical and problematic way on the functional and cultural reasons behind the emergence of the earliest ceramic technology. Were the initial motivations which led to the invention of pottery potentially different on the African continent compared and in the Near East? Did the economic and social context where such innovation took place influence and, in a certain sense, determine the choices – technological, aesthetic and functional – adopted by the first potters?

Sai Island in northern Upper Nubia (northern Sudan)

The context of the case studies for this research is of exceptional importance for both its location and cultural history – Sai Island, in the Middle Nile Valley. This island is located in a strategic geographic position, on the border between ancient Upper and Lower Nubia, and is a key setting to understand the cultural and political dynamics that led to the emergence of the two most important North African cultures: the Pharaonic kingdom of Egypt and the Nubian kingdom of Kerma.

Sai is also one of the largest islands in the River Nile. Its landscape, varying from a barren, inhospitable interior to a relatively fertile perimeter lapped by the Nile River, appears extraordinarily spectacular and those, like myself, who have had the fortune to go there, could have the feeling they were diving into a past environment that still looks real and vibrant. All its geographic features and natural elements, from water to sand, and to the silt of the Nile, are particularly tangible and the island offers geologically an almost unlimited supply of raw materials.

Sai Island has a very long and continuous occupational history. It was first occupied during the Early Stone Age (*c.* 220,000-150,000 years ago) through recent prehistory and into the Kerma period (*c.* 2450-1480 BC). During the Early New Kingdom (*c.* 1550 BC), it was of prime importance for Egyptian southern expansion, becoming one of the main Pharaonic centres in Upper Nubia (Budka 2014). The island was still occupied in medieval times and the Ottoman period (end of the sixteenth-beginning of the nineteenth century AD) until the present day.

The time frame considered in this study covers about five millennia of later prehistory, extending from the initial appearance of ceramic technology made by sedentary hunters-fisher-gatherer communities of the Early Holocene (local Khartoum Variant horizon: *c.*7600–4800 BC), through the emergence, during the Middle Holocene, of the first 'Neolithic' pastoral societies (Abkan horizon: *c.* 5550-3700 BC), into the flourishing Pre-Kerma Nubian culture (*c.* 3600-2500 BC). The uniqueness of the geographic context and its exceptionally long sequence of occupation provide a first-rate opportunity to widely investigate the cultural and economic significance that the invention of pottery had in the African continent from its earliest appearance within the 'Pre-Neolithic' ('pre-pastoral') communities of the Early Holocene to the first proto-historical cultures of the Late Holocene.

The ceramic sample consists of over 3,000 sherds selected from four sites excavated on Sai Island during successive field seasons starting from January 2009 until February 2013. A thorough analysis and discussion of these ceramic finds is followed by a comparison across a broader geographical area, including the regions of Lower and Upper Nubia, Central Sudan and the Egyptian Western Desert. The final result of this study suggests an original synthesis and interpretation of the ceramic traditions in Nubia and Sudan.

Methodological approach and research questions: style, traditions and change

The approach adopted for the analysis of the pottery sample was essentially technological and primarily inspired by the concepts of *chaîne opératoire* (Leroi-Gourhan 1964, 1965; Cresswell 1972) and 'technological style' (Lechtman 1977).

The concept of style was for a long time almost solely related to the aesthetic characteristics of an object (i.e., the decorative style of a pot) and so the earliest studies on Saharan and Sudanese pottery were mainly dedicated to the analysis of decoration, elaborating typologies of decorative motifs, styles and design structures (e.g., Camps Fabrer 1966; Bailloud 1969; Nordström 1972). Later, Caneva (1983, 1988, 1995; Caneva and Marks 1990; Caneva *et al.* 1993) developed these earliest typologies into a new system of classification which first linked the decorations with the techniques and tools used by the potters.

Style is however not just an aesthetic concept which relates to 'a specific and characteristic manner of doing something that is always peculiar to a specific time and place' (Sackett 1977: 370). It is above all a powerful communication tool through which people can mark boundaries or, otherwise, convey information between different groups (Wobst 1977).

For this reason, David and Kramer (2001:219) have defined style as 'a relational quality, the potential for which resides in those formal characteristics of an artifact that are acquired in the course of manufacture as the consequence of the exercise of cultural choice'. Central to this definition is the concept of 'technological style', firstly developed by Lechtman (1977), and the insight that the style potentially resides in every phase of the manufacturing sequence (or chaîne opératoire) (Sillar and Tite 2000: 8).

Keeping this in mind, I have chosen to combine a stylistic assessment of the ceramics with a technological analysis, aimed at outlining all stages of the manufacturing processes, i.e., preparation, production, finishing, use, and discard. The whole analysis was structured in two steps: beside a through macroscopic observation of all the potsherds, a selected number of sherds were also submitted to archaeometric petrographic, mineralogical and chemical analyses in order to better investigate specific aspects of the chaîne opératoire.

Archaeometric and technological studies have already been conducted on Nubian and Sudanese prehistoric ceramic assemblages (e.g., Dal Sasso *et al.* 2014; Francaviglia and Palmieri 1983; Hays and Hassan 1974; Khabir 1991; Klein *et al.* 2004; Nordström 1972), however this is the first time that this analysis covers such a remarkably long chronological sequence across different phases which, both economically and culturally, may be considered decisive not just for the history of Sudan but for the whole human past.

The main purpose of this book consists in fact not in the study of the ceramic assemblages from Sai Island and not even in the definition of the evolution of the stylistic and technological traits of the Nubian and Sudanese pottery traditions. I am rather interested in understanding the cultural and functional meaning that ceramics had for ancient people when they first made them and, further, in the perception that those people had of their past environment and of the other populations living nearby.

What were the functions and the social meanings of the earliest Nubian and Sudanese ceramic productions? Did this meaning eventually change over time in the transition from the first hunter-fisher-gatherers communities of the Early Holocene to the Middle Holocene cultures with a productive economy? Did the development of stylistic and decorative aspects of pottery correspond to the technological innovations or were they independent variables? How did people choose the raw materials, technologies and tools for making their pottery? And, moreover, what was the cultural significance of these choices?

This book aims at seeking a meaningful answer to such questions by establishing an ongoing dialogue between the material evidence and the socio-cultural context, starting from the analysis of the ceramic assemblages of Sai Island and ending with a broader evaluation of contemporary Nubian and Sudanese ceramic complexes. Finally, this research wants to offer its contribution to the study of pottery, being of interest not only for Africanists but potentially to scholars also working in the Near East as well as in other cultural and geographical contexts.

Structure of the research

The book consists of six chapters as follows:

The first chapter describes the geographical and cultural area covered by the research project and provides a detailed review of the chronological events which led to the development of Nubian and Sudanese societies in the Nile Valley and in the neighbouring deserts, with particular attention to the development of the local Khartoum Variant, Abkan and Pre-Kerma cultures.

The second chapter is devoted to the specific context of Sai Island and describes the research, the environment and the geomorphological features of the island and of the four sites where the ceramic assemblages originated.

Chapters 3 deals with the classification and with the statistical processing of the ceramic data pertaining to each context which was carried out on a relational data base. Chapter 4 presents the results of the archaeometric characterization conducted on a large sample of artefacts and sediments from the four sites.

Chapter 5 deals with interpretation and discussion. A comparison is made between macroscopic observations and the evidence resulting from archaeometric analyses. In this chapter, the data pertaining to the different ceramic assemblages, which had previously been discussed separately, are compared in terms of style and technology in order to outline a possible chronological variability (or else continuity) across the different local cultural horizons and also within the same cultural complex.

Finally, Chapter 6 compares the evidence collected from Sai Island with the evidence available for the Nubian and Sudanese contexts which are discussed in the opening chapter. In this chapter an attempt is made to provide both a synthesis and an interpretation of the development of ceramic traditions in the Nubian and Sudanese cultures throughout the Holocene period, with reference to the questions raised at the outset.

1. Nubia and its cultural sequences between the 8th and the 3rd millennium BC: Khartoum Variant, Abkan and Pre-Kerma

Introduction

The region of Nubia encompasses a broad geographical territory, extending along the Nile River from the First Cataract, in modern Egypt, to the Sixth Cataract, in modern Sudan and including the desert and semi-desert areas east and west of the Valley. This territory comprehends different ecological niches and cultural entities, which, depending on times, climate and environmental conditions have been variously related to each other, building more or less solid and durable networks.

This chapter describes the chronological development and the geographical distribution of the Early to Late Holocene cultures in Nubia (Middle Nile Valley), locally named Khartoum Variant, Abkan and Pre-Kerma, with particular regard to the evidence from Sai Island.

First, a general description of climatic and environmental conditions in the Nile Valley and in the Egyptian Western Desert and an analysis of the different land occupation strategies is provided for each period, from the Early (*c.* 8000–5000 BC) through the Middle (*c.* 5000–3500 BC) and into the Late Holocene (*c.* 3500–2500 BC). Then follows a discussion of the specific cultural complexes in Nubia (Khartoum Variant, Abkan and Pre-Kerma) and a definition of their chronology and settlement evidence.

In retracing this sequence of developments, the local Khartoum Variant, Abkan and Pre-Kerma cultures have been compared with the other cultures concurrently emerging in Central Sudan, Upper Nubia and in the Egyptian Western Desert. The intention has been to try, as far as possible, to establish a relationship among those regional entities in time and space.

For each chronological horizon, the various cultural complexes are referred to by the names that have become customary in discussion of the region (i.e., Khartoum Variant in Lower and northern Upper Nubia, Mesolithic in the rest of Upper Nubia, Early Khartoum in Central Sudan, and Early Neolithic in the Egyptian Western Desert). A synopsis of the chronological sequences in the different regions during the Holocene is showed in **Figure 1.1**.

Radiocarbon dating for the different cultural contexts is expressed throughout the text in cal BC. When calibrated dates were not available, calibrations were obtained by the author using OxCal v. 4.2.4 Bronk Ramsey (2013), IntCal13 atmospheric curve (Reimer *et al.* 2013).

Climate, environmental conditions and human occupation during the Early Holocene along the Nile Valley and in the Egyptian Western Desert

From about 8300 BC on, a more or less simultaneous return to more humid climatic conditions has been recorded in the different parts of the eastern Sahara from the latitude of Khartoum (around 15°N) up to Egypt **(Figure 1.2)**. This rather unexpected phenomenon, which initiated before the start of the Holocene and lasted until the Late Holocene (12,800–3500 BC), is known as the African Humid Period (AHP) (deMenocal *et al.* 2000; Gasse 2000; McGee *et al.* 2013; Tierney and deMenocal 2013) or the time of the 'Green Sahara' (Drake *et al.* 2011; Dunne *et al.* 2012; Sereno *et al.* 2008), although it underwent a number of dry intervals.

The AHP can be attributed primarily to an increased extension of the monsoon palaeosystem driving the Intertropical Convergence Zone (ITCZ) northwards with a resulting gradual shifting of the summer

Ceramic manufacturing techniques and cultural traditions in Nubia

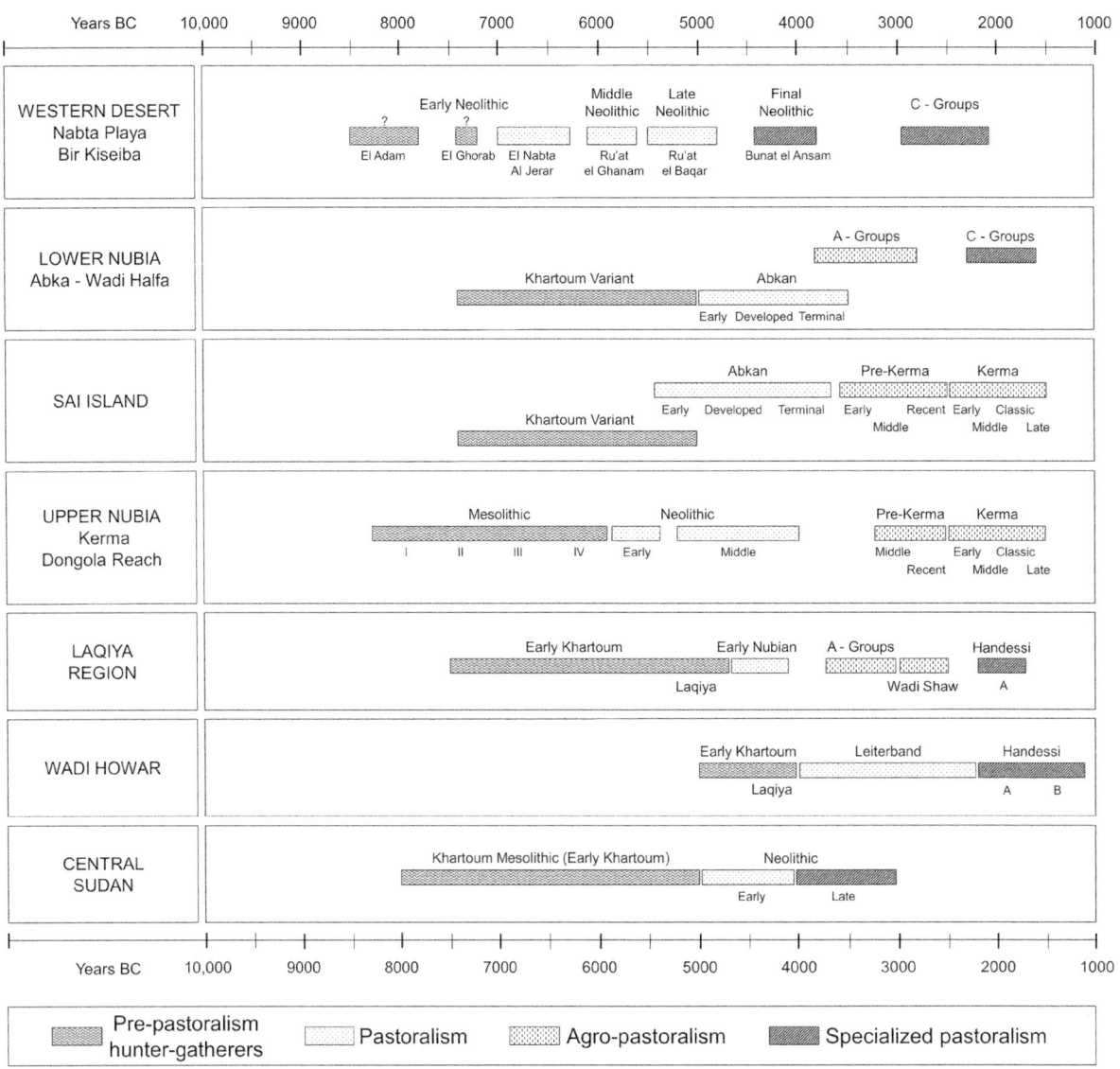

Figure 1.1 A synopsis of the chronological sequences during the Holocene in the regions of Egypt and Sudan mentioned in the text (figure by G. D'Ercole).

tropical rain front towards northern latitudes (Gasse 2000; Maley 1991). The Sahara consequently came out of a hyper-arid phase and began slowly to be re-populated (Manning and Timpson 2014), so that even the Egyptian Western Desert, for a long time an uninhabited and inhospitable area, became suitable for human settlements thanks to the presence of several playa basins fed by seasonal rains (Kuper and Kröpelin 2006; Nicoll 2004).

In spite of these general climatic conditions, during the Early Holocene, the Sudanese and Egyptian Nile Valley appeared as a 'mosaic of habitats' (Florenzano *et al.* 2016: 3) including diverse complementary environments with specific local or regional ecological conditions corresponding to the different latitudes but also to the specific hydro-geological and sedimentological history of the various sectors of the river (see also Honegger and Williams 2015).

Central Sudan at a latitude of 17° North was at that time an area of swamps and lakes (Pachur *et al.* 1990), with shallow lakes reaching as far as west of the Gebel Tageru Highlands (Williams and Adamson

1980). The Nile was directly connected to these western areas of the Erg Ennedi and Gebel Tageru and from those to the Sahara by a network of channels and tributaries including the Wadi Howar, a river, now dried up, which flowed for over 1000km from the Ennedi Mountains to Old Dongola (Kröpelin 2007) **(Figure 1.2)**. These regions on the southern border of the Sahel received plentiful constant rainfall, which supported the growth of wooded savannah vegetation and enabled the survival of several animal species including large mammals and different kinds of fish and shellfish (Gabriel 1976, 1986; Haynes 1985; Neumann 1989; Pachur et al. 1990). Remains of Celtis integrifolia fruits have been found along the central Nile Valley in the site of Khartoum Hospital together with the remains of gastropods like Limicolaria caillaudi, indicating the presence of a humid climate and of rainfall estimated at over 400mm/y, twice as much as the present values for that area (Arkell 1949).

In this part of the Nile Valley, permanent or semi-permanent settlements were created from approx. 8000 BC on by communities of hunter-fisher-gatherers who manufactured pottery and belong to the 'Early Khartoum' (or 'Khartoum Mesolithic') tradition (c. 8000-5000 BC) (Arkell 1949; Caneva 1983; Caneva et al. 1993).

Further north, along the northern Dongola reach of the Nile (19°N), in Upper Nubia **(Figure 1.2)**, recent archaeological and geological studies, supported by a consistent sequence of OSL and radiocarbon dates (Macklin et al. 2013, 2015; Welsby et al. 2001, 2002; Woodward et al. 2001), testified to the presence during the Early Holocene of extensive alluvial plains, terraces and paleochannels (see also Williams et al. 2010).

In Upper Nubia the first evidence of human occupation, at the onset of the Holocene, dates from c. 8300 BC and come from the region of Kerma, 40km north of Dongola, on the east bank of the Nile (Honegger 2014; Honegger and Williams 2015). It refers to hunter-gatherer groups producing pottery who, as their southern neighbours in the Khartoum region, also lived in semi-permanent or permanent villages exploiting the various resources offered by the Nilotic environment.

At the latitude of Sai Island (20° 42′ 30″ N), between the Second and the Third Cataract of the Nile **(Figure 1.2)**, the reconstruction of pollen spectra describes 'a regional plant landscape characterized by an open (desert) savannah with a local fresh water habitat covered with riverine vegetation' (Florenzano et al. 2016: 12). The pollen data indicate at c. 6400-6200 BC (8200 cal BP) a dry environment with only xerophilous vegetation together with seasonal species. The presence of tropical Acacia and Limicolaria gastropods are testimony approx. one millennium later (c. 5050-4800 BC) of a wet, warm phase occurring after the 'cooling and dry 8200 cal BP event' (Florenzano et al. 2016).

In this region of northern Upper Nubia, the first Holocene hunter-gatherer settlements with ceramic assemblages are attributied to the Khartoum Variant cultural complex (see below) and appeared slightly later compared to the region of Kerma. The earliest dates from Sai Island are between c. 7600-4800 BC and indicate a continuous occupation throughout the 8200 cal years BP dry event (Garcea et al. 2016a; Florenzano et al. 2016).

As we proceed further north along the Nile, near the present Egyptian border, a drier environment can be imagined, although wet conditions endured in the vicinity of the Nile or in the hinterland where seasonal rains and underground springs offered ecological refuges permitting the establishing of human settlements (Kuper and Kröpelin 2006).

According to Kuper and Kröpelin (2006), this stage corresponds to the 'Early Holocene reoccupation phase' of the Eastern Sahara (8500-7000 BC) and coincides with the beginning of the Holocene humid optimum. During this phase, groups of hunter-gatherers started to settle around the many playas and hilly areas of the Western Desert (Kuper 2002; Wendorf and Schild 2001) **(Figure 1.2)**. In contrast, the

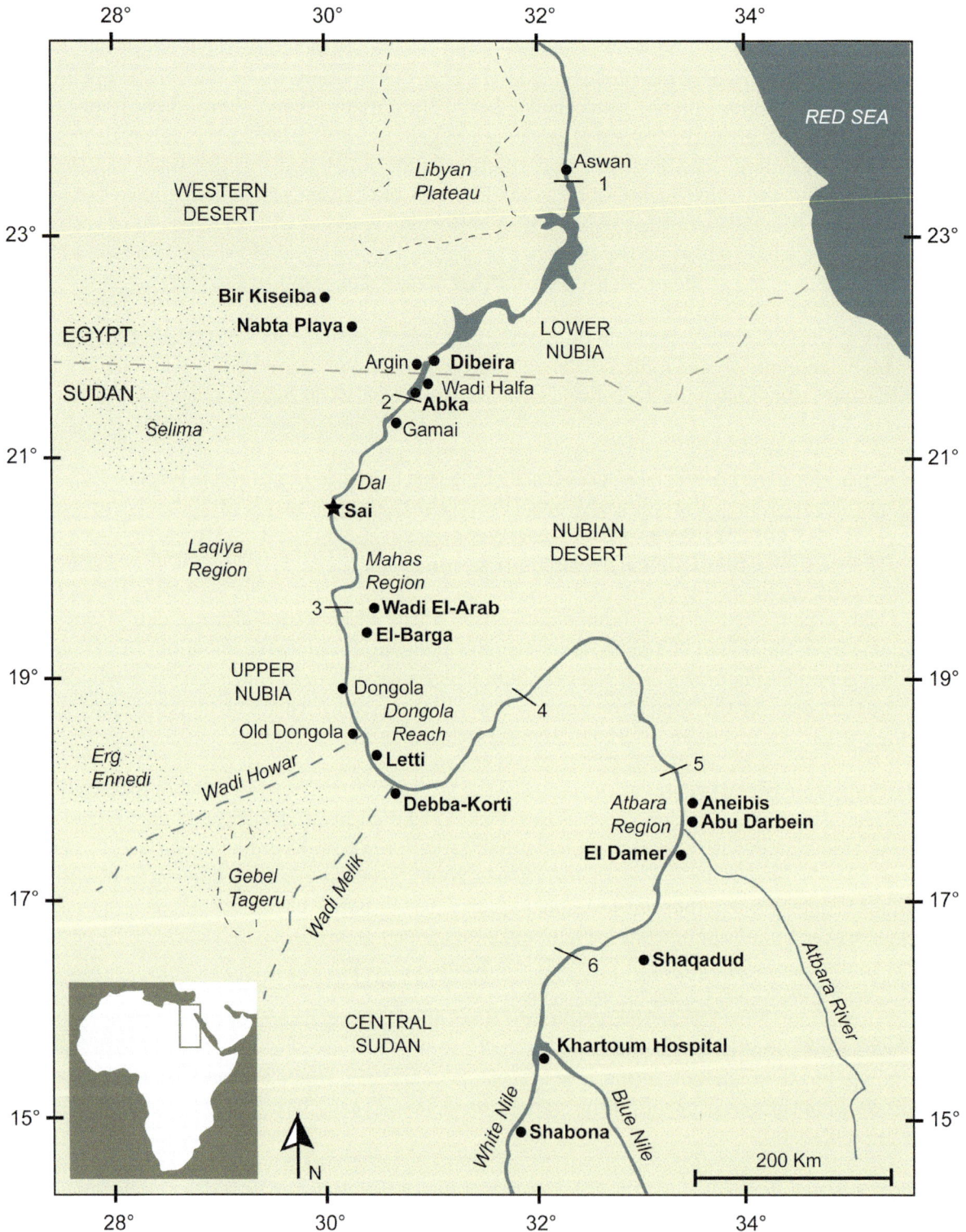

Figure 1.2 Map of Egypt and Sudan during the Early Holocene (c. 8000-5000 BC) with geographical features and sites cited in the text (figure by G. D'Ercole).

Egyptian Nile Valley at that time was almost completely lacking in human settlements (Vermeesch 2002). A similar scenario characterizes also the following period ('Mid-Holocene formation phase') (7000-5300 BC). Human groups became well established and increasingly sedentary at Nabta Playa as in the northern oases (Kuper and Kröpelin 2006), while in the Egyptian Nile Valley only a few sites have been documented (Vermeesch 2002).

The intense and repeated flooding registered during the Early-Mid Holocene in the Egyptian sector of the Nile Valley most probably limited the human capacity to exploit the territory being responsible for a prolonged hiatus in occupation, something which is not recorded in the Sudanese Nile Valley. Further, the frequent floods of the Nile possibly obliterated part of the evidence of the Early Holocene settlements in Egypt (Honegger and Williams 2015; see also Vermeesch 2002).

The Khartoum Variant culture: definition, chronology and settlements

The definition of the Khartoum Variant culture dates back to the 1960's when, less than ten years after the publication of Arkell's volume 'Early Khartoum' (1949), Myers (1958, 1960) excavated a group of sites in the vicinity of Abka – close to the border between modern Egypt and Sudan **(Figure 1.2)**. These sites appeared to have some aspects of material culture in common with the sites in the central Nile Valley, which form part of the Early Khartoum tradition.

A few years later, Shiner (1968a) attributed the archaeological evidence found in the Abka – Wadi Halfa region to a new cultural phase, which he called 'Khartoum Variant'. According to the author, the Khartoum Variant was the result of technological knowledge and of cultural options introduced from outside (Shiner 1968a). In the opinion of Shiner (1968a), this culture appeared however less developed and less rich in terms of the production of material objects than the Early Khartoum culture in the central Nile Valley, hence its definition as a varying northern development (i.e., 'Variant') of the original Early Khartoum (see also Garcea and Hildebrand 2009).

Similarly, Nordström (1972: 10-11) described the Khartoum Variant horizon as being 'a secondary phase of the Khartoum techno-complex', suggesting for this culture of Lower Nubia a chronological position younger than the Early Khartoum, that is between *c.* 5500 and 4500 BC, whereas the Early Khartoum is dated between *c.* 8000-5000 BC (Arkell 1949; Caneva 1983; Caneva *et al.* 1993).

During the 1960's, a number of intensive surveys and excavations were carried out on the occasion of the rescue operations in Lower Nubia, from the First to the Second Cataract, connected with the construction of the Aswan High Dam. The results were published by the Combined Prehistoric Expedition (CPE) (Wendorf 1968) and by the Scandinavian Joint Expedition (SJE) (Nordström 1972).

The CPE investigated eight Khartoum Variant sites, five in the vicinity of Abka (277, 1022, 2016, 2006, 1045), one (Dibeira West 5) north of Argin and two (626 and 628) in the desert, at about one kilometer north-west of Wadi Halfa (Shiner 1968a). The SJE identified four additional sites, two in the vicinity of the Second Cataract (423 and 428 = CPE 1045) and two north of the Egyptian border (89 and 18A) (Nordström 1972). Between 1963 and 1966, new settlements attributable to the same culture were discovered in the Abka district between Gamai and Dal (Mills 1965, 1968; Mills and Nordström 1966) and in the portion of the valley south of the Dal Cataract (Vila 1975-1979) **(Figure 1.2)**.

Khartoum Variant sites in the region of Abka – Wadi Halfa (Lower Nubia)

The sites located around the Second Cataract in Lower Nubia represent the core area of distribution of the Khartoum Variant culture. Shiner (1968a) and Nordström (1972) describe them as small 'open-air camps' mostly located a few hundred metres from the river, not directly on the river bank but rather upon Precambrian hills or on gentle slopes, probably to ensure protection from the frequent floods. These open-air sites have been for the most part destroyed by erosion and have left shallow deposits containing clusters of lithic and ceramic artefacts.

Based on material data, it was suggested that these settlements might have been used only seasonally either as workshops or as stopover stations by mobile groups which seasonally engaged in gathering, hunting and fishing activities (cf., Shiner 1968a).

Two cave shelters attributed to this cultural horizon have also been identified in the Abka district: sites Abka V and Abka IX (Myers 1958, 1960). These shelters are located at different levels along the river: the 'Upper Site' and the 'Low Site' to be used when the Nile flow was lower (Myers 1960).

None of these sites around the Second Cataract provided radiocarbon dates except for Abka IX, which provided a date from level 6 (M-804) of 8260 ± 400 uncal years bp (c. 8291-6396 BC) (Myers 1960), and Dibeira West 5 which provided a date of 6540±110 uncal years bp (c. 5667-5306 BC) (Hays 1984; Shiner 1960) **(Table 1.1)**. Further, older dates come from this area from the two 'Shamarkian' sites of Dibeira West 51 and 53 (Schild *et al.* 1968), which are nearly contemporary with Abka level 7 (cf., Usai 2004: 23) **(Table 1.1)**.

To-day it is possible to extend the sphere of influence of the Khartoum Variant horizon farther from its original core area around the Second Cataract. Its social and economic identity is being more clearly defined as well as its position in the chronology of the Early-Mid Holocene cultures.

Khartoum Variant sites at Sai Island (northern Upper Nubia)

New evidence pertaining to the Khartoum Variant complex has been detected on Sai Island **(Figure 1.2)**. On the island were identified several Khartoum Variant sites of which two (sites 8-B-10C and 8-B-76) were extensively excavated (Garcea 2006-2007, 2011-2012, 2016a; Garcea and Hildebrand 2009; Garcea *et al.* 2016a).

Both sites 8-B-10C and 8-B-76 are located in a formerly vegetated floodplain habitat (Garcea 2016b) and both provided a long stratified sequence of occupation indicating a repeated use of the site by human groups at different times (for a detailed description of these sites, see Chapter 2).

At site 8-B-10C, level 1 is dated between *c.* 5050 and 4800 BC, while level 2 revealed an earlier phase of occupation, dating between *c.* 7600 and 7200 BC (Garcea *et al.* 2016a) **(Table 1.2)**.

Site 8-B-76 provided both horizontal and vertical stratigraphy (Garcea 2016a). The Khartoum Variant materials appear concentrated in the northeasternmost units – towards the inland – while the south-

Site	Period	# Sample	Context	Material	14C yr bp	References	Years BC 95.4% Probability range
Dibeira West 5	KV	TX–1155	settlement	charcoal	6540±110	Hays 1984	5667–5306
Dibeira West 51	SHK	SMU–585	settlement	charcoal	8860±90	Schild et al. 1968; Wendorf et al. 1979	8256–7684
		WSU–176		charcoal	7700±120		7023–6263
Dibeira West 53	SHK	SMU–4	settlement	charcoal	7910±120	Schild et al. 1968	7083–6483
Abka IX Level 6, bottom	KV	M–804	settlement	shell	8260±400	Myers 1960	8291–6396

Table 1.1 Radiocarbon dates from Khartoum Variant (KV) and Shamarkian (SHK) sites in the Abka-Wadi Halfa region (Lower Nubia). Calibrations were obtained by the author using OxCal v. 4.2.4 Bronk Ramsey (2013); IntCal13 atmospheric curve (Reimer et al. 2013).

1. Nubia and its cultural sequences between the 8th and the 3rd millennium BC: Khartoum Variant, Abkan and Pre-Kerma

Site	Period	# Sample	Context	Material	14C yr bp	Cal BP	Years BC	References
8-B-10C								Garcea et al. 2016a
Level 2, 102N/106E	KV	ISGS–A2745	settlement	Pila shell	8505±25	9515±15	7550–7580	
Level 2, 95N/106E		ISGS–A2744		ostrich eggshell	8280±25	9275±120	7205–7450	
Level 1, Hearth 1D		KIA-24463		charcoal	6080±35	6810±90	4945–5040	
Level 1, Hearth 1D		ISGS–A2737		charcoal	6030±25	6865±60	4855–4975	
Level 1, Hearth 1A		ISGS–A2736		charcoal	5995±25	6835±45	4840–4930	
Level 1, Hearth 1A		KIA-24464		charcoal	5980±40	6935±85	4800–4930	
8-B-76								
Level 2, 11-13m	KV	ISGS–A2308	settlement	charcoal	7460±30	8275±65	6260–6390	

Table 1.2 Radiocarbon dates from Khartoum Variant (KV) sites on Sai Island (northern Upper Nubia). Calibrations in Garcea et al. (2016a) based on Riemer et al. (2013).

westernmost unit – closer to the present course of the Nile – only includes later Abkan materials. In the vertical stratigraphy, the Khartoum Variant occupation lies below the Abkan deposit with a date to 6260-6390 BC (Garcea *et al.* 2016a) **(Table 1.2)**.

Overall, this suggests a more stable and structured occupation strategy in comparison with the Khartoum Variant sites around the Second Cataract. Both the Khartoum Variant people in the region of Abka – Wadi Halfa and those settled on Sai Island were Nilotic hunter-fisher-gathers with a similar economic strategy ('delayed-return foraging') (Garcea 2016b; see also Garcea 2006a; Woodburn 1982, 1988) and comparable ceramic and lithic assemblages. However, the settlement pattern of the former ('small-open air sites' with absence of stratigraphic deposit, except for Abka IX, and scanty evidences of domestic features) hint at a mobile or semi-mobile occupation strategy, with possible short-distance seasonal displacement of part of the group or of the entire community. In contrast, the archaeological evidence from Sai Island (sites with thick stratigraphic sequence, hut floors and other domestic features) indicates a sedentary or near-sedentary occupation strategy, with sites that were 'either permanently occupied or systematically reoccupied in the same spot' (Garcea 2016b: 40). If we consider the geographical distribution of the Khartoum Variant culture, from the Second towards the Third Cataract, these differences in settlement pattern could be explained taking into account the different morphological and hydro-geological history of the various sectors of the river (i.e., severe erosion at the sites around Abka – Wadi Halfa) as well as the peculiarity of each habitat, either island or mainland (cf., Garcea 2016b).

From a chronological perspective, the earliest radiocarbon dates from Sai Island (*c*. 7600 BC) would appear to overlap with the dates of the Early Khartoum sites in the central Nile Valley. This indicates the possibility that these two cultures developed simultaneously and independently of each other (cf., Garcea and Hildebrand 2009). The Khartoum Variant and Early Khartoum cultures also differ in their ceramic traditions (Garcea and Hildebrand 2009; Jesse 2002; see for details Chapter 6).

On the other hand, a review of the lithic (Usai 2004, 2005) and ceramic (Gatto 2002a, 2006a) assemblages has provided elements for a comparison between the Khartoum Variant in the Nile Valley and the Early

Site	Period	# Sample	Context	Material	14C yr bp	References	Years BC 95.4% Probability range
Nabta Playa E-75-6	Early Neolithic El Nabta phase	Gd–6260	settlement	charcoal	8260±100	Schild & Wendorf 2001	7516–7069
		Gd–6257		charcoal	7770±110		7028–6434
Nabta Playa E-91-1	Early Neolithic Al Jerar phase	DRI–3526	settlement	charcoal	7740±115		7027–6399
		Gd–12188		charcoal	7360±90		6414–6059
Nabta Playa E-75-6	Early Neolithic Al Jerar phase	Gd–6507	settlement	charcoal	7610±120		6742–6219
		Gd–6510		charcoal	7330±100		6400–6020

Table 1.3 Radiocarbon dates from Early Neolithic (El Nabta/Al Jerar phases) sites in the Nabta-Kiseiba region (Western Desert). For periods when several dates are available, only the earliest and the most recent are provided. Calibrations were obtained by the author using OxCal v. 4.2.4 Bronk Ramsey (2013); IntCal13 atmospheric curve (Reimer et al. 2013).

Neolithic cultures in the Egyptian Western Desert (Nabta Playa and Bir Kiseiba region) **(Figure 1.2)**. Both the Khartoum Variant ceramics and the Early Neolithic production from Nabta-Kiseiba are decorated with typical rocker impressions mostly in bands of dotted wavy lines of the type 'short waves' (Jesse 2002: 80). Another peculiar feature which these productions have in common are some decorations on the vessels rims, known as 'milled' and 'notched impressions' (Gatto 2002a). At Nabta Playa, dotted wavy line decorations firstly appeared within the El Nabta/Al Jerar phase (c. 7100-6200 BC) (Nelson 2002a) **(Table 1.3)**. In Lower Nubia, a dotted wavy line sherd was found in the Khartoum Variant site of Abka IX (level 6) in association with a date (8260±400 uncal bp or 8291-6396 BC) which corresponds to the El Nabta sequence (Gatto 2006a: 65). At Sai Island ceramics decorated with dotted wavy line motifs first appeared at site 8-B-10C, level 2 (see Chapters 3 and 6).

Additional comparisons were also made between the Khartoum Variant and the Early Holocene hunter-gatherer cultures from the region of Kerma (northern Dongola Reach) (Garcea 2011-2012; Garcea and Hildebrand 2009). These groups, ascribable to the Mesolithic of Sudan (Honegger 2014; Honegger and Williams 2015), lived in semi-permanent villages and produced ceramics in a style comparable with the Khartoum Variant tradition. The sites of this period show indistinctly the same distribution being all located outside of the alluvial plain, safe from the Nile floods (Honegger and Williams 2015). This topographical choice found a comparison with the Khartoum Variant sites around Abka – Wadi Halfa (Lower Nubia); however, these sites in the Kerma region are more extended and, similarly to sites 8-B-10C and 8-B-76 at Sai Island, yielded thick stratigraphic deposits. The two most relevant are El-Barga and Wadi El-Arab **(Figure 1.2)**.

The site of El-Barga consists of a habitation structure, with traces of a semi-underground floor dug directly into the Nubian sandstone, which dated from c. 7500 BC (Honegger 2006, 2010, 2014; Honegger and Williams 2015). Further, several graves have been identified both within and close to the hut. Overall, approximately 50 inhumations were detected with dates between c. 7800-7000 BC (Honegger 2004a, 2005, 2014; Honegger and Williams 2015). No grave goods were found with the exception of two graves with bivalve Nile shells (Honegger 2004a, 2005, 2014) **(Table 1.4)**.

The site of Wadi el-Arab extends for more than four hectares with a stratigraphic sequence dated between c. 8300 and 6300 BC **(Table 1.4)**. The excavations yielded one oval habitation structure similar to the one from El-Barga and at least three other structures of which two were also dug into the ground and one, pertaining to a second phase, delimited by stones used to sustain posts for a superstructure (Honegger 2014). In the site were also discovered hearths, pits and 'some ten tombs disseminated within the habitations' (Honegger and Williams 2015: 6).

Site	Period	# Sample	Context	Material	14C yr bp	Cal BP	Years BC	References
Wadi El-Arab	Mesolithic I	ETH–31788	settlement	ostrich eggshell	8990±65	10,250–9910	8300–7960	Honegger & Williams 2015
		ETH–40941		ostrich eggshell	8820±40	10,500–9700	8200–7750	
	Mesolithic II	ETH–40949	settlement	ostrich eggshell	8795±40	10,300–9630	8180–7680	
		ETH–31787		ostrich eggshell	8140±65	9300–8790	7350–6840	
	Mesolithic III	ETH–40946	settlement	ostrich eggshell	8135±40	9250–9000	7300–7050	
		ETH–40524		ostrich eggshell	7590±40	8450–8340	6500–6390	
	Mesolithic IV	ETH–40526	settlement	ostrich eggshell	7400±40	8340–8070	6390–6120	
		ETH–51610		ostrich eggshell	7365±35	8310–8050	6360–6100	
El-Barga	Mesolithic II	ETH–27205	settlement	ostrich eggshell	8730±70	10,20–9540	8170–7590	
		ETH–31779		ostrich eggshell	8180±65	9400–9000	7450–7050	
	Mesolithic III	ETH–27206	grave 33	Nile shell	8020±65	9080–8640	7130–6690	

Table 1.4 Radiocarbon dates from Mesolithic sites in the Kerma region (Upper Nubia). For periods when several dates are available, only the earliest and the most recent are provided. Calibrations in Honegger and Williams (2015) based on Riemer et al. (2013).

Finally, settlements dated to the Early Holocene and possibly related to the Khartoum Variant complex have also recently been discovered in the region of Mahas – Third Cataract on both banks of the Nile (Herbst 2008; Edwards and Osman 2012) **(Figure 1.2)**. The distribution of these sites, a few kilometres from the present course of the Nile, reflects the hydrological regime of the river during the Early-to-Mid Holocene so that the sites may be expected to lie above the high flood level of the period (cf., Edwards *et al.* 2012). They did not yield domestic features but are characterized by clusters of lithic artefacts, shells, bones and ceramic sherds whose style can be attributed, in my opinion, to the Khartoum Variant tradition (see Edwards and Osman 2012: 42-44).

A comparison between the Khartoum Variant ceramics from Sai Island and the other contemporary sites mentioned above from Lower Nubia, Upper Nubia and the Western Desert is provided in detail in Chapter 6.

Climate, environmental conditions and human occupation during the Middle Holocene along the Nile Valley and in the Egyptian Western Desert

Climatic and environmental conditions in northern Sudan and southern Egypt began to deteriorate in the second half of the Holocene. From approx. 5300 BC, levels of humidity dropped considerably and seasonal rainfall became less frequent over a large part of the region, turning the climate to semi-arid and marking the return of drier conditions especially in the Western Desert (Kuper and Kröpelin 2006). This process followed different patterns in the Egyptian and Sudanese portions of the valley and in the hinterland, east and west of the Nile.

Botanical and faunal data from the Khartoum region provide a picture not unlike the one reconstructed for the Early Holocene sites. Seeds of *Celtis integrifolia* were found in the Neolithic site of esh-Shaheinab **(Figure 1.3)** together with remains of gastropods of *Limicolaria caillaudi* and remains of thirty-two species of mammals, of which buffalo, giraffe and hippopotamus were the most plentifully represented among the wild animals (Arkell 1953: 12). This indicates the persistence of a humid climate and rainfall levels above 400mm/y (Wickens 1975).

The esh-Shaheinab Neolithic (*c.* 4800-4500 BC), represented by a number of sites located on the west bank of the Nile around Khartoum and its immediate hinterland, is essentially still a Nile based culture, with a mixed hunting-fishing-gathering and herding economy (Arkell 1953; Garcea 2006b; Haaland 1987). In contrast, due to dissimilar geomorphologic conditions, in the sites east of Khartoum on the opposite bank of the Nile, herding almost totally replaced the earlier hunting-fishing-gathering economy (Garcea 2006b; Peters 1986). The site of Umm Direiwa, on the eastern bank **(Figure 1.3)**, provided some of the earliest evidence for domestic cattle from this area with dates from around 5050-4790 BC (Haaland 1987).

A hiatus in human occupation to be possibly linked to the decline of the Nile flow at this time is registered in the region of Kerma (Upper Nubia) between approx. 5500 and 5000 BC (Macklin *et al.* 2015). From 5300 BC on, due to the drier conditions, the region is re-occupied with new settlements located on the alluvial plain (Honegger and Williams 2015). This phase corresponds to the development of Neolithic pastoral societies (Chaix and Honegger 2015; Honegger 2014).

At Sai Island **(Figure 1.3)**, local seasonally wet conditions persist between 5000 and 4300 BC as witnessed by the presence of the terrestrial gastropod *Limicolaria* sp. (Florenzano *et al.* 2016). Here, there are no gaps in the human occupation and the new pastoral Abkan groups (see below) settle on the same site as the previous Khartoum Variant foragers (Garcea 2016a).

Farther from the Nile Valley, the environment must have been less hospitable. According to Nicoll (2004), the Sudanese/Sahelian vegetation typical of the Selima region, west of Sai Island **(Figure 1.3)**, did not change until approx. 5000 BC. There followed a gradual desiccation which can be reconstructed based on the levels of the local lake. At first (6000-5000 BC), levels started to fall as water evaporated and eventually the lake dried up completely around 3000 BC (Ritchie and Haynes 1987).

The desiccation process was perhaps more rapid and more intense in the Egyptian territory. With the exception of a few sheltered areas, all other locations in the Western Desert started to become depopulated. This process began around approx. 5000 BC just as the series of humid phases came to an end and local playa basins disappeared, often silted up by the new phenomenon of wind-blown sand (Nicoll 2004).

This period corresponds to the 'Mid-Holocene regionalization phase' of human occupation in the Eastern Sahara (5300 to 3500 BC) (Kuper and Kröpelin 2006). It is associated with the establishment of specialized fully pastoral societies with cattle, of either local or Near East origin (Gautier 2001, 2002) in addition to domesticated caprines imported from the Near East (cf. Gifford-Gonzalez and Hanotte 2011; Linseele 2010; Linseele *et al.* 2014).

In the same period (toward the end of the 6th millennium BC), Near East domesticated cereals firstly appeared in Nubia (Garcea *et al.* 2016b). The earliest evidence comes from cemetery R12, south of Kadruka, in Upper Nubia **(Figure 1.3)**, and consist of phytoliths of *Hordeum* sp. and/or *Triticum* sp., (Madella *et al.* 2014). These cereals were possibly used as both food but also as a funerary offering or they may have served for some ritual function within the burials (cf. Out *et al.* 2016). In spite of this initial evidence, regular crop cultivation would appear in Nubia only much later, around 2700 BC, that is during the Pre-Kerma horizon (Hildebrand 2006-2007) (see below).

1. Nubia and its cultural sequences between the 8th and the 3rd millennium BC: Khartoum Variant, Abkan and Pre-Kerma

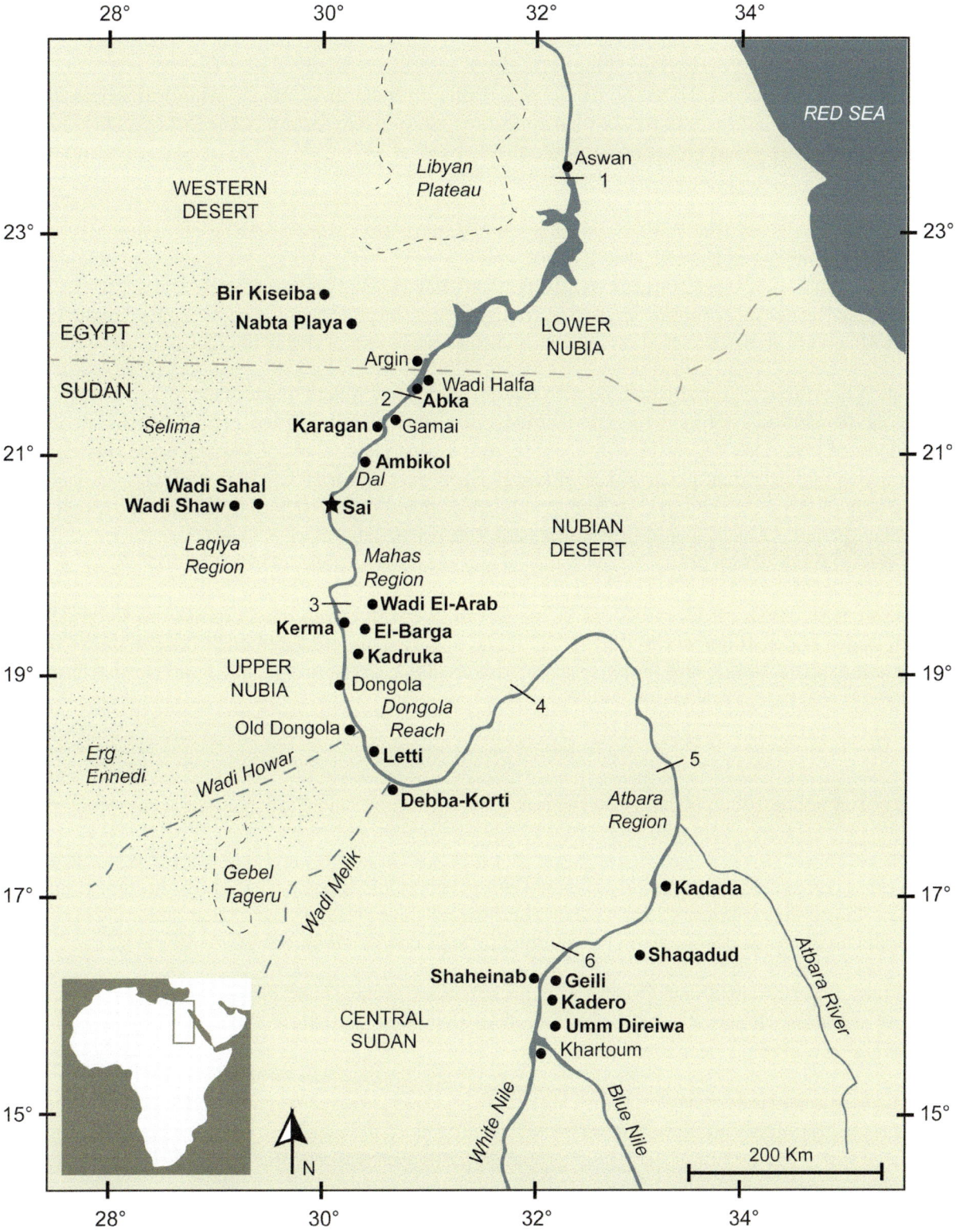

Figure 1.3 Map of Egypt and Sudan during the Middle Holocene (c. 5000-3500 BC) with geographical features and sites cited in the text (figure by G. D'Ercole).

The Abkan culture: definition, chronology and settlements

The term 'Abkan' was proposed by Shiner (1968b) to indicate a group of sites (no. 94, 604, 629, 1001, 1029, 2002 and 2007) concentrated along the Nile Valley in the area of Abka – Wadi Halfa. These sites are distinguished by a lithic industry, which appears to have developed out of the local Qadan tradition (Wendorf 1968), but by ceramics which are completely innovative in style and technology, in comparison with Khartoum Variant ceramics (Nordström 1972: 80).

Shiner (1968b) believed that Abkan and Khartoum Variant sites were partially contemporary – and this is the case indeed of the two sites 8-B-10C and 8-B-76 investigated on Sai Island (see below). The material production of these cultures is different, yet they overlap geographically and share the same settlement patterns and economic strategies, though Shiner implied that it was still to be demonstrated that the Abkan group had adopted certain characteristics of a pastoral way of life.

The Abkan horizon was initially divided into two phases: 'Early Abkan' (5000-4500 BC) and 'Developed Abkan' (4500-4000 BC) (Shiner 1968b). Nordström (1972) added a third phase: 'Terminal Abkan' (4000-3500 BC). He also suggested a revision of the entire chronological sequence for this culture. According to him, the final phase of the Abkan horizon overlaps with the first evidence available for the A-Group ('Early A-Group'), known from northern locations in Lower Nubia, which it had direct links with. The Terminal Abkan would thus constitute 'the starting point of a trajectory which led to the emergence of the A-Group proper in the southern part of Lower Nubia' (Nordström 2006: 37).

Several sites, which were described as being Abkan, were discovered in the Second Cataract and Batn-el-Hagar region **(Figure 1.3)** during the salvage work promoted by UNESCO in response to the construction of the Aswan dam (Adams and Nordström 1963; Carlson 1966; Mills and Nordström 1966; Nordström 1962, 1972; Verwers 1962; Williams 1989).

Abkan sites in the region of Abka – Wadi Halfa (Lower Nubia)

Similarly, to the Khartoum Variant, also the Abkan sites in Lower Nubia are open-air settlements, which have been considerably damaged as a result of severe erosion and possess very few typical structural features except for clusters of burned stones sometimes associated with layers of ash and concentrations of lithic and ceramic artefacts. The location of Abkan settlements is however slightly different from that of earlier Khartoum Variant sites. Rather than choosing hillsides or slopes, the settlers generally preferred the floodplain, river terraces or even the mouth of former, now dried up, tributaries of the Nile.

At site 1001, described by Shiner (1968b), copious fish remains have been found together with a smaller quantity of mammal bones. It would seem, therefore, that it was actually a fishing camp even though no formal fishing tools have been identified.

Site 5-S-25, on the island of Shagir – in the centre of the Second Cataract – is another fishing camp possibly belonging to the Abkan horizon[1]. The site yielded remains of different fish species over a layer of ash associated with a few ceramic sherds. In all probability, this area was repeatedly used by human groups at particular times of year possibly to smoke fish caught in the rapids close by (Adams and Nordström 1963).

Nordström (1972) dated CPE sites 1029, 2002 and SJE 414 in the district of Abka to the Early Abkan phase (*c.* 5000-4500 BC). Sites SJE 365, 369 and 371 as well as sites CPE 604 and 629 are attributed to the Developed Abkan (*c.* 4500-4000 BC) and are only slightly larger, without any trace of huts.

---

[1] This site, together with Site 5-T-38 on the island of Matuga (Second Cataract), is mentioned among the *'Early A-Group'* contexts of the region although both sites possibly belong to an earlier cultural phase. According to the authors, these sites *'presumably form an older part of a continuous development during this Nubian period'* (Adams and Nordström 1963: 18 footnote 18).

Levels 5 and 4 at the Abka IX shelter might also belong to the Abkan horizon **(Table 1.5)**. The ceramics found in those levels are different from those from level 6 and are described by Myers (1960: 176-177) as being 'made of sandy Nile mud ware with the surface very crudely combed or perhaps wiped with grass'. Gatto (2006a: 64) has included however the pottery of Abka IX, level 5 ('Ware C') still within the scope of the Khartoum Variant tradition.

South of the Second Cataract, in the locality of Wadi Karagàn **(Figure 1.3)**, site 11-I-16 provided one of the most recent dates for this area: 4935 ± 130 uncal bp (4032-3377 BC) (Carlson 1966) **(Table 1.5)**. Two further dates come from Ambikol East **(Figure 1.3)** site 16-S-10 to 5730 ± 160 uncal bp (4991-4263 BC) and 5330 ± 80 uncal BP (4332-3991 BC) (Hays 1984) **(Table 1.5)**.

Abkan sites at Sai Island (northern Upper Nubia)

Evidence of the Abkan culture has also been recently found on Sai Island (Geus 1998, 2000, 2002, Garcea 2006-2007, 2011-2012, 2016a; Garcea *et al.* 2016a) **(Figure 1.3)**.

These sites are located in the vicinity of the area occupied by the earlier Khartoum Variant settlements that is, near the former course of the Nile River, along the pediment, which at that time it should also correspond to the maximum extension of the island (Geus 2000) (for a detailed description of these sites, see Chapter 2).

At site 8-B-76, including both Khartoum Variant and Abkan materials, the Abkan occupation was closer to the river, following the accretion of sediment around the island. Material from the Abkan deposit – where ceramics analysed in this work were collected (see Chapter 3 and 4) – has been attributed to several dates stretching from approx. 5500 to 3700 BC (Garcea et al. 2016a) **(Table 1.6)**.

Another Abkan site, which includes exclusively Abkan artefacts, is site 8-B-81. This site is located on the same western side of the island, north of 8-B-76, and dates to between approx. 5000 and 4300 BC (Garcea *et al.* 2016a) **(Table 1.6)**.

Site	Period	# Sample	Context	Material	14C yr bp	References	Years BC 95.4% Probability range
Dibeira West 4	Post-SHK	WUS–103	settlement	charcoal	5220±50	Schild et al. 1968; Wendorf et al. 1979	4230–3955
Dibeira West 50	Post-SHK	SMU–2	settlement	charcoal	5880±150	Schild et al. 1968; Wendorf et al. 1979	5207–4406
		SMU–1		charcoal	5410±150		4586–3944
Abka IX							
Level 5	ABK	M–803	settlement	shell	5960±400	Myers 1960; Stuckenrath & Ralph 1965	5666–3995
Level 4		M–802		ostrich eggshell	4470±300		3946–2351
Level 4		M–801		charcoal	4500±350		4041–2235
CPE 605	ABK	WSU–190	settlement	charcoal	6430±200	Hays 1984	5734–4911
Wadi Karagan 11-I-16	Terminal ABK	GXO.423	settlement	charcoal	4935±130	Carlson 1966	4032–3377
Ambikol East 16-S-10		U–820		charcoal	5730±160		4991–4263
		U–2490		charcoal	5330±80		4332–3991

Table 1.5 Radiocarbon dates from Abkan (ABK) and Post-Shamarkian (Post-SHK) sites in the Abka-Wadi Halfa region (Lower Nubia). For periods when several dates are available, only the earliest and the most recent are provided. Calibrations were obtained by the author using OxCal v. 4.2.4 Bronk Ramsey (2013); IntCal13 atmospheric curve (Reimer et al. 2013).

Site	Period	# Sample	Context	Material	14C yr bp	Cal BP	Years BC	References
8-B-76								Garcea et al. 2016a
Level 1, 13-14m		ISGS–A2750		pottery	6500±20	7425±10	5465–5485	
Level 1, 11-13m		ISGS–A2749		pottery	6345±20	7275±20	5305–5345	
Level 1, 9-11m		ISGS–A2748		pottery	6210±20	7100±70	5080–5220	
Level 2, 25-27m	ABK	ISGS–A2751	settlement	pottery	6205±20	7095±70	5075–5215	
Level 9, TU 1-2		ISGS–A2752		pottery	6195±25	7095±70	5075–5215	
Level 4, 25-27m		ISGS–A2309		charcoal	5710±25	6490±40	4500–4580	
Level 7, TU 1-2		ISGS–A2307		charcoal	5395±25	6230±40	4240–4320	
Level 6, TU 1-2		ISGS–A2306		charcoal	5005±25	5755±90	3715–3895	
8-B-81								
Level 3, 1-2m		ISGS–A2755		charcoal	6060±25	6920±35	4935–5005	
Level 2, 3-4m	ABK	ISGS–A2759	settlement	charcoal	6020±25	6845±50	4845–4945	
Level 2, 1-2m		ISGS–A2754		charcoal	5475±25	6260±40	4270–4350	

Table 1.6 Radiocarbon dates from Abkan (ABK) sites on Sai Island (northern Upper Nubia). Calibrations in Garcea et al. (2016a) based on Riemer et al. (2013).

'Abkan-related' sites with shared technological and stylistic options – especially in pottery production – were also observed in the Laqiya region **(Figure 1.3)**, south-west of Sai Island ('Early Nubian' sites at Wadi Shaw and Wadi Sahal) (Lange and Nordström 2006) and in the Egyptian Western Desert (Middle-Late Neolithic sites in the Nabta-Kiseiba region) (Nelson 2002a) **(Tables 1.7 and 1.8)**.

Along the Nile Valley, south of Sai Island, ceramic traditions which show an affinity with Abkan productions have been detected in the region of Kerma, (Honegger 2006, 2014), in the Wadi el-Khowi area (Reinold 2000, 2001, 2004) and as far south as Debba and Korti in the Dongola Reach, close to the Fourth Cataract. The latter belong to the so-called 'Karat Group' (Marks and Ferring 1971) where Gatto (2002b) has observed the presence of a ceramic style with affinities with Abkan sites. Similarly fashioned ceramic sherds were also found in the Letti Basin (Usai 1998) **(Figure 1.3)**.

Site	Period	# Sample	Context	Material	14C yr bp	Years BC	References
Wadi Shaw							Lange & Nordström 2006
82/82-2		KN–3080		charcoal	5730±160	4587±170	
	Early Nubian	KN–3877	settlement	charcoal	5680±130	4535±139	
82/66		KN–3331		charcoal	5530±180	4354±208	
		KN–3180		charcoal	5410±65	4218±98	

Table 1.7 Radiocarbon dates from Early Nubian sites in the Laqiya region (Northwest Sudan). For periods when several dates are available, only the earliest and the most recent are provided. Calibrations in Lange and Nordström (2006).

1. Nubia and its cultural sequences between the 8th and the 3rd millennium BC: Khartoum Variant, Abkan and Pre-Kerma

Site	Period	# Sample	Context	Material	14C yr bp	References	Years BC 95.4% Probability range
Nabta Playa E-75-8	El Ghanam Middle Neolithic	SMU–2745	settlement	charcoal	7220±75	Schild & Wendorf 2001	6236–5928
		SMU–368		charcoal	6500±90		5618–5316
Nabta Playa E-75-8	Ru'at El Baqar Late Neolithic	SMU–487	settlement	charcoal	6550±80		5633–5362
		SMU–473		charcoal	5810±80		4842–4466

Table 1.8 Radiocarbon dates from Middle and Late Neolithic (El Ghanam, Ru'at El Baqar phases) sites in the Nabta-Kiseiba region (Western Desert). For periods when several dates are available, only the earliest and the most recent are provided. Calibrations were obtained by the author using OxCal v. 4.2.4 Bronk Ramsey (2013); IntCal13 atmospheric curve (Reimer et al. 2013).

The earliest evidence of an Early Neolithic necropolis (site El-Barga, dating between *c.* 6000 and 5500 BC) comes from Kerma (Honegger 2006, 2014) **(Table 1.9)**. This necropolis provided also the first indication of a domesticated bovine from the region, whose skull was deposited on top of a tomb (Chaix and Honegger 2015; Honegger 2014).

After the abandonment of the El-Barga necropolis in about 5500 BC, the occupations moved to the alluvial plain. Beneath the Eastern Cemetery of Kerma, a large Middle Neolithic settlement dating between approx. 5000 and 4000 BC was discovered (Honegger and Williams 2015) **(Table 1.9)**. Several post holes have been identified at this site, marking the position of oval huts, together with outdoor hearths and fire pits. Remains of fences and palisades indicate the location of livestock enclosures within the perimeter of the settlement. The faunal remains, except for a few fish vertebrae, are all from domesticated cattle and goats (Honegger 2006, Honegger 2014). The general impression is of a large settlement, which was not only used on a seasonal basis but was probably occupied by a numerous population who lived by herding and fishing (Honegger 2003).

Lange and Nordström have proposed the definition of 'Abkan Culture Group' to describe 'different regional cultures related in different ways to the Abkan, combining clear affinities as regards pottery, lithic traits, settlement pattern and economic structure' (Lange and Nordström 2006: 310). This seems to be an effective way to reflect the idea of a cultural complex with a Nubian matrix inherently consistent, yet split into increasingly limited regional areas.

A comparison between the Abkan ceramics from Sai Island and the other contemporary sites mentioned above from Lower Nubia, Upper Nubia and the Western Desert is provided in detail in Chapter 6.

Site	Period	# Sample	Context	Material	14C yr bp	Cal BP	Years BC	References
Wadi El-Arab	Neolithic I	ETH–51609	settlement	ostrich eggshell	7006±36	7940–7750	5990–5800	Honegger & Williams 2015
		ETH–47148		ostrich eggshell	6526±33	7510–7330	5560–5380	
El-Barga	Early Neolithic	ETH–27208	grave 16	ostrich eggshell	7045±70	8000–7720	6050–5770	
		ETH–28405	grave 70	ostrich eggshell	6605±60	7580–7430	5630–5480	
Cemetery of Kerma	Middle Neolithic	ETH–18827	settlement	charcoal	5815±60	6770–6470	4820–4520	
		ETH–51603		charcoal	5458±30	6300–6210	4350–4260	

Table 1.9 Radiocarbon dates from Neolithic sites in the Kerma region (Upper Nubia). For periods when several dates are available, only the earliest and the most recent are provided. Calibrations in Honegger and Williams (2015) based on Riemer et al. (2013).

Climate, environmental conditions and human occupation during the Late Holocene along the Nile Valley and in the Egyptian Western Desert

The end of the African Humid Period (c. 3500 BC) marks definitively the shift to arid and hyper-arid conditions in Egypt and northern Sudan. Kuper and Kröpelin (2006) define the period between 3500 and 1500 BC as a phase of marginalization of human occupation ('Late Holocene marginalization phase'). During that period, the human groups in the Eastern Sahara were forced to leave the inhospitable interior and to 'migrate' either towards the Nile Valley or to the few areas that could still offer a more hospitable environment and sufficient resources.

By the Late Mid-Holocene, along the length of the 'Saharan Nile', the vegetation dwindled progressively and aeolian sands advanced (Woodward *et al.* 2015). Rainfall decreased or ceased altogether even in areas which until a few centuries earlier had provided ecological 'refuges' and had been selected by groups of humans as favoured settlement sites (Kuper and Kröpelin 2006).

Between *c*. 3000 and 2500 BC, the lake of Selima, west of Sai Island, and, further south, the depressions of El 'Atrun and Gebel Tageru **(Figure 1.4)** in rapid succession dried up completely (Pachur *et al.* 1990; Ritchie and Haynes 1987). As the climate became drier, also most of the tributary wadis of the Nile (such as Wadi Howar and Wadi Melik), which were active during the AHP, progressively dried up (Nicoll 2004; cf. also Woodward *et al.* 2001, 2015).

The strip of land along the Nile was at that time the only green area. At the latitude of Khartoum, rainfall stabilized at 100-200mm/y (approx. half of the value recorded during the Early Holocene) and a vegetation typical of the Sahelian band replaced the former wooded savannah (Sadig 2010).

In Central Sudan, a hiatus in occupation of more than two millennia followed the end of the Neolithic period (around 3500 BC) (Honegger and Williams 2015; Usai 2014). During this phase, the necropolis at Kadada **(Figure 1.4)** is the only site known from the area (Reinold 2007; see also Salvatori and Usai 2006-2007).

A similar picture comes from Kerma and most of the regions in Upper Nubia. Within the shift to the drier climate of the Late Mid-Holocene, the local Neolithic cultures collapsed and no new sites were established in the area before the emergence, at *c*. 3400 BC, of the Pre-Kerma culture (Honegger 2014; Honegger and Williams 2015).

In the region of Batn-el-Hagar, 30km north of Sai Island, a drop of the Nile flow was recorded *c*. 4200 BC (de Heinzelin 1968). These episodes of drops in river levels were responsible for narrowing the floodplains on Sai's western margins. However, large floodplains and productive environments endured on the eastern side of the island (cf. Garcea and Hildebrand 2009).

Between *c*. 3600 and 2500 BC, different sites ascribable to the Pre-Kerma culture are documented on Sai Island (Garcea and Hildebrand 2009; Geus 1998; Hildebrand 2006-2007; Hildebrand and Shilling 2016) (see below). This period corresponds to the development in Nubia of agro-pastoral societies.

North of Sai Island, in the Egyptian Western Desert, the cultural sequence of Nabta-Kiseiba, terminates around 3800 BC at the end of the last wet phase of the 'Bunat El Ansam Final Neolithic.' The following centuries are characterized by growing aridity interrupted briefly during the 3rd millennium BC by wetter periods, when areas associated with the pastoral culture of the C-Groups were occupied on a temporary basis (Applegate and Zedeňo 2001). In the Egyptian sector of the Nile Valley, the period between *c*. 3800 and 3000 BC, corresponds to the flowering of the Pre-Dynastic and A-Groups cultures (Gatto 2006b).

1. Nubia and its cultural sequences between the 8th and the 3rd millennium BC: Khartoum Variant, Abkan and Pre-Kerma

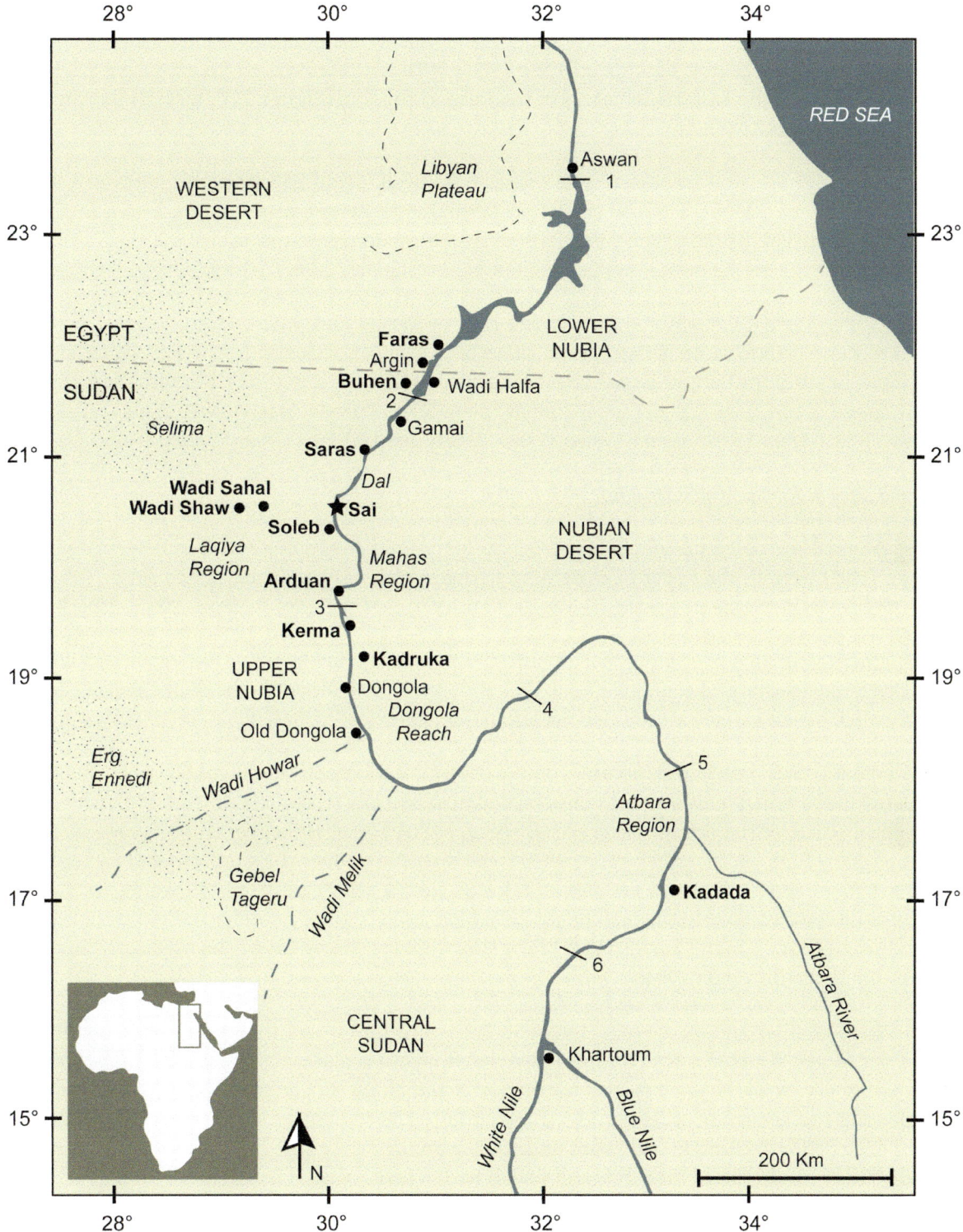

Figure 1.4 Map of Egypt and Sudan during the Late Holocene (c. 3500-2500 BC) with geographical features and sites cited in the text (figure by G. D'Ercole).

Overall, from Central Sudan to Upper Nubia up to the Egyptian territory, the Late Holocene period sees a cohabitation between fully nomadic pastoralists (*sensu* Garcea 2016a), in the desert, and emerging 'state-level' agro-pastoral or farmer societies, along the Nile Valley.

The Pre-Kerma culture: definition, chronology and settlements

This cultural horizon was first described in the 1980's with the discovery of an extensive settlement area in the vicinity of the Eastern Cemetery of the Kerma kingdom **(Figure 1.4)**, directly beneath later Kerma burials (Bonnet 1988).

In chronological terms, two separate cultural phases have been identified within this horizon: a 'Middle phase' dated to approx. 3000 BC (Eastern cemetery of Kerma) and a 'Recent phase' (approx. between 2700 and 2600 BC) represented at the site of Boucharia II (Honegger 2014; Honegger and Williams 2015) **(Table 1.10)**. The earliest phase has not yet been fully documented, but presumably dates to the mid-4th millennium BC.

The 'Middle Pre-Kerma' is mainly known in the area of Kerma and in the Third Cataract while the 'Recent Pre-Kerma' shows a larger geographical extent, from the Second to the Fourth Cataract. In Lower Nubia, north of the Batn-el-Hagar, the Pre-Kerma horizon appears to overlap with the last phases of development of the A-Groups. Ceramics similar to Pre-Kerma products have been found at Saras and Faras as well as at the site of Buhen (Honegger 2004b) **(Figure 1.4)**.

Pre-Kerma sites at Kerma (Upper Nubia)

The most extensive and best documented site pertaining to the Pre-Kerma culture is the settlement discovered by Bonnet (1988) beneath the Eastern Cemetery of Kerma, which has been investigated since 1995 by the 'Mission Archéologique Suisse au Soudan' (Honegger 1995, 1997, 1999, 2003, 2004b, 2004c, 2004d, 2006c).

This settlement is located over an earlier Neolithic site, on the alluvial plain (Honegger and Williams 2015). It covered a surface of some five to six hectares stretching along the bank of the Nile, and has been excavated over a surface area of two hectares (Honegger 2004c, 2014).

The settlement shows a certain structural complexity and a coherent, articulated layout. The storage area located in the north-western part consists of almost 300 pits (originally at least 500), which were most probably used for the storage of cereals (Honegger 2004c, 2014, Honegger and Williams 2015). The remains of many huts, on the other hand, seem to be distributed over the southern and eastern parts of the settlement and constitute a residential nucleus clearly separated from the storage pits. Some smaller pits discovered close to the huts might have been used to store grain for individual households whereas the larger silos, in the western part of the site, would probably have been used communally for different functions (Honegger 2004c).

The huts were circular and on average 4m wide. It has been proposed that a different function was reserved for larger structures which might have been meeting places, the huts of leading members of the community, workshops or stables for domesticated animals (Honegger 2004c). The size of the post holes and the remnants of fences identified close to the structures suggest a rather sturdy and durable construction technique. The walls and roofs were probably supported by a frame of wooden poles covered with interlaced branches and sealed with a layer of mud (Honegger 2006). Rectangular buildings, probably intended for purposes other than domestic use, were also discovered, together with large fortifications, with several parallel rows of fences (Honegger 2004c, 2014).

This settlement, in its internal layout and the characteristics of the structures it contained, is reminiscent of villages of certain agro-pastoral peoples such as the Teso of Uganda or the Zulu in South Africa (cf., Honegger 2003).

Only four graves have been discovered in the vicinity of this site (the only Pre-Kerma burials known so far). Two of them have yielded a few grave goods. These objects are associated with the burial of a female and are similar to typical grave goods from A-Group cemeteries (Honegger 2004d).

1. Nubia and its cultural sequences between the 8th and the 3rd millennium BC: Khartoum Variant, Abkan and Pre-Kerma

Site	Period	# Sample	Context	Material	14C yr bp	Cal BP	Years BC	References
Busharia II	Middle PK	LY–1662–(OXA)	settlement	charcoal	4470±45	5300–4900	3350–2950	Honegger & Williams 2015
	Recent PK	LY–11146	settlement	charcoal	4345±65	5280–4830	3330–2880	
		ETH–20839		charcoal	4085±50	4820–4440	2870–2490	
Cemetery of Kerma	Middle PK	ETH–188828	settlement	charcoal	4400±55	5280–4850	3330–2900	
		ETH–188829		charcoal	4365±55	5270–4840	3320–2890	
	Recent PK	ETH–47154	grave 463	blades of grass	4079±31	4810–4440	2860–2490	
		ETH–47157	grave 442	blades of grass	3920±31	4440–4240	2490–2290	
	Early Kerma I	ETH–27203	grave 321	ostrich eggshell	4050±55	4810–4420	2860–2470	
		ETH–40519	chapel	charcoal	3810±35	4400–4090	2450–2140	

Table 1.10 Radiocarbon dates from Pre-Kerma (PK)/Kerma sites in the Kerma region (Upper Nubia). For periods when several dates are available, only the earliest and the most recent are provided. Calibrations in Honegger and Williams (2015) based on Riemer et al. (2013).

Pre-Kerma sites at Sai Island (northern Upper Nubia)

On Sai Island, the Pre-Kerma culture is represented by sites 8-B-10A and 8-B-52A, which are respectively a habitation site (8-B-10A), and a storage site (8-B-52A) with no indication of habitation, but over 100 pits used as storage facilities for local wild plants and Asian domesticated cereals (Geus 1998; Hildebrand 2006-2007; Hildebrand and Schilling 2016) (for a detailed description of these sites, see Chapter 2). Site 8-B-52A has provided a sequence of radiocarbon dates spanning the period between c. 3600 and 2500 BC, which demonstrates an initial phase of construction and use of the site earlier than at the sites around Kerma, already during the 'Early Pre-Kerma' phase; while preliminary dating at site 8-B-10A indicates a later phase of occupation (Hildebrand and Schilling 2016) **(Table 1.11)**. Both sites yielded ceramics typical of the Pre-Kerma tradition (Garcea and Hildebrand 2009) (see Chapters 3 and 6).

Further evidence for the Pre-Kerma culture was provided by surveys and excavations carried out on the island of Arduan, in the region of Mahas, in the Third Cataract (Edwards and Osman 1992, 1993, 2000, 2012) **(Figure 1.4)**. The most important find in this region is site ARD001, identified in 2000 on the island of Arduan, a few kilometers south of the village of Arduan. Nine pits of more or less circular shape were identified in an area of approx. 0.5 hectares. Four of the five pits excavated were larger and are thought to have been used for storage (Edwards and Osman 2012).

Outside the valley, in the regions west of the Nile, a continuity in cultural traditions can be observed, as shown by the technological and stylistic characteristics of pottery from the so-called 'dune habitats' in the Lower Wadi Howar dating from the Middle Holocene (c. 4000-2200 BC) (Jesse 2006, 2008). In the region of Laqiya at site 82/52 in the Wadi Shaw **(Figure 1.4)**, dating to c. 2500 BC (Lange 2006), the ceramics have many common features with the pottery traditions of the A-Group culture, but have additional characteristics, which indicate also possible links to the Pre-Kerma complex **(Table 1.12)**.

A comparison between the Pre-Kerma ceramics from Sai Island and the other contemporary sites mentioned above from Lower Nubia, Upper Nubia and the Western Desert is provided in detail in Chapter 6.

Site	Period	# Sample	Context	Material	14C yr bp	Years BC	References
8-B-52A							Geus 1998; Hildebrand & Schilling 2016
silo 70, daub seal		ISGS A1741		chaff	4865±20	3696–3637	
silo 26, daub seal		ISGS A1764		chaff	4730±20	3632–3380	
silo 10, daub seal	PK	ISGS A1742	grain storage site	chaff	4300±25	3010–2881	
silo 11		UtC 5295		barley	4151±44	2881–2587	
silo 1		UtC 5294		barley	4142±48	2878–2581	
silo 39, daub seal		ISGS A1763		charcoal	4140±20	2871–2626	
silo 11, daub seal		ISGS A1743		charcoal	4070±20	2836–2496	
8-B-10A							
bag 94, lower deposits		ISGS A1740		charcoal	3485±20	1182–1748	
bag 64, middle deposits	PK - Kerma	ISGS A1739	settlement	charcoal	3420±20	1861–1659	
bag 31, upper deposits		ISGS A1738		charcoal	3465±20	1880–1697	

Table 1.11 Radiocarbon dates from Pre-Kerma (PK)/Kerma sites on Sai Island (northern Upper Nubia). Calibrations in Hildebrand and Schilling (2016) by OxCal v. 4.2.4 Bronk Ramsey (2013); IntCal13 atmospheric curve (Reimer et al. 2013).

Site	Period	# Sample	Context	Material	14C yr bp	Years BC	References
Wadi Sahal	A-Group						Lange 2003
82/38-1		KN–3013	settlement	charcoal	5000±170	3801±171	
82/38-2		KN–3014		bone	4350±320	2973±423	
82/38-3		KN–3177		charcoal	4390±160	3074±224	
82/38-4		KN–3144		charcoal	4990±150	3794±148	
82/38-6		KN–3083		charcoal	4470±50	3173±120	
Wadi Shaw	A-Group		settlement				
82/33-41		KN–3091		charcoal	4320±60	2960±67	
82/59		KN–3145		charcoal	4410±130	3110±82	
83/120		KN–3415		charcoal	4400±400	3013±511	

Table 1.12 Radiocarbon dates from A-Groups sites in the Laqiya region (Northwest Sudan). Calibrations in Lange (2003).

2. Sai Island: archaeological research and cultural sequence

Introduction

Sai Island (about 12km long x 5.5km wide) is one of the largest islands in northern Upper Nubia, between the Second and the Third Cataracts of the Nile. It occupies a strategic geographic position, being related to both Lower Nubia, in Egypt, and Upper Nubia, in Sudan, and in contact with places on either side of the valley: eastwards into the Eastern Nubian Desert and westwards into the Western Egyptian Desert **(Figure 2.1)**.

Along the course of the Nile, the island is enclosed by two imposing rocky uplands: the outcrop of Batn-el-Hagar to the north, and the highlands of the Nubian Sandstone Formation to the south (Maley 1970; Whiteman 1971). As a result of this geographic configuration, Sai could act as a proper 'strategic stronghold' (Garcea and Hillebrand 2009: 311), protected naturally on both its northern and southern approaches by the presence of these two uplands from possible incursions or external threats. However, the local geography has never represented a real barrier to stop groups of humans from reaching the island; and similarly, foodstuffs and other material products as well as ideas and cultural traditions were free to move along the Nile and east-west between the valley and the deserts. Viewed in this light, Sai takes on rather the appearance of a border crossing point for groups of different origins and cultures.

The island has been occupied from the Early Stone Age (Van Peer *et al.* 2003, Van Peer and Herman 2006) up until the Ottoman and medieval period (Geus 2004a; Hafsaas-Tsakos and Tsakos 2012; Tsakos and Hafsaas-Tsakos 2014), and indeed into the present day. Since the beginning of the Holocene, it has always been at the centre of cultural currents in the region, registering, as climatic conditions changed, significant alterations both in the economic choices and in the settlement patterns of the human groups occupying it. Over the course of millennia, these people transformed their way of life, moving from an economy of hunting, gathering and fishing – Khartoum Variant horizon (*c.* 7600–4800 BC) – to the gradual emergence of a production economy based on animal husbandry – Akban (*c.* 5500–3700 BC) – and then to the development of a complex agro-pastoral society – Pre-Kerma horizon (*c.* 3600–2500 BC).

Geological setting and paleo-climatic reconstruction

Geologically, Sai Island is located in the territory of the Saharan Metacraton in Sudan, whose eastern boundary is a contact with the Arabian-Nubian Shield (ANS) (Guiraud *et al.* 2005). This region is composed of two main lithological domains: the older Precambrian Basement – consisting mostly of migmatites and granitites, and minor gabbroic rocks, amphibolites and meta-sedimentary sequences (Shang *et al.* 2010) – and the younger sedimentary unit of the Nubian Sandstone, which lays on top of it, being deposited during the Cretaceous period or possibly slightly later (Gardiner 2010; Klemm *et al.* 2001; Prasad *et al.* 1986) **(Figure 2.1)**. Younger alluvial sediments, related to the Nile drainage system, are exposed along the Nile Valley. Coarse sediments (gravels) characterize the higher Nile terraces, while the finer sediments are typical of the lower terraces and occur often within beds of aeolian sand, as a consequence of desertification (D'Ercole *et al.* 2015; cf., also Pachur and Hoelzmann 2000).

Sai Island well represents the geological history of this region **(Figure 2.2)**. Nowadays, its landscape varies from an arid, inhospitable, gravelly interior **(Figure 2.3)**, to a perimeter where sediments have been deposited abundantly by the river, and which for this reason is preferred by the current population as being better suited to cultivation and settlement (Geus 2000). During the Early Holocene, ancient Sai was limited to just the pediment around Jebel Adu, an isolated inselberg of Nubian sandstone protruding above the eroded surface of the Precambrian Basement, which is today visible in only a few parts of the island, for the most part in association with outcrops of schist (metamorphic rock) and quartz (Geus 2000) **(Figure 2.2)**.

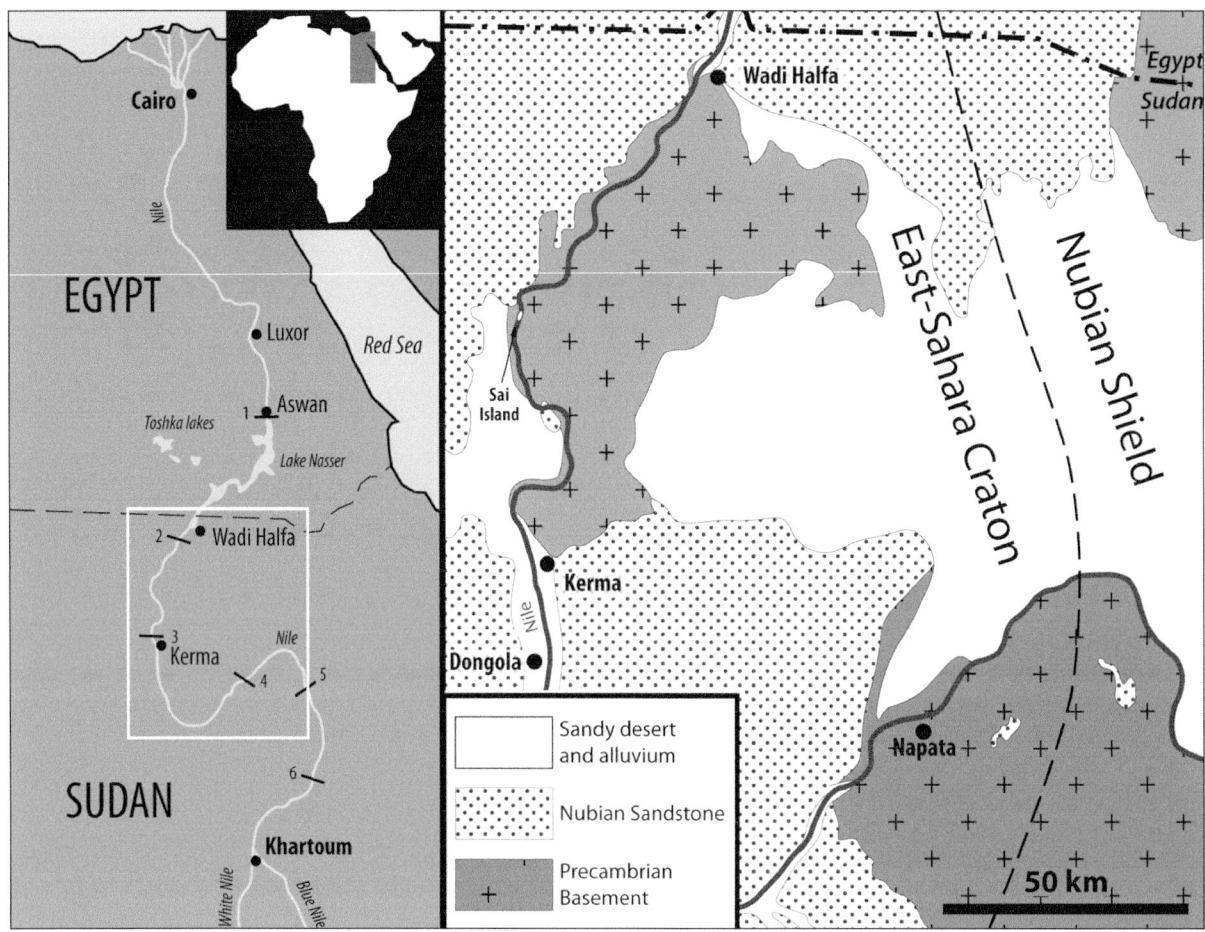

Figure 2.1 Geographical location of Sai Island and schematic geological and morphological map of northern Sudan (modified after D'Ercole et al. 2015).

Climatic conditions at that time were relatively wet, with seasonal rainfall, enough to support the growth of vegetation even in the island's interior (Garcea and Hildebrand 2009; cf., also Florenzano *et al.* 2016). Only after the fifth millennium BC, following a progressive deterioration of the climate (Küper and Kröpelin 2006; Nicoll 2004) and as a direct consequence of a drop in water levels in the Nile, did Sai take on its present morphology. Wide floodplains formed in the south-eastern part of the island, which then remained suitable for human habitation even in later periods that were marked by increasing aridity (Garcea and Hildebrand 2009). However, the two riverbanks differ from one another. The west bank is distinguished by the presence of abundant river sediment, stretching uninterruptedly from north to south, and by a landscape dominated by sand and gravel bars; while along the eastern side, the deposits of sand and gravel are more limited and in some areas completely absent, because the more intense action of the Nile has brought about greater erosion (Geus 2000) **(Figures 2.2 and 2.4)**. Finally, the northern point of Sai was formed only in a more recent period, probably after the first millennium BC, attaching itself to the main part of the island when an ancient channel of the Nile was blocked by the formation of a sand dune.

The archaeological sites on the island

The first archaeological excavations on Sai were carried out in the 1950s and 1970s (Vercoutter 1958, 1972, 1985) but they were limited to an exploration of the more recent material, principally from sites

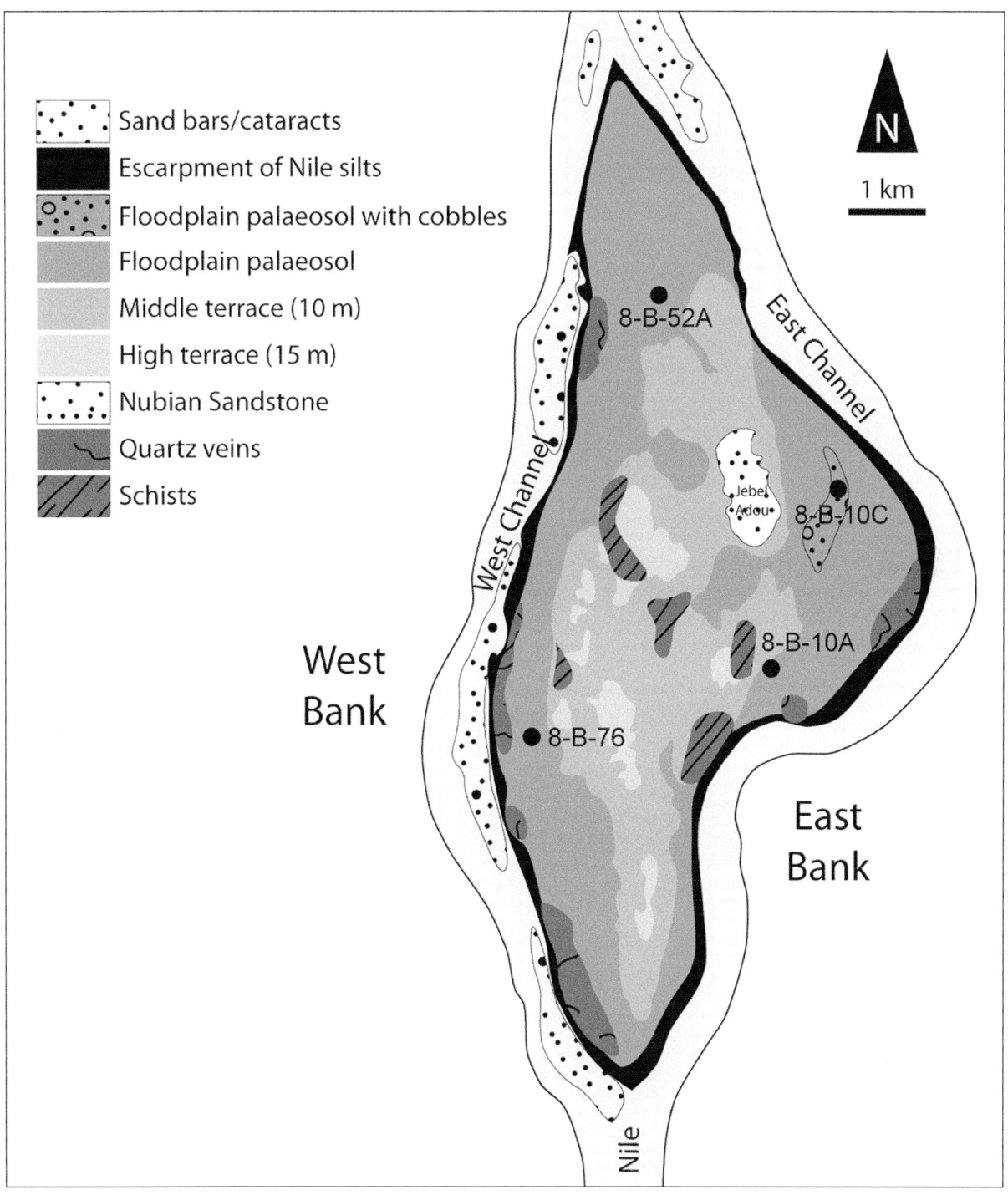

Figure 2.2 Geological map of Sai Island with the location of the four sites from which the pottery sample come (modified after D'Ercole et al. 2015).

dating to the third and to the second millennium BC, such as the extensive cemetery attributed to the Kerma culture (Gratien 1978, 1986) and the Pharaonic town (Azim 1975; Azim and Carlotti 2011-2012; Budka 2014, 2015; Budka and Doyen 2013; Doyen 2014; Geus 2004a).

Figure 2.3 View of the interior landscape of Sai Island with the profile of the Jebel Abri in the background (photo by G. D'Ercole).

Figure 2.4 View of the western riverbank of Sai Island with young alluvial sediments and typical riverine vegetation (photo by G. D'Ercole).

Only from 1993 on did the team led by Francis Geus set in motion a large scale research project aimed at reconstructing the different phases of occupation of the island from the earliest times onward (Geus 1994, 1995a, 1995b, 1996, 1998, 2000, 2002, 2004b; Goosssens *et al.* 1997; Hesse 1996). This project combined traditional archaeological research with a geomorphological survey of the entire area of the island, with the aim of identifying and defining where human groups had chosen to settle, and linking archaeological sites to different geomorphological features. A first geological map of Sai was produced as a result, including numerous archaeological sites that were attributed for the most part to the Early and Middle Holocene, and which had been identified by means of surface surveys, remote sensing, and aerial photographs taken from a kite (Hesse and Chagny 1994).

The Early Middle Stone Age sites were located exclusively in the interior of the surface of the pediment, at the foot of the outcrop of Jebel Adu, on a terrace dating from the Middle Pleistocene. Two settlements from the Later Stone Age are situated on the silt deposits near the Pharaonic town (Geus 2000; Van Peer and Herman 2006). Early and Middle Holocene settlements are for the most part found along the perimeter of the pediment, near what must have been, at least until the fifth millennium BC, the original course of the Nile (Geus 2000). The sites of the Kerma cemetery and of the Pharaonic settlement are also contained within the ancient nucleus of the island, on the northern and eastern edges respectively. The Christian cathedral and more recent occupation are found on the northern point of the island, which was formed during the first millennium BC. All the sites identified during the course of this research were numbered according to the conventions of the Archaeological Map of Sudan produced by Hinkell in 1977 (Geus 1995b; Goossens *et al.* 1997).

In 2004, E. A. A. Garcea had been entrusted with the direction of the Research Unit for the Later Prehistory of the island, while E. Hildebrand had been responsible for the archeobotanical research. A new cycle of excavations and geomorphological research was inaugurated and all the settlements from the Early and Middle Holocene, which had been identified previously, have been relocated and georeferenced in order to create a small-scale map of the later prehistoric occurrences (Garcea 2006–2007).

Altogether there were initially recognized eight sites dating from the Khartoum Variant horizon (8-B-10C, 8-B-10D, 8-B-49, 8-B-50, 8-B-77, 8-B-78, 8-G-46A, and 8-G-46B) and three Abkan sites (8-B-10B, 8-B-10E and 8-B-76) (Garcea 2006–2007). Recently, at site 8-B-76 excavations led by Garcea exposed surfaces and levels with a succession of Khartoum Variant and Abkan deposits (see below). In 2013, Garcea also excavated the Abkan site 8-B-81, on the western side of the island (Garcea 2016a; Garcea *et al.* 2016a). Two Pre-Kerma sites (8-B-52A and 8-B-53) had been identified and marked on the archaeological map by F. Geus, in the district of Adu (north-eastern sector of the island). Another site which can be assigned to this cultural complex (site 8-B-10A) has been brought to light recently in the eastern part of the island, a few kilometres south of the Khartoum Variant settlement 8-B-10C (Garcea and Hildebrand 2009). This site was previously called SKP1 (Geus 1998, 2002).

The pottery samples analysed in this work come from four of the sites mentioned above: 8-B-10C, 8-B-76, 8-B-52A and 8-B-10A, and belong, respectively, to the Khartoum Variant, Abkan, and Pre-Kerma (8-B-52A and 8-B-10A) horizons. An analytical description of these four sites is provided below **(Figure 2.2)**.

Selected sites

Site 8-B-10C

Site 8-B-10C is located on the eastern sector of the island, south of the modern village of Adu, in a locality called Eshak Nirki. The settlement lies on a prominent gravel bar above a fluvial terrace that was formed during the Early Holocene, probably about 8000 BC, which stretches for over 2km along the south-eastern edge of the island (Garcea and Hildebrand 2009) **(Figure 2.2)**. At that time, because of the seasonal floods of the Nile, a swampy environment characterized the site and the gravel bars periodically became separated islands (Garcea 2016a; Geus 1998, 2000; Hesse 1996).

In 2004, an area of 35m² of the site was excavated. This was later increased to 105m² (7 x 15m), divided internally into a grid of 1 x 1m squares, with a north-south orientation. The excavation, concluded in January 2011 under the direction of E. A. A. Garcea, brought to light a complete stratigraphic sequence, showing a continuous and repeated use of the site by human groups. It could be divided into a sequence of five levels of deposits, including two different phases of occupation which showed evidence of settlement structures (levels 1 and 2) (Garcea 2006-2007, 2011-2012; Garcea and Hildebrand 2009) **(Figure 2.5)**.

Level 1 consists of a layer of loose sandy sediment with nodules of calcium carbonate (*kankar*), probably formed in some post-depositional event. The layer is at most 13cm thick and contains a particularly high concentration of lithic and ceramic artefacts. At the bottom of this level, a habitation surface was identified, marked by a complex of different structures, which indicate a phase of use as a place of permanent or semi-permanent settlement. Seven hut floors have been identified (A-G), three hearths (1A, 1D and 1F), three rubbish pits (1B, 1C, and 1E) and 100 postholes (Garcea 2006-2007) **(Figure 2.6)**.

The hut floors are either oval in shape or sub-circular with a maximum width of 4.5m, and can be as much as 30cm deep. Frequently the structures are concentric, one hut floor formed inside another, with the internal surface covered by just a rather thin layer of sediment (Garcea 2006-2007). The hearths and the rubbish pits are located near the huts, but for the most part outside the area they actually occupy. Broken burned stones were found in the hearths as well as the remains of stone and pottery artefacts,

Figure 2.5 Excavation at site 8-B-10C (photo by R. Ceccacci).

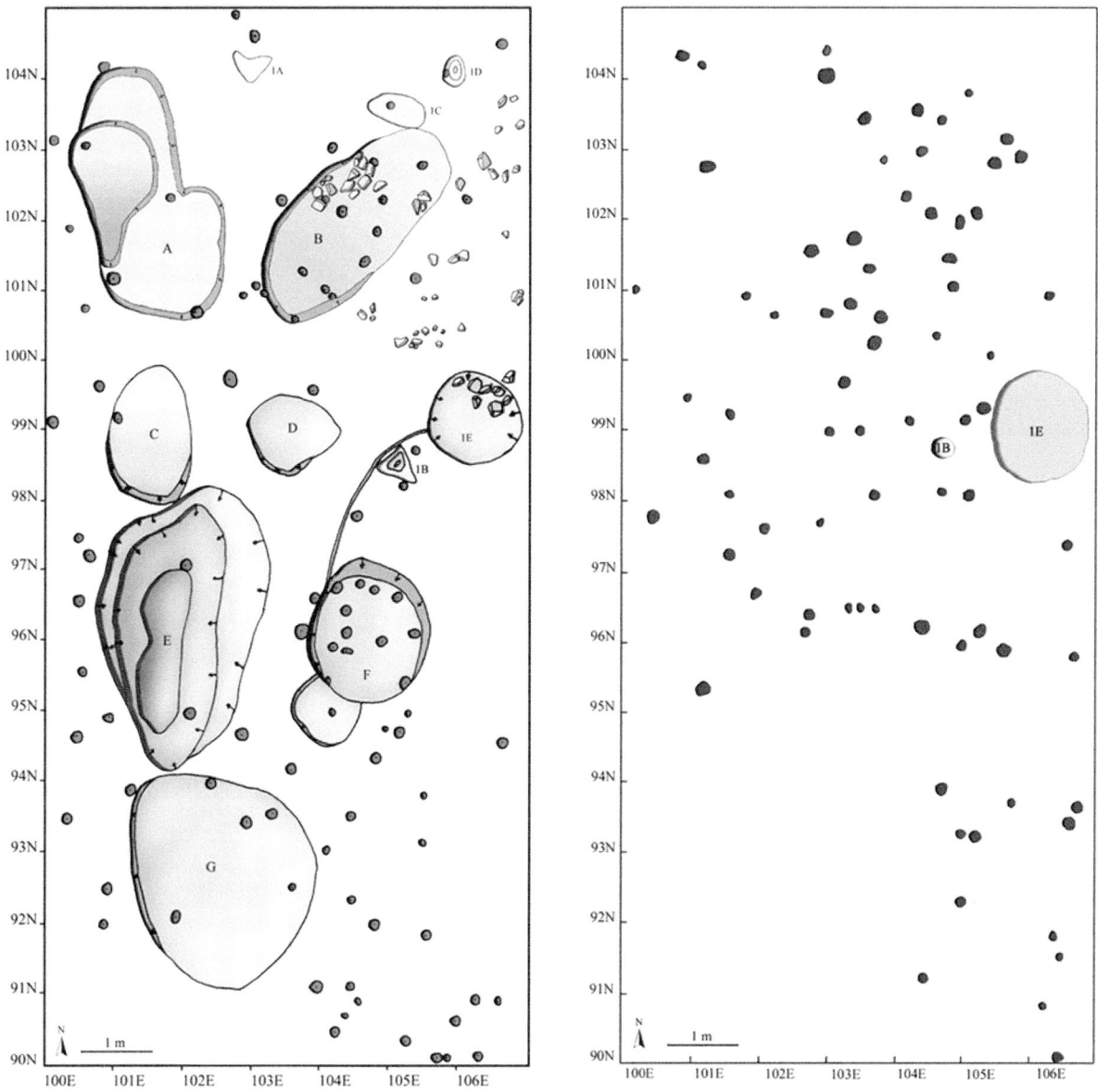

Figure 2.6 Plan of levels 1 (on the left) and 2 (on the right) of site 8-B-10C showing hut floors, features and post holes (modified after Garcea 2011-2012).

showing that they could also function if necessary as places for the disposal of refuse. Some areas of the surface near pit 1E, the largest, show signs of oxidisation, suggesting – in this part of the site – the original presence of smaller fireplaces (Garcea 2006-2007). The many postholes found in this first level seem to be evenly distributed over the whole surface, though tending to be concentrated more outside the area of the hut floors. The few found inside the structures were probably intended for poles used to support hut walls. Others, located outside the huts and forming straight lines, have been interpreted as being part of possible pens and/or windbreaks. It is probable that some of the smaller poles were placed close by larger posts to reinforce them (Garcea 2006-2007). Four radiocarbon dates come from Level 1. The first two are between 4945-5040 cal BC (6080 ± 35 BP) and 4800-4930 cal BC (5980 ± 40 BP) (Garcea 2006-2007). Two new dates of 4855-4975 cal BC (6030 ± 25 BP) and 4840-4930 cal BC (5995 ± 25 BP) are consistent with the first dates (Garcea 2016a; Garcea *et al.* 2016a). This range corresponds with the final stage of development of the Khartoum Variant, following the chronological sequence revised by Gatto (2006a).

The excavation of level 2 – a deposit of dark silt, cemented and pedogenized, rich in calcareous concretions – exposed a second habitation surface. Unlike the surface of level 1, this one has no hut floors, but does have 75 post holes, some of which are laid out in a circular pattern outlining further huts, deriving from a more ancient phase of occupation (Garcea 2011-2012) **(Figure 2.6)**. This level provided two radiocarbon dates, which are much older than level 1: 7550-7580 cal BC (8505 ± 25 BP) and 7205-7450 cal BC (8280 ± 25 BP) (Garcea *et al.* 2016a).

Level 3 consists of a compact sedimentary deposit lying just above the archaeologically sterile pale surface, containing rounded pebbles. There is no sign of possible structures, but the significant number of stone artefacts recovered suggest a phase of repeated occupation, though probably more occasional than in levels 1 and 2. Only 16 pottery fragments have been found in level 3. The stratigraphic sequence ends, after 12cm of especially compacted sandy sediment (level 4), with the white saprolithic level of the surface of the bedrock (level 5) (Garcea and Hildebrand 2009).

Faunal remains from site 8-B-10C consist exclusively of wild species, predominantly aquatic (*Pila cf. wernei and Lanistes carinatus*) and terrestrial gastropods (*Limicolaria sp.*), and few ostrich eggshells (Garcea 2016a; Garcea *et al.* 2016a). The pollen record from the site refers mainly to tropical species, such as *Acacia*, *Barleria* and *Cadaba*, as well as from plants of the families of the *Cyperaceae* and *Poaceae*, possibly transported by the river during the seasonal floodings. Cereal pollen belonging to the *Avena/Triticum* group were also found at the site, although their presence cannot be considered indicative of local occurrence of domesticated plants (Florenzano *et al.* 2016).

Site 8-B-76

Site 8-B-76, originally identified by F. Geus, and attributed by him to the Abkan horizon, is located near the village of Arodin in the south-western part of the island **(Figure 2.2)**. It lies on a sloping transition from a high, Pleistocene, surface to a younger and currently inactive floodplain surface, and borders to the north and northeast with metamorphic rock outcrops (D'Ercole *et al.* 2015).

In 2008, a test excavation of 2 x 1m (TU 1-2) was conducted under the direction of E. A. A. Garcea (Garcea 2016a; Garcea and Hildebrand 2009). The ceramic assemblage analysed in this study comes from surface collection and from the excavation of these first test units and includes exclusively Abkan material. In 2011, a further transect of 27 x 1m was opened along the slop from NE to SW. The transect was excavated in some of its units and revealed horizontal and vertical stratigraphies with surfaces and levels containing Khartoum Variant and Abkan materials (Garcea 2016a). The Khartoum Variant deposit was dated to 6260-6390 cal BC (7460 ± 30 BP), while the Abkan deposits provided eight radiocarbon dates from 5465-5485 cal BC (6500 ± 20 BP) to 3715-3895 cal BC (5005 ± 25 BP) (Garcea 2016a; Garcea *et al.* 2016a) **(Figure 2.7)**.

A fragment of lower molar of cattle was recovered from the Abkan level and other faunal remains from both the Khartoum Variant and Abkan layers refer to gastropods (Garcea *et al.* 2016a). The pollen samples indicate the presence of arboreal plants, including *Acacia*, like at site 8-B-10C. However, pollen of *Artemisia* and *Cornulaca* suggest an occupation phase during a relatively drier phase (Florenzano *et al.* 2016).

Site 8-B-52A

Site 8-B-52A is located in the northern part of the island of Sai, just a few kilometres north of Jebel Adu, near the Pharaonic town and areas occupied by modern villages and farms. It lies about 9m above the present floodplain on a low hill bordered on the west by the former western branch of the Nile, and on the south by a minor tributary – Khor Hamar – which is now dried up (Geus 2004b) **(Figure 2.2)**.

Figure 2.7 Excavation at site 8-B-76 (photo by E. A. A. Garcea).

F. Geus discovered the site in 1996 (Geus 1998). Since then a first surface collection was carried out on an area of 10 x 10m, and later on an additional 27 squares, each measuring 5 x 5m. In the following years, the excavation was continued on only the original 10 x 10m square and on eight of the smaller squares (Geus 2000, 2004b). In 2006, E. Hildebrand took over responsibility for completing publication of research and the spatial documentation of the site (Garcea and Hildebrand 2009; Hildebrand 2006-2007; Hildebrand and Schilling 2016).

The original extent of the Pre-Kerma site is now difficult to determine because its northern and western edges were disturbed by the later Meroitic burial ground (site 8-B-52B) but it may be that it covered a total area of about a hectare (Geus 2004b) **(Figure 2.8)**.

No evidence has emerged from the excavation of structures suitable for habitation, but >100 circular or sub-circular pits were found, of which 72 were excavated by Geus team (Hildebrand and Schilling 2016). It was thought at first that these pits were possibly graves, but they are in fact grain storage pits dug directly into the ground – compacted acidic sediment rich in calcium carbonate – chosen perhaps as especially suitable for storing grain and other foodstuffs precisely because of its natural characteristics.

The pits are of varying size and appearance. There are single and double pits, some found isolated, others grouped together in distinct clusters, and spread out over the area in such a way that access to the different silos was never impeded (Geus 2004b; Hildebrand 2006-2007). Hildebrand and Schilling (2016) identified four groups of pits among those excavated on the site. These groups can be interpreted

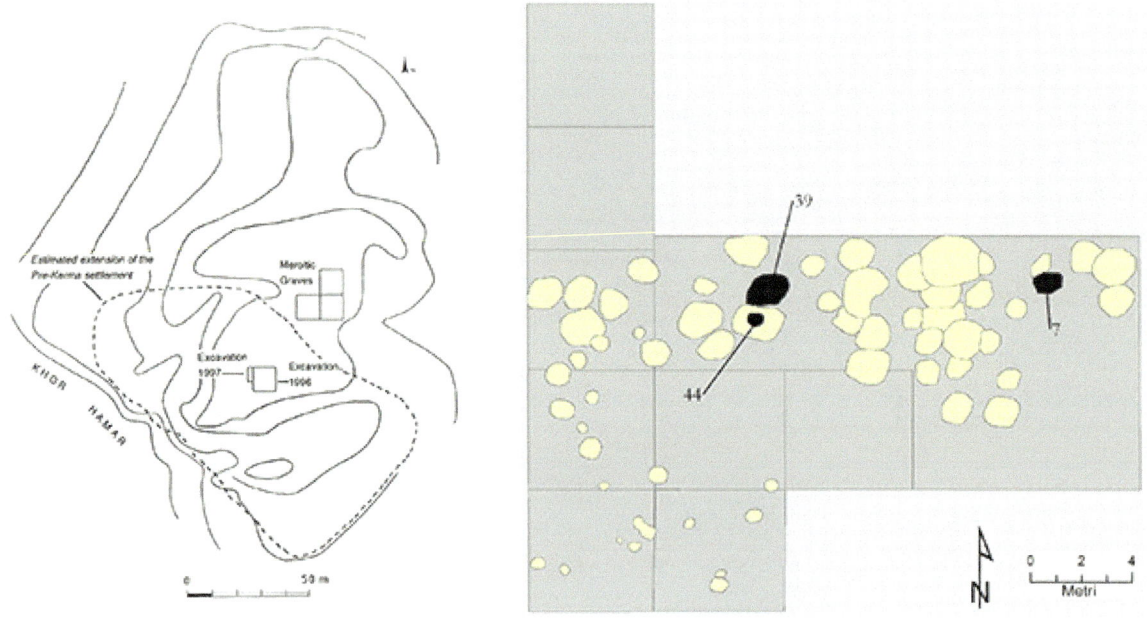

Figure 2.8 Estimated extension of site 8-B-52A (on the left) and schematic map (on the right) showing the three silos from which the ceramic sample analysed come (map modified after Hildebrand and Shilling 2016).

in different ways: either as multiple pits built and used concurrently by households or kin groups, or as groups of pits built in different times, that is 'if that the pit collapsed or become unusable, they repaired it or dug a new one adjacent' (Hildebrand and Schilling 2016: 10). In this latter case, 'households or kin groups may have used only a single pit at any one time' (Hildebrand and Schilling 2016: 10).

Most pits were classified by F. Geus (2000) as 'simple units' as they present a single chamber. These pits are either bell shaped, cylindrical, or basin shaped (Hildebrand and Schilling 2016). Some others ('two level units') (Geus 2000) showed a more complex structure, made up of a first chamber – moderately deep with a wide opening – inside which another smaller chamber has been dug out, deeper and with only a narrow opening into it, sealed sometimes with mortar or with one or two slabs of schist (cf., also Hildebrand and Schilling 2016) **(Figure 2.9)**.

Some of the pits turned out to be completely empty, as though they had been carefully emptied and sealed in response to some external event and then never used again during the site's later occupation (Geus 1998). Most of them however yielded up abundant lithic and ceramic material, as well as animal skin and bones, beads and pieces of ostrich eggshell, and above all, copious macrobotanical remains. These vegetable remains included pieces of wood, both burned and not, leaves and twigs, fruit seeds and different types of dried cereals, among which were grains of domesticated barley (*Hordeum vulgare*) and wheat (*Triticum dicoccum*) but also two different types of wild millet, and local species such as *Citrullus*, *Zizyphus*, and Cucurbitaceae (Geus 1998).

F. Geus (1998) dated the barley to 4142±48 BP (UTC-5294) and 4151±44 BP (UTC-5295) or between 2872 and 2612 BC. Further, five more radiocarbon dates were obtained from chaff and charcoal remains from daub seals of different silos (Hildebrand and Schilling 2016). The new dates indicate that the site's life was about 800 years from c. 3600 BC to c. 2500 BC, which is from the Early Pre-Kerma, contemporary to Predynastic in Egypt, to the Middle and Late Pre-Kerma (cf., Honegger 2004b). This means that the construction of the silos took place not one-off but occurred over a prolonged period. During this time, also the original function of the site might have been changed. It is reasonable, however, that coping

Figure 2.9 Site 8-B-52A: detail of a 'two level units' pit with the slab of schist used to seal the opening of the pit (photo by G. D'Ercole).

with 'seasonal or long-term fluctuations in resource ability' was the initial motive to build the site at its early construction at *c.* 3600 BC, in correspondence with the onset of dry conditions at Sai Island (Hildebrand and Schilling 2016: 11).

The ceramic assemblage analysed in this study comes from silos 7, 39 and 44 **(Figure 2.8)**.

Site 8-B-10A

Site 8-B-10A is located in the south-eastern part of the island, a few kilometres north-east of the Kerma cemetery, on a floodplain at the foot of the escarpment of Jebel Adu **(Figure 2.2 and Figure 2.10)**. This is a multi-component habitation site with Pre-Kerma to Kerma Classic pottery on the surfaces, preliminary radiocarbon dated between 1882-1748 and 1861-1659 BC (Hildebrand and Schilling 2016).

Pre-Kerma ceramics were first discovered at the site in 2006 near a trench dug in 1990. In 2008, the excavation of the site started with a test excavation of 2 x 1m (TU 1 and 2) which showed an intact stratigraphic sequence consisting of successive levels of deposits containing lithic and ceramic artefacts, charcoal and numerous animal remains (Garcea and Hildebrand 2009). Investigation was taken up again by E. Hildebrand in the course of the 2009 season, opening up two new trial excavations, each of 2 x 1m, one to the east (Nord 20/Est 48-49) and one to the west (Nord 20/Est 20-21) of the original test units (now renamed Nord 20/Est 35-36). The excavation was later extended over a total area of 12m², starting from the two western units, and then continued in the squares N20/E20-21 and in the eastern trial trench (squares N20/E48-49). The ceramic assemblage analysed in this study comes from surface collection and from the excavation of the 2009 test units.

Particularly significant data came out of a study of the faunal evidence from the site (Chaix 2011). The fauna consists above all of domesticated species, mainly sheep with occasional goats (*Ovis aries* L. and *Capra hircus* L.) and only to a lesser extent cattle (*Bos taurus* L.). Wild species, in contrast, such as fish and molluscs, are only rarely represented, suggesting that their contribution to the settlement's economy was not very important (Chaix 2011: 212, Tab. 1).

Figure 2.10 Excavation at site 8-B-10A (photo by R. Ceccacci).

3. Ceramic productions on Sai Island: analysis of the macroscopic data

Materials and methods

The ceramic sample is made up of 3887 potsherds, of which 905 come from site 8-B-10C, 39 from the test excavation carried out at site 8-B-76, and 448 from site 8-B-52A (silos 7, 39 and 44). Finally, 2495 sherds were collected in the course of the 2009 season at site 8-B-10A **(Table 3.1)**.

The ceramic data were entered on a relational database realized on a Microsoft Access® platform. The database, designed by E. A. A. Garcea, has been used for previous studies of ceramics from Sai Island (Garcea 2011-2012, 2012; Garcea and Hildebrand 2009; D'Ercole 2015) and for the analysis of pottery coming from other Saharan, Sudanese and Sahelian contexts (Garcea 2005, 2006b, 2008, 2013; Garcea and Caputo 2004). It collects information on the entire sequence of manufacture of pottery or *chaîne opératoire*, covering the phases of preparation, production and finishing, and considering features such as the thickness of the vessel, the texture of the pastes, the frequency of the non-plastic inclusions, their size and type, and finally, the colour and treatment of a vessel's internal and external surfaces. For the question of decoration, the database uses the system of classification developed by Caneva (Caneva 1983, 1987; Caneva and Marks 1990). This system, starting with the technique (e.g., rocker stamp, alternately pivoting stamp, simple impression, incision, rouletting) and the implement used (e.g., comb with evenly/unevenly serrated edge, plain edged implement, double pronged implement, stylus, roulette) distinguishes in succession, in hierarchical order, the element (e.g., dots, dashes, lines), the motif (e.g., straight, curved, packed or spaced zig-zags, single, paired, parallel or dotted wavy lines), and the structure of the decoration (e.g., continuous, banded, wavy banded, panelled, criss-cross, herring-bone, geometrical).

Textures were categorized by the dimensions of their inclusions: 'fine', with small inclusions, that is, <1mm; 'medium', with inclusions >1mm and <2mm; and 'coarse', with inclusions >2mm. For this type of classification, the frequency of inclusions is also significant. Consequently, a paste can be defined as medium if it has a high frequency of inclusions even when those inclusions are smaller than 1mm; and similarly, pastes are considered to be coarse when the inclusions are only medium sized but make up a high proportion of the fabric. When merely the frequency of the inclusions is referred to, inclusions are described as 'frequent' (a high proportion of the total fabric), 'common' (medium), or 'rare' (a low proportion).

The levels of sphericity ('low', 'medium' and 'high') and of angularity ('angular', 'rounded') used to describe the mineral inclusions follows the criteria established in the evaluation charts of Orton *et al.* (1993: 239). The presence of inclusions with a medium or high level of sphericity and a rounded shape is commonly attributed in the literature (Orton *et al.* 1993; Velde and Druc 1999) to the use of secondary clays, gathered from lake or river deposits, in which the mineral component – often affected by erosion

	8-B-10C		8-B-76		8-B-52A		8-B-10A	
	No.	%	No.	%	No.	%	No.	%
Classifiable	889	98.2	39	100.0	441	98.4	1044	41.8
Unclassifiable	16	1.8	0	0.0	7	1.6	1451	58.2
Total	905	100.0	39	100.0	448	100.0	2495	100.0

Table 3.1 Distribution of classifiable and unclassifiable sherds, by site.

and made up of more or less rounded grains – shows the effect of the natural process of transport which the sediment has undergone since leaving the mother rock. If rounded grains of sand are found in a single sherd together with angular, irregularly shaped inclusions, it is possible that some mineral tempers were added to the paste by the potter during preparation, and that it is for this reason that it shows different characteristics from a paste of entirely natural origin (Orton *et al.* 1993: 115).

The inclusions are then also classified by type, whether they are mineral or organic. If not otherwise specified, a paste contains only mineral inclusions. When the inclusions are organic, a distinction is made between 'flat' and 'tubular' fibres; flat fibres indicate the use of grass leaves and culm as tempering material, whereas tubular fibres suggest the preparation of intentionally chopped straw or chaff, and possibly also herbivore dung (cf., Livingstone Smith 2001). In those samples where both types of organic inclusions are found, the more frequent type is the one indicated.

Surfaces can be burnished internally and/or externally, or else polished. Burnishing is made by rubbing the leather-hard vessels with a hard tool, such as a smooth pebble; this treatment enhances the colour of the fabric and produces glossy surfaces with irregular lustre. It is different from polishing, which produces a more significant and uniform level of lustre and very glossy surfaces (Orton *et al.* 1993; Sinopoli 1991). The colour of the surfaces is defined using Munsell's charts. To simplify the compilation of the database and subsequent processing of the data, the entry 'red' refers to the following tonalities: 2.5 YR 3/4, 4/1, 4/2, 4/3, 4/4, 4/6, 5/3, 5/4, 5/6, and 6/6; the entry 'brown' indicates these tonalities: 5 YR 4/2, 4/3, 4/4, 5/1, 5/2, 5/3, 5/4, 6/4, 7/4 and 7.5 YR 4/1, 5/3, 5/4, 6/3, 6/4, 6/6, 7/4, 7/6, 7/8.

In order to classify the different pot shapes, after an initial sorting of the sherds into walls, rims and bottoms, an attempt is made, where possible, to identify the different types of pot. These include: open vessels with straight or slightly convex walls, globular bowls with more curved sides, usually hemispherical or saucer-shaped; vessels with narrower openings, ranging from the cylindrical, with straight walls, to the more globular, usually referring to closed forms, slightly ovoid in profile, such as jars.

A systematic description of technological and stylistic aspects of the ceramics, examined through the different headings used in the database, is provided below. A more thorough discussion and a comparison between the assemblages of the different sites will be put off until Chapter 5.

Site 8-B-10C

State of preservation

The assemblage from site 8-B-10C is made up of 905 pottery sherds, of which 101 come from the surface and 746 from level 1. Levels 2 and 3 have provided only a limited number of artefacts, 27 and 16 samples respectively **(Table 3.2)**. The materials coming from the hearth 1A (a single piece) and from the rubbish pit 1E (14 sherds) have been classified independently, discussing these two features as distinct stratigraphic units **(Table 3.2)**. The pottery from the surface and level 1 has been studied previously by Garcea (Garcea 2011-2012, 2012; Garcea and Hildebrand 2009). These results are presented and discussed here together with the data from levels 2 and 3 in order to analyse the pottery production of the entire sequence.

Out of a total of 905 sherds, 98.2% were classified **(Table 3.2)**. The category of unclassifiable sherds refers to fragments with badly preserved surfaces, or which are very small (less than 1cm), that is, the part of the sample that shows the damage produced by environmental or taphonomic processes or by human agency. On the whole, the state of preservation of the ceramic material is good, even though the site has been subject to cyclic flooding from the yearly high waters of the Nile, which

8-B-10C	S		L1		L2		L3		1A		1E		Total	
	No.	%	No.	%	No.	%	No.	%	No.	%	No.	%	No.	%
Classifiable	100	11.0	731	80.8	27	3.0	16	1.8	1	0.1	14	1.5	889	98.2
Unclassifiable	1	0.1	15	1.7	0	0.0	0	0.0	0	0.0	0	0.0	16	1.8
Total	101	11.2	746	82.4	27	3.0	16	1.8	1	0.1	14	1.5	905	100.0

Table 3.2 Stratigraphic distribution of classifiable and unclassifiable sherds from site 8-B-10C.

brought about repeated deposition of silt in a moist, and possibly stagnant, low energy environmental system of a type associated with floodplains (see Chapter 2). The relatively good state of preservation of most of the ceramics from the site (leaving aside the significantly high level of breakage) can be attributed to technological choices made during production: the presence of mineral tempers in the pastes, contributed to producing a 'hard' ceramic, characterised by low porosity and a 'rough' to 'harsh' texture (Orton *et al.* 1993: 70). These products would probably be more resistant to damage resulting from environmental or taphonomic action than some more recent pottery with – as we will see below – its 'soft' consistency (cf., also Peacock 1977: 30) and the high porosity associated with pastes that are rich in organic inclusions.

Preparation: clay processing and addition of non-plastic inclusions

The ceramics of site 8-B-10C vary in colour between red and light brown; dark grey to black cores are rare and present only in the few samples. Irregularly coloured zones, occasionally visible on the side margins of the sherds, may be related to taphonomic conditions and result from post-depositional causes (D'Ercole *et al.* 2015).

Textures are gritty because of the high content of mineral inclusions in the pastes. Over 80% of the samples have a paste made with inclusions that are exclusively mineral, while only a reduced number of sherds contain also organic matter. 124 pieces, or 13.9% of the total, contain flat organic fibres, and only one sherd has been found with tubular fibres **(Table 3.3)**.

Mineral inclusions were classified as angular in 85.8% of cases (763 fragments) while rounded inclusions are found in only 13.4% of the sample (119 fragments) **(Figure 3.1) (Plate 3.23: 1)**. Ceramics with common, angular inclusions which have a low or medium degree of sphericity (33.4% and 19.2% respectively) predominate, together with a coarser product also characterised by angular inclusions, and present in relatively large quantities. A small proportion of pottery, exclusively from the surface and level 1, shows a finer texture, with rare inclusions, which are either angular with a low degree of sphericity or rounded with a different degree of sphericity **(Table 3.4)**.

8-B-10C	S		L1		L2		L3		1A		1E		Total	
	No.	%	No.	%	No.	%	No.	%	No.	%	No.	%	No.	%
Organic tubular	0	0.0	0	0.0	0	0.0	0	0.0	0	0.0	1	0.1	1	0.1
Organic flat	14	1.6	107	12.0	0	0.0	2	0.2	0	0.0	1	0.1	124	13.9
Only mineral	86	9.7	624	70.2	27	3.0	14	1.6	1	0.1	12	1.3	764	85.9
Total	100	11.2	731	82.2	27	3.0	16	1.8	1	0.1	14	1.6	889	100.0

Table 3.3 Stratigraphic distribution of types of inclusions from site 8-B-10C.

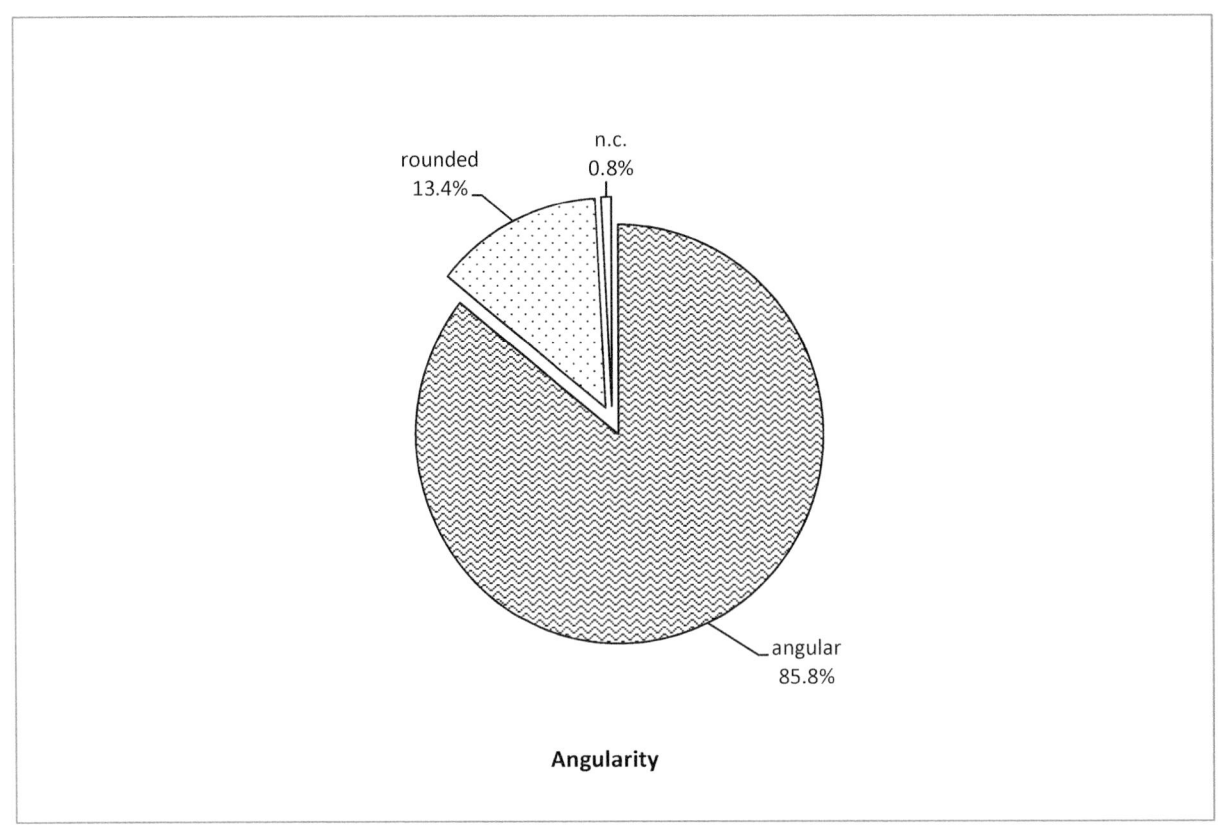

Figure 3.1 Angularity of mineral inclusions from site 8-B-10C.

8-B-10C			S		L1		L2		L3		1A		1E		Total	
Inclusions	Sphericity	Angularity	No.	%	No.	%	No.	%	No.	%	No.	%	No.	%	No.	%
N.c.	N.c.	N.c.			4	0.4			2	0.2					6	0.7
N.c.	Low	Angular			1	0.1									1	0.1
Common	N.c.	N.c.			1	0.1									1	0.1
Common	N.c.	Angular			1	0.1									1	0.1
Common	High	Rounded	11	1.2	18	2.0									29	3.3
Common	Low	Angular	17	1.9	272	30.6							8	0.9	297	33.4
Common	Low	Rounded			1	0.1									1	0.1
Common	Medium	Angular	7	0.8	145	16.3	11	1.2	5	0.6			3	0.3	171	19.2
Common	Medium	Rounded	14	1.6	50	5.6	1	0.1			1	0.1	1	0.1	67	7.5
Frequent	N.c.	Angular			1	0.1									1	0.1
Frequent	High	Rounded											1	0.1	1	0.1
Frequent	Low	Angular	27	3.0	152	17.1									179	20.1
Frequent	Low	Angular	11	1.2	6	0.7	15	1.7	8	0.9					40	4.5
Frequent	Medium	Rounded	6	0.7	3	0.3			1	0.1			1	0.1	11	1.2
Rare	High	Rounded	1	0.1											1	0.1
Rare	Low	Angular	5	0.6	68	7.6									73	8.2
Rare	Medium	Rounded	1	0.1	8	0.9									9	1.0
Total			100	11.2	731	82.2	27	3.0	16	1.8	1	0.1	14	1.6	889	100.0

Table 3.4 Frequency, sphericity and angularity of inclusions from site 8-B-10C.

ANALYSIS OF THE MACROSCOPIC DATA

8-B-10C	S		L1		L2		L3		1A		1E		Total	
	No.	%	No.	%	No.	%	No.	%	No.	%	No.	%	No.	%
Fine	19	2.1	185	20.8	9	1.0	5	0.6	1	0.1	7	0.8	226	25.4
Medium	57	6.4	471	53.0	15	1.7	7	0.8	0	0.0	4	0.4	554	62.3
Coarse	24	2.7	70	7.9	3	0.3	3	0.3	0	0.0	3	0.3	103	11.6
N.c.	0	0.0	5	0.6	0	0.0	1	0.1	0	0.0	0	0.0	6	0.7
Total	100	11.2	731	82.2	27	3.0	16	1.8	1	0.1	14	1.6	889	100.0

Table 3.5 Stratigraphic distribution of types of textures from site 8-B-10C.

Overall, the frequency data, combined with observations on the dimensions of the inclusions, show a dominating presence of ceramics made of textures of the medium type (62.3%), and a lower proportion of fragments with fine (25.4%) or coarse texture (11.6%) **(Table 3.5)**. If the samples from the surface and from level 1 are compared, a level of variation in the percentages of fine and coarse ceramics can be observed: on the surface 2.7% of the fragments are of coarse and 2.1% of fine texture, but level 1 contains a higher proportion of fine (20.8%) and coarse pastes (7.9%) **(Table 3.5)**. On levels 2 and 3 the medium pastes dominate statistically, followed by fine and coarse types **(Table 3.5)**.

When the data on the different types of texture and on the levels of angularity of the inclusions are combined, there is a distinct association of fine ceramics with rounded inclusions. As can be seen from the graph below **(Figure 3.2)**, fine texture in the pottery correlates in 46.2% of cases with the presence of rounded inclusions. This value is halved for angular inclusions, which are found in only 22.3% of fine pastes, but in 65.1% of artefacts with medium texture.

A similar conclusion can be drawn from the following graph **(Figure 3.3)**, where the data on the type of texture is plotted against the data on the type of inclusions: more than 60% of the sherds made from pastes containing organic matter, whether flat or tubular, have a fine texture. In conclusion, fine

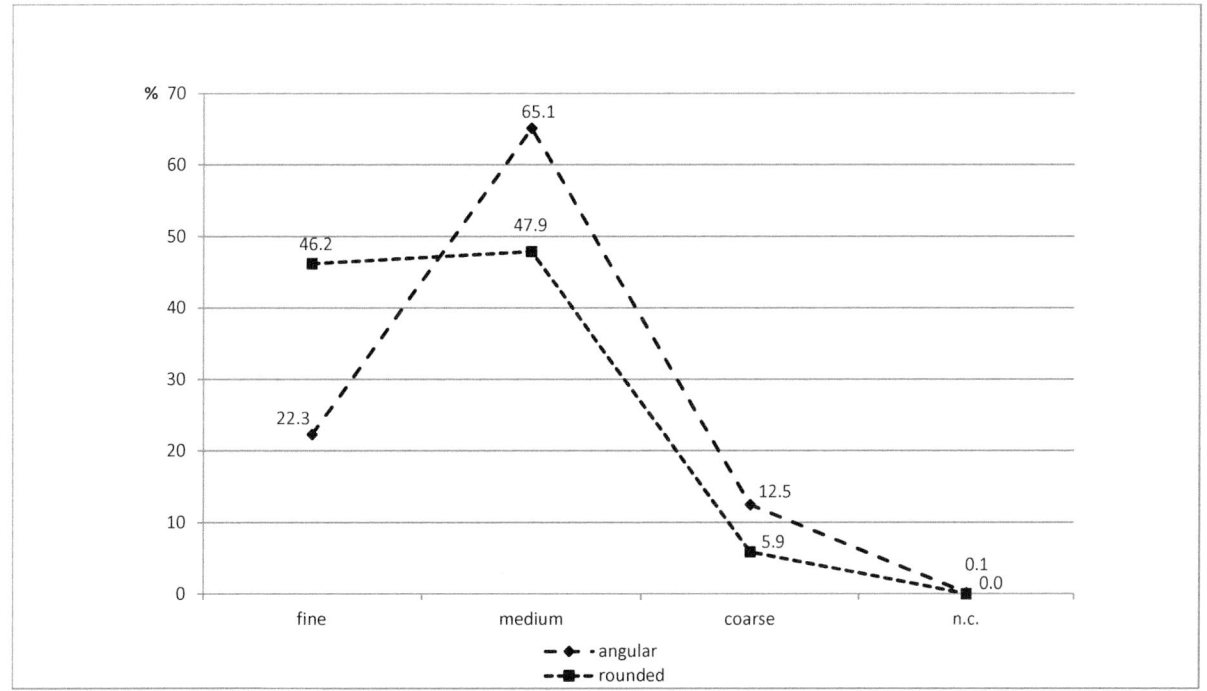

Figure 3.2 Distribution of angular and rounded inclusions in relation to the different types of textures from site 8-B-10C.

Ceramic manufacturing techniques and cultural traditions in Nubia

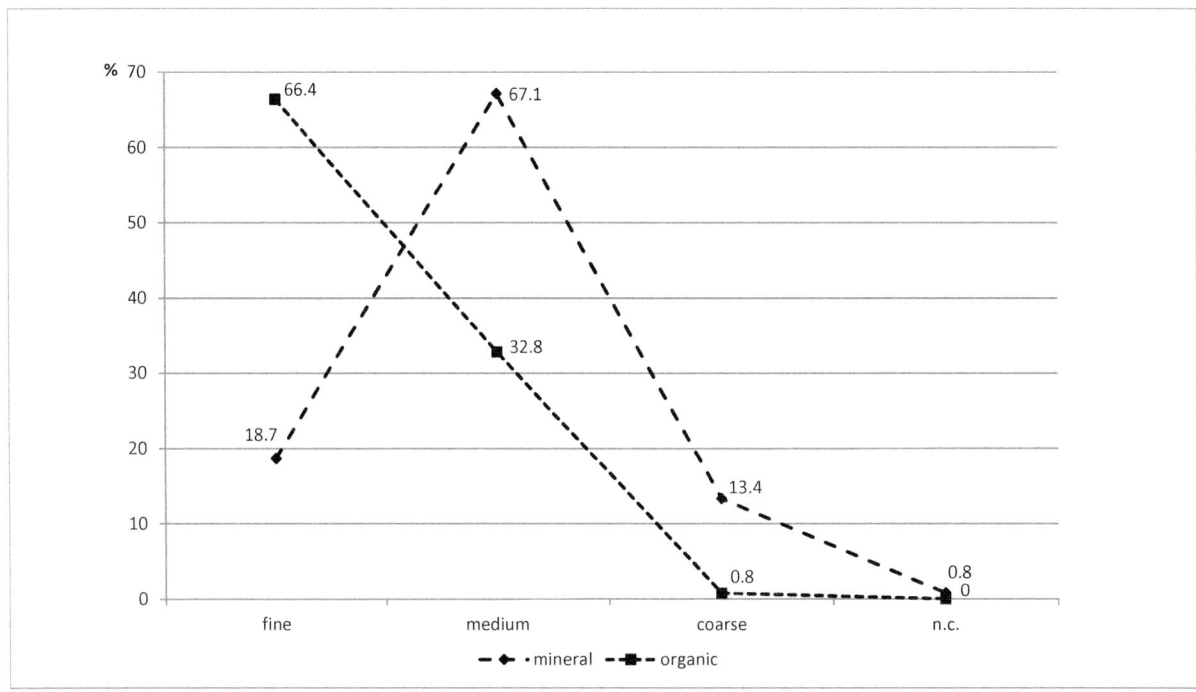

Figure 3.3 Distribution of mineral and organic inclusions in relation to the different types of textures from site 8-B-10C.

pottery seems to be associated with a particular type of paste, characterised by mineral inclusions which are more often rounded and by the presence of organic matter; whereas the pottery with a medium texture, the most common in all stratigraphic levels, and the coarse ware are made of pastes dominated by mineral inclusions with an irregular, angular morphology.

Production: shaping

At site 8-B-10C, only 4.4% of the fragments (39 fr.) are rims and 5.8% (52 fr.) are bottoms. The rest of the samples are walls **(Table 3.6)**.

The rimsherds are usually too small to allow a calculation of the diameter of the vessel, but their level of curvature suggests both large open bowls and jars. The average wall thickness is between 6 and 10mm. Some pieces, usually bottoms, are thicker (from 11 to 14mm), and one or two pieces stand out by being more than 14mm thick (between 16 and 26mm) **(Figure 3.4)**.

Production: decoration techniques, implements, elements, motifs and structure of the decoration

Only 24% of these ceramics are decorated (213 fr.) **(Plates 3.12 and 3.13)**. The highest proportion of decorated sherds (21.1%) is found in level 1 **(Table 3.7)**.

8-B-10C	S		L1		L2		L3		1A		1E		Total	
	No.	%	No.	%	No.	%	No.	%	No.	%	No.	%	No.	%
Walls	87	9.8	653	73.5	27	3.0	16	1.8	1	0.1	14	1.6	798	89.8
Bottoms	7	0.8	45	5.1	0	0.0	0	0.0	0	0.0	0	0.0	52	5.8
Rims	6	0.7	33	3.7	0	0.0	0	0.0	0	0.0	0	0.0	39	4.4
Total	100	11.2	731	82.2	27	3.0	16	1.8	1	0.1	14	1.6	889	100.0

Table 3.6 Stratigraphic distribution of body parts of the vessel from site 8-B-10C.

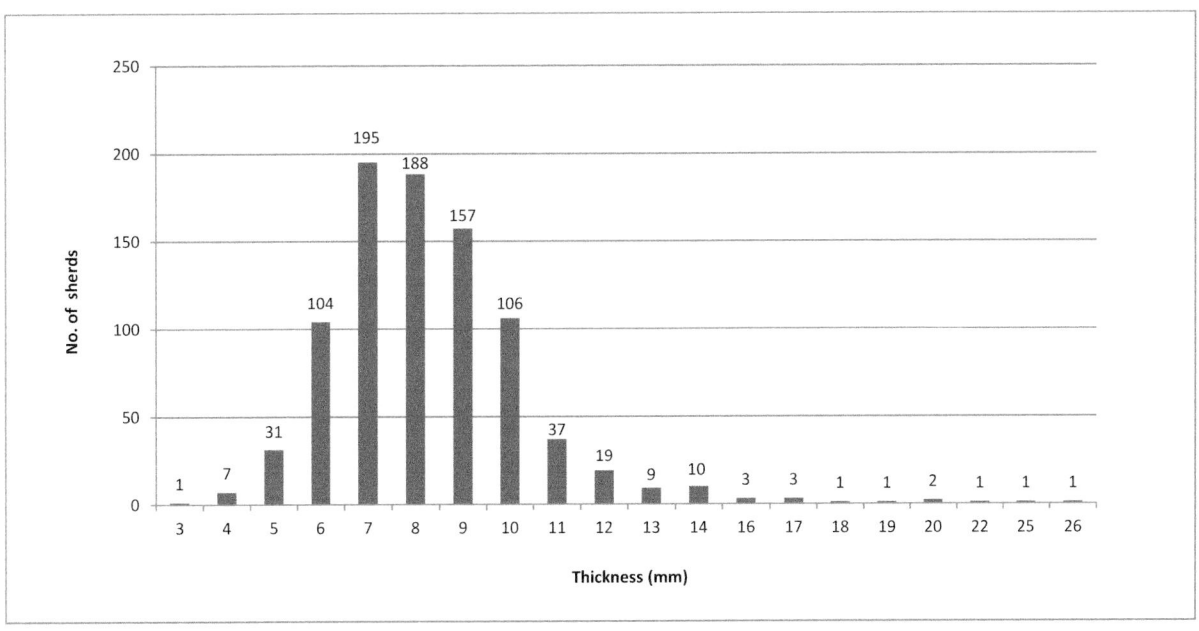

Figure 3.4 Sherd thicknesses from site 8-B-10C.

8-B-10C	S		L1		L2		L3		1A		1E		Total	
	No.	%	No.	%	No.	%	No.	%	No.	%	No.	%	No.	%
Decorated	15	1.7	188	21.1	2	0.2	2	0.2	1	0.1	5	0.6	213	24.0
Undecorated	85	9.6	543	61.1	25	2.8	14	1.6	0	0.0	9	1.0	676	76.0
Total	100	11.2	731	82.2	27	3.0	16	1.8	1	0.1	14	1.6	889	100.0

Table 3.7 Stratigraphic distribution of decorated and undecorated sherds from site 8-B-10C.

Rim decorations, when they occur, often consist of simple impressions, of varying depth, either milled or notched, similar to the decorations described by Gatto (2002a: 70-71) at Nabta Playa in the Nabta/Al Jerar cultural complexes **(Plate 3.12: 1)**.

The rocker stamp is the most frequently used technique on the body of the vessel: 184 sherds (86.4%) show decorations made with variants of this technique **(Table 3.8)**. Pottery decorated using the alternately pivoting stamp technique or by incision is not common; altogether these two methods occur in less than 1% of the samples. Finally, 7%, or 15 fragments are decorated with simple impressions **(Table 3.8)**.

8-B-10C	S		L1		L2		L3		1A		1E		Total	
	No.	%	No.	%	No.	%	No.	%	No.	%	No.	%	No.	%
APS	0	0.0	0	0.0	1	0.5	0	0.0	0	0.0	1	0.5	2	0.9
Incision	0	0.0	1	0.5	0	0.0	0	0.0	0	0.0	0	0.0	1	0.5
Rocker stamp	13	6.1	166	77.9	1	0.5	0	0.0	0	0.0	4	1.9	184	86.4
Simple impression	0	0.0	14	6.6	0	0.0	0	0.0	1	0.5	0	0.0	15	7.0
N.c	2	0.9	7	3.3	0	0.0	2	0.9	0	0.0	0	0.0	11	5.2
Total	15	7.0	188	88.3	2	0.9	2	0.9	1	0.5	5	2.3	213	100.0

Table 3.8 Stratigraphic distribution of decorative techniques from site 8-B-10C.

8-B-10C	S		L1		L2		1E		Total	
	No.	%	No.	%	No.	%	No.	%	No.	%
Evenly serrated edge	13	7.1	153	83.2	1	0.5	4	2.2	171	92.9
Plain edge	0	0.0	1	0.5	0	0.0	0	0.0	1	0.5
Roulette	0	0.0	9	4.9	0	0.0	0	0.0	9	4.9
n.c.	0	0.0	3	1.6	0	0.0	0	0.0	3	1.6
Total	13	7.1	166	90.2	1	0.5	4	2.2	184	100.0

Table 3.9 Stratigraphic distribution of tools used for rocker stamping from site 8-B-10C.

The rocker stamp technique is usually (in 92.9% of cases) applied using a comb with an evenly serrated edge **(Plate 3.12: 2)**. In only one case has an implement with a plain edge been used, and there are no examples of use of a comb with an unevenly serrated edge **(Table 3.9)**. Table 3.9 shows that in nine cases the rocker stamp motifs seem to have been produced not by using a comb but with an implement known in the literature as roulette (Soper 1985).

This cylindrical or almost cylindrical instrument, consisting of a rigid or a flexible core around which one or more pieces of cord have been wrapped – a 'cord-wrapped stick' or a 'cord-wrapped cord' (MacDonald and Manning 2010; Soper 1985) – corresponds to the 'peigne fileté rigide' and the 'peigne fileté souple', respectively, described in Camps-Fabrer (1966) and Camps (1969). An original classification of the roulette as a type of comb however is incorrect. Even though it is used to produce zigzag patterns using the rocker stamp technique (and not the technique known as rouletting), similar to designs that could be made with a comb, the roulette is a different tool. In this variant of rocker stamp technique, the zigzags appear packed very close to each other, and are made up of dashes rather than dots, in which at times the structure of the organic fibres of the tool can be seen. By employing cords of different thickness, and wrapping them in different ways around the core of the implement, zigzag patterns can be produced using dashes of varying sizes and depths, which can be straight or curved depending on whether the roulette is rigid or flexible **(Plate 3.13: 1-6)**.

Packed zigzags of dots or dashes are the most common pattern (61% of the samples). Spaced zigzags are found on only a few sherds **(Table 3.10)**.

11.3% of the potsherds (24 pieces) are decorated with dotted wavy lines **(Table 3.10)**. These wave patterns can be short or long, but more commonly short, run parallel to the rim, and often two lines are placed symmetrically, forming semicircles that mirror one another **(Plate 3.12: 4-7)**. In a few cases this decoration appears in a variant made up of arch-shaped motifs **(Plate 3.12: 3)**. Implements with a plain edge, making simple impressions, produce designs consisting of parallel lines (1.9%) or, more often, straight zigzags (3.8%) **(Table 3.10)**. Paired lines made up of dots are produced by a double pronged implement, using either the alternately pivoting stamp technique or by simple impression **(Table 3.10)**.

The combs used to make dotted wavy lines must have been quite small, from 7 to 11mm long, with an average of 9.4mm. Packed zigzags were made with a wider instrument (from 7 to 25mm, on average 12.9mm long) **(Figure 3.5)**. Parallel lines were a similar distance apart, the tool used being between 10 and 20mm across, on average 13.3mm. Paired lines and straight zigzags were produced by slightly smaller instruments (from 4 to 9mm and 5 to 9mm respectively, with averages of 6.5mm and 6.8mm) **(Figure 3.5)**.

A similar variety can be found in the number of teeth in a comb. From the evidence provided by dotted wavy lines and by packed zigzags, it seems that the tools used in the first case had only two or three

8-B-10C		S		L1		L2		L3		1A		1E		Total	
Tools	Motifs	No.	%	No.	%	No.	%	No.	%	No.	%	No.	%	No.	%
Double pronged	Paired lines					1	0.5					1	0.5	2	0.9
Stylus	Single lines			1	0.5									1	0.5
N.c.	Packed zig-zags			1	0.5									1	0.5
Evenly serrated edge	N.c.	1	0.5	11	5.2									12	5.6
Evenly serrated edge	Dotted wavy lines	1	0.5	20	9.4	1	0.5					2	0.9	24	11.3
Evenly serrated edge	Packed zig-zags	11	5.2	117	54.9							2	0.9	130	61.0
Evenly serrated edge	Straight zig-zags			1	0.5									1	0.5
Evenly serrated edge	Zig-zags with spaced dashes			1	0.5									1	0.5
Evenly serrated edge	Zig-zags with spaced dots			3	1.4									3	1.4
Plain edge	Straight zig-zags (rocker)			1	0.5									1	0.5
Roulette	Packed zig-zags			9	4.2									9	4.2
Double pronged	Dotted wavy lines			1	0.5									1	0.5
Double pronged	Paired lines			1	0.5									1	0.5
Plain edge	Parallel lines (simple impression)			4	1.9									4	1.9
Plain edge	Straight zig-zags (simple impression)			8	3.8									8	3.8
N.c.	N.c.	2	0.9	9	4.2			2	0.9	1	0.5			14	6.6
Total		15	7.0	188	88.3	2	0.9	2	0.9	1	0.5	5	2.3	213	100.0

Table 3.10 Stratigraphic distribution of tools and decorative motifs from site 8-B-10C.

teeth, and that for the packed zigzags the combs were larger – as was noted above – with between three and 12 teeth **(Figure 3.6)**.

The structure of the decoration, when it can be determined, is usually composed of bands: simple (17.8%), straight bands (3.3%), and wavy bands (2.3%); while 16.9% of the decorative motifs show a continuous structure **(Table. 3.11)**. Only rarely do the ceramics from this site display a more complex structure: criss-cross (0.5%), or panels (0.9%) **(Table 3.11)**.

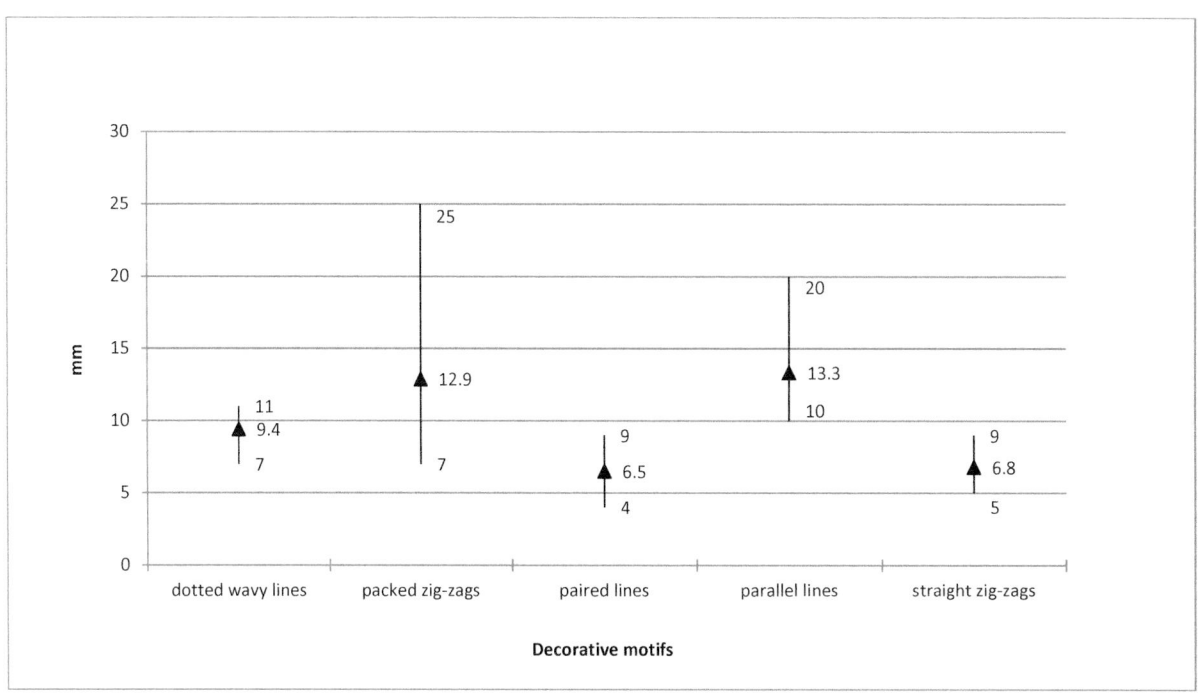

Figure 3.5 Ranges of the lengths of decorative motifs from site 8-B-10C (minimum, maximum and average).

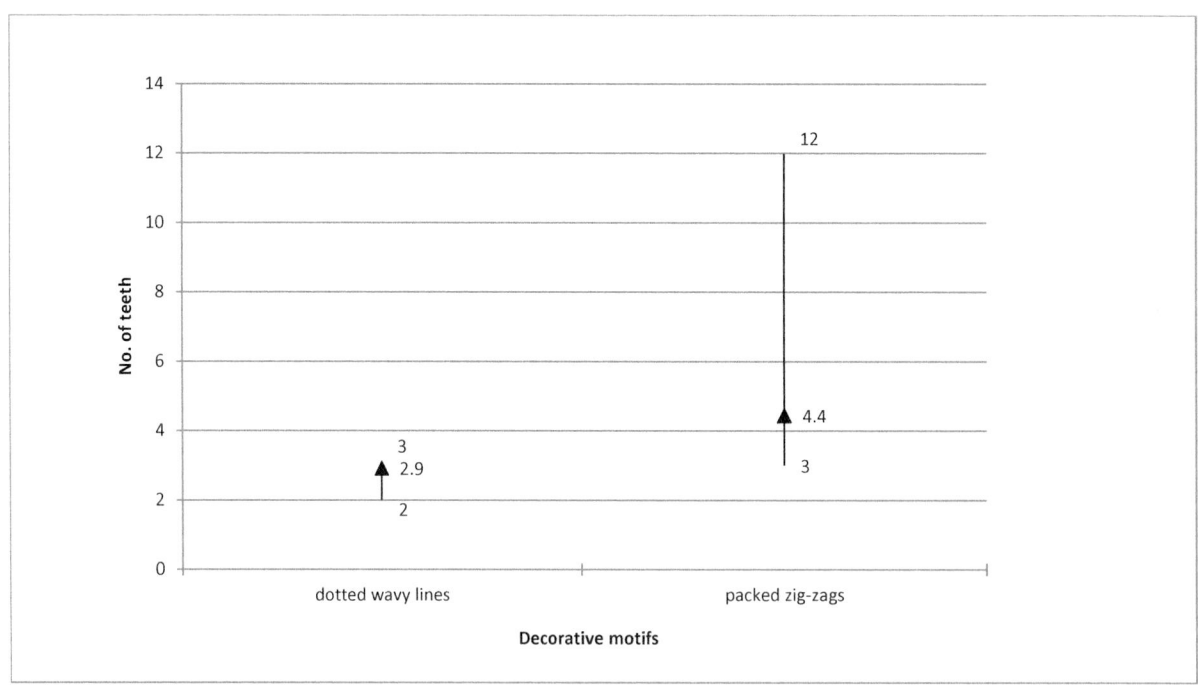

Figure 3.6 Ranges of the teeth numbers of combs with evenly serrated edges from site 8-B-10C (minimum, maximum and average).

Production: surface treatment

Most (97%) of the analysed potsherds from the site had plain or smoothed surfaces; however a small percentage of ceramics were also burnished or polished. Burnishing, if present, is limited in most cases to the inside of the pots (1.7%) **(Figure 3.7)**. It is always associated with brown or black surfaces (internal and/or external) **(Table 3.12)**.

ANALYSIS OF THE MACROSCOPIC DATA

8-B-10C	S		L1		L2		L3		1A		1E		Total	
Structure	No.	%	No.	%	No.	%	No.	%	No.	%	No.	%	No.	%
Banded	1	0.5	36	16.9	0	0.0	0	0.0	0	0.0	1	0.5	38	17.8
Straight bands	1	0.5	6	2.8	0	0.0	0	0.0	0	0.0	0	0.0	7	3.3
Wavy bands	1	0.5	3	1.4	0	0.0	0	0.0	0	0.0	1	0.5	5	2.3
Continuous	0	0.0	32	15.0	2	0.9	0	0.0	0	0.0	2	0.9	36	16.9
Criss-cross	0	0.0	1	0.5	0	0.0	0	0.0	0	0.0	0	0.0	1	0.5
Panelled	0	0.0	2	0.9	0	0.0	0	0.0	0	0.0	0	0.0	2	0.9
N.c.	12	5.6	108	50.7	0	0.0	2	0.9	1	0.5	1	0.5	124	58.2
Total	15	7.0	188	88.3	2	0.9	2	0.9	1	0.5	5	2.3	213	100.0

Table 3.11 Stratigraphic distribution of decorative structures from site 8-B-10C.

8-B-10C		S		L1		L2		L3		1A		1E		Total	
Burnishing in	Burnishing out	No.	%	No.	%	No.	%	No.	%	No.	%	No.	%	No.	%
N.p.	N.p.	100	11.2	724	81.4	18	2.0	5	0.6	1	0.1	14	1.6	862	97.0
N.p.	Red	0	0.0	4	0.4	0	0.0	0	0.0	0	0.0	0	0.0	4	0.4
Black	N.p.	0	0.0	3	0.3	8	0.9	2	0.2	0	0.0	0	0.0	13	1.5
Black	Black	0	0.0	0	0.0	0	0.0	4	0.4	0	0.0	0	0.0	4	0.4
Brown	N.p.	0	0.0	0	0.0	1	0.1	1	0.1	0	0.0	0	0.0	2	0.2
Brown	Black	0	0.0	0	0.0	0	0.0	1	0.1	0	0.0	0	0.0	1	0.1
Brown	Brown	0	0.0	0	0.0	0	0.0	3	0.3	0	0.0	0	0.0	3	0.3
Total		100	11.2	731	82.2	27	3.0	16	1.8	1	0.1	14	1.6	889	100.0

Table 3.12 Stratigraphic distribution of inside and outside burnishing from site 8-B-10C.

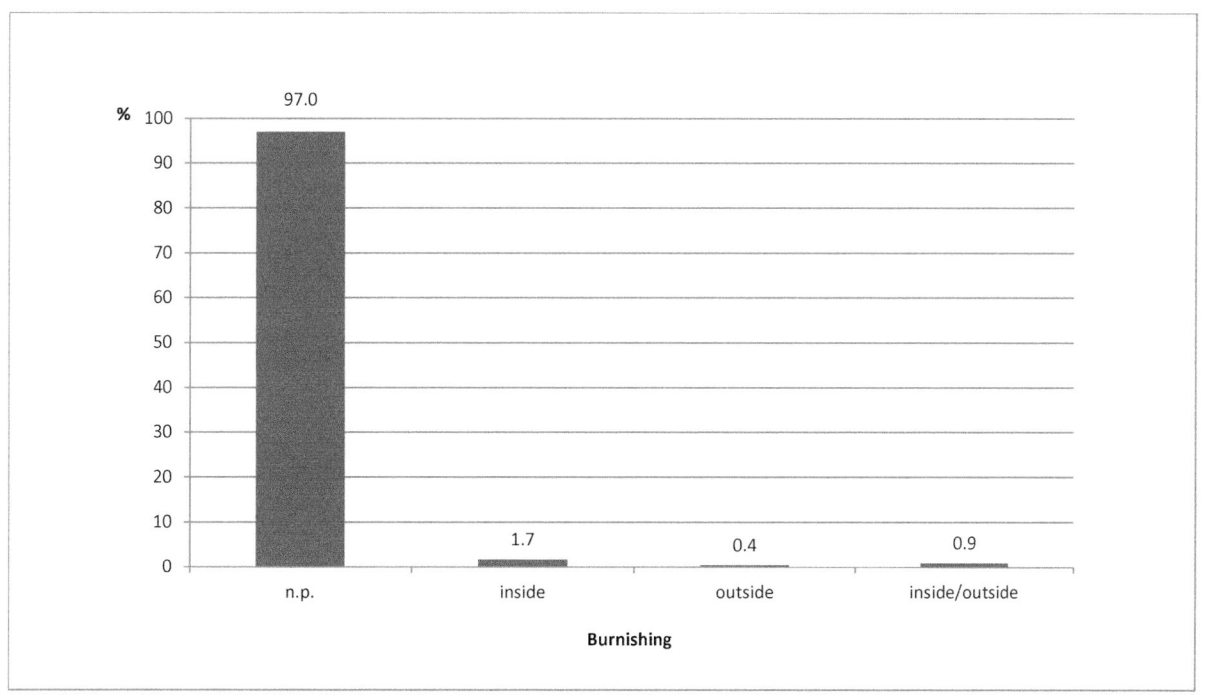

Figure 3.7 Percentages of burnishing from site 8-B-10C.

8-B-10C	S		L1		L2		L3		1A		1E		Total	
	No.	%	No.	%	No.	%	No.	%	No.	%	No.	%	No.	%
N.p.	89	10.0	585	65.8	27	3.0	16	1.8	1	0.1	10	1.1	728	81.9
Inside	9	1.0	101	11.4	0	0.0	0	0.0	0	0.0	2	0.2	112	12.6
Outside	0	0.0	18	2.0	0	0.0	0	0.0	0	0.0	2	0.2	20	2.2
Inside/outside	2	0.2	27	3.0	0	0.0	0	0.0	0	0.0	0	0.0	29	3.3
Total	100	11.2	731	82.2	27	3.0	16	1.8	1	0.1	14	1.6	889	100.0

Table 3.13 Stratigraphic distribution of polishing from site 8-B-10C.

In the same way, polishing, when observed, was found mostly on inner surfaces (12.6%) **(Table 3.13)**.

This fact might suggest that polishing was chosen for functional rather than aesthetic reasons. No relationship is found between this treatment and the presence of decoration either, since polishing, whether internal or external, or on both surfaces, is for the most part associated with undecorated vessels (50.9%) **(Table 3.14)**. Similarly, polishing does not seem to be characteristic of the site's fine pottery (if present, polishing is found more often on ceramics with medium pastes: 10.9%) **(Table 3.15)**. On some potsherds, the surface seems to be covered with a thin layer of clay – a self-slip – identical in composition with the clay of the rest of the pot.

Spatial distribution

For every level of the site, the spatial distribution of artefacts has been recorded on maps of the site, and shown in relation to its various structures, as they occur on the different levels. As can be seen from the maps below, produced using Surfer 9.0 software, ceramic artefacts have been found mainly in the east of the site. This is particularly the case with level 1, where the greatest concentration of sherds (orange and red symbols: respectively from 11 to 20 and 21 to 45 sherds per square) is found outside the area occupied by the huts, in squares 90N/105-106E and 104N/105-106E **(Figure 3.8)**. In level 2, the artefacts are again concentrated mainly in the east of the site, though in this case the number of samples is very much lower (blue symbols: 1 to 5 sherds per square) **(Figure 3.9)**. On the surface **(Figure 3.8)** and in level 3 **(Figure 3.9)** the artefacts are spread over the north-eastern part of the area of excavation: blue symbols (1 to 5 sherds per square), green (6 to 10 sherds per square), and orange (11 to 20 sherds per square). In addition, **Figure 3.10** shows the spatial distribution in level 1 of the different types of pastes employed. From this further analysis it emerges that the coarse ware (blue symbols) is found for the most part away from the huts, while the fine or medium wares (red and green symbols, respectively) are found both inside and outside domestic structures.

8-B-10C	Decorated		Undecorated		Total	
	No.	%	No.	%	No.	%
Inside	30	18.6	82	50.9	112	69.6
Outside	4	2.5	16	9.9	20	12.4
Inside/outside	5	3.1	24	14.9	29	18.0
Total	39	24.2	122	75.8	161	100.0

Table 3.14 Distribution of polishing in relation to decorated and undecorated sherds from site 8-B-10C.

8-B-10C	Present		N.p.		Total	
	No.	%	No.	%	No.	%
Fine	46	5.2	180	20.2	226	25.4
Medium	97	10.9	457	51.4	554	62.3
Coarse	18	2.0	85	9.6	103	11.6
N.c.	0	0.0	6	0.7	6	0.7
Total	161	18.1	728	81.9	889	100.0

Table 3.15 Distribution of textures in relation to the presence /absence of polishing from site 8-B-10C.

ANALYSIS OF THE MACROSCOPIC DATA

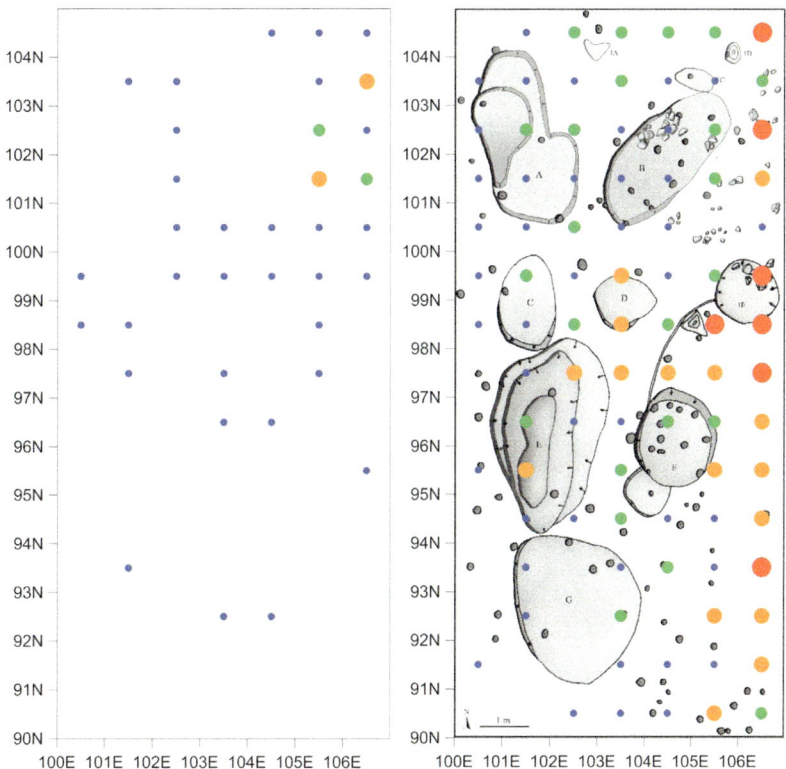

Figure 3.8 Spatial distribution of ceramic sherds from the surface (on the left) and from level 1 (on the right) at site 8-B-10C. Blue symbols (from 1 to 5 sherds/square); green symbols (from 6 to 10 sherds/square); orange symbols (from 11 to 20 sherds/square); red symbols (from 21 to 45 sherds/square) (figure by G. D'Ercole).

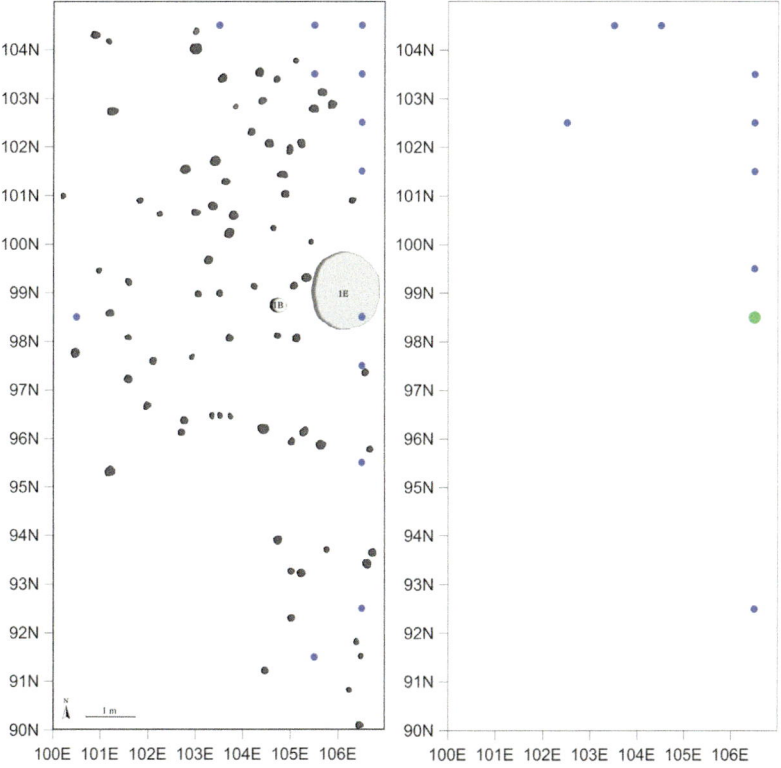

Figure 3.9 Spatial distribution of ceramic sherds from levels 2 (on the left) and 3 (on the right) at site 8-B-10C. Blue symbols (from 1 to 5 sherds/square); green symbols (from 6 to 10 sherds/square) (figure by G. D'Ercole).

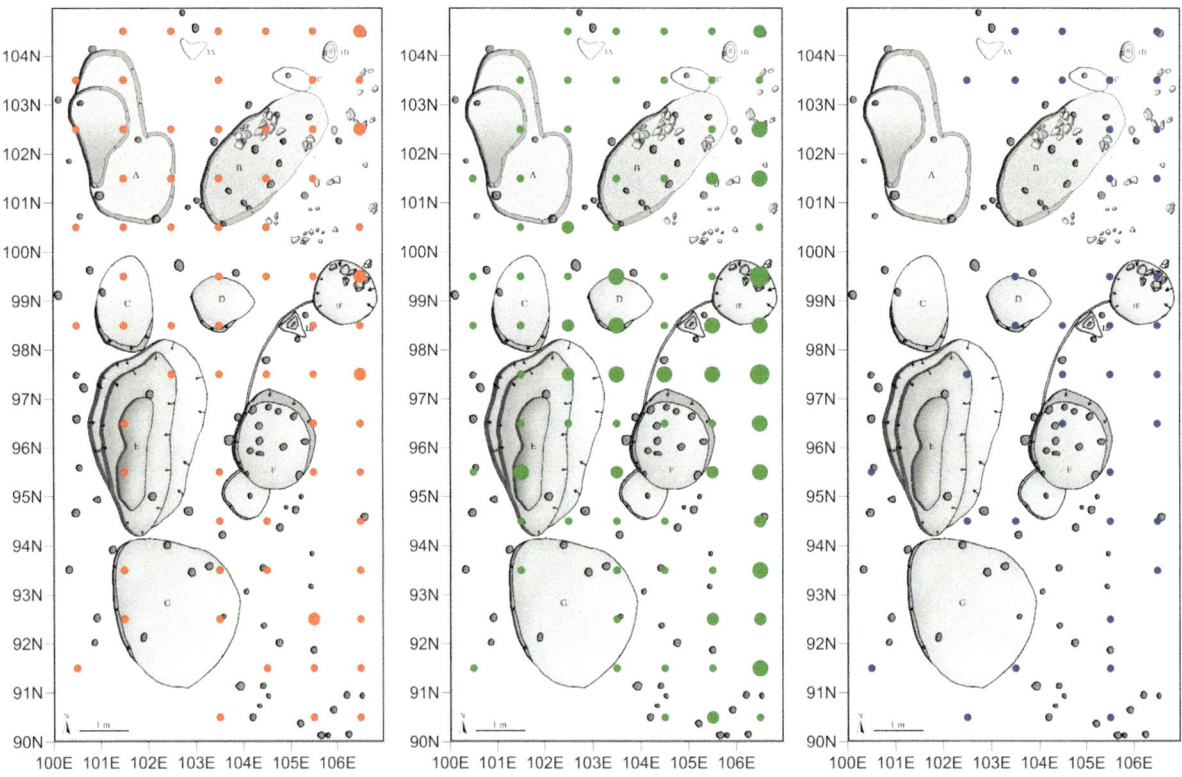

Figure 3.10 Spatial distribution of ceramic sherds with fine (in red on the left), medium (in green in the middle) and coarse texture (in blue on the right) at site 8-B-10C (figure by G. D'Ercole).

Site 8-B-76

State of preservation

The sample analysed from site 8-B-76 consists of 39 potsherds. They come from the surface and from levels 1, 4, 6 and 9 of Test Unit 1 (TU1) and 2 (TU2). No ceramics were found in levels 2 or 3. Two single fragments from Test Pit 1 (STP1) have been classified independently.

Preparation: clay processing and addition of non-plastic inclusions

The ceramics from site 8-B-76 are made from sandy pastes, brown, dark grey or greyish-brown in colour, even completely black. The sample analysed has commonly a medium (43.6%) or fine (53.8%) textures **(Table 3.16)**. Mineral inclusions were classified as angular in 61.5% of cases while rounded inclusions are found in only 38.5% of the sample **(Figure 3.11)**. Inclusions of angular appearance and a medium level of sphericity are the 56.4% with dimensions <1mm or >1 and <2mm. Pastes with common inclusions that are rounded and have a high (23.1%) or medium level of sphericity (10.3%) are found less often **(Table 3.17)**.

8-B-76	S		L1		L4		L6		L9		STP1		Total	
	No.	%	No.	%	No.	%	No.	%	No.	%	No.	%	No.	%
Fine	7	17.9	12	30.8	0	0.0	1	2.6	1	2.6	0	0.0	21	53.8
Medium	4	10.3	8	20.5	1	2.6	1	2.6	1	2.6	2	5.1	17	43.6
Coarse	0	0.0	0	0.0	0	0.0	0	0.0	1	2.6	0	0.0	1	2.6
Total	11	28.2	20	51.3	1	2.6	2	5.1	3	7.7	2	5.1	39	100.0

Table 3.16 Stratigraphic distribution of types of textures from site 8-B-76.

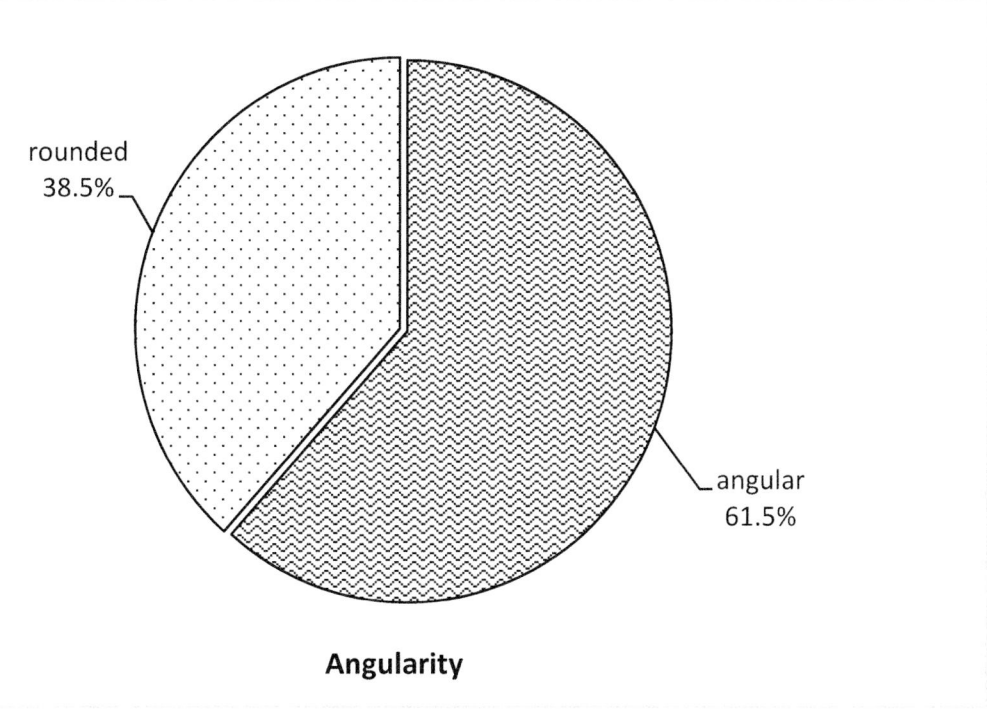

Figure 3.11 Angularity of mineral inclusions from site 8-B-76.

8-B-76			S		L1		L4		L6		L9		STP1		Total	
Inclusions	Sphericity	Angularity	No.	%	No.	%	No.	%	No.	%	No.	%	No.	%	No.	%
Rare	High	Rounded	0	0.0	0	0.0	0	0.0	0	0.0	1	2.6	0	0.0	1	2.6
Common	High	Angular	0	0.0	1	2.6	0	0.0	0	0.0	0	0.0	0	0.0	1	2.6
Common	High	Rounded	7	17.9	2	5.1	0	0.0	0	0.0	0	0.0	0	0.0	9	23.1
Common	Medium	Angular	2	5.1	15	38.5	1	2.6	1	2.6	1	2.6	2	5.1	22	56.4
Common	Medium	Rounded	2	5.1	1	2.6	0	0.0	1	2.6	0	0.0	0	0.0	4	10.3
Frequent	High	Angular	0	0.0	0	0.0	0	0.0	0	0.0	1	2.6	0	0.0	1	2.6
Frequent	High	Rounded	0	0.0	1	2.6	0	0.0	0	0.0	0	0.0	0	0.0	1	2.6
Total			11	28.2	20	51.3	1	2.6	2	5.1	3	7.7	2	5.1	39	100.0

Table 3.17 Frequency, sphericity and angularity of inclusions from site 8-B-76.

All the ceramics contain organic inclusions, responsible for the characteristic dark coloration of the pastes. These organics are tubular in 79.5% of cases (31 fr.) **(Table 3.18)**.

Production: shaping

The sample from site 8-B-76 contains eight rimsherds (20.5% of the total), five of which are decorated **(Table 3.19)**. The identifiable vessel shapes are globular jars **(Plate 3.1: 1–2)** or straight walled jars **(Plate 3.1: 3)**. The former have simple or inverted rims, that are flat **(Plate 3.1: 1)** or rounded **(Plate 3.1: 2)** at the lip. The straight walled jar has a rim in the form of a collar that is bevelled on the outside **(Plate 3.1: 3)**. The thickness of the walls varies from 3 to 8mm, though for the most part between 4 and 6mm. Just one piece stands out by being 11mm thick **(Figure 3.12)**.

8-B-76	S		L1		L4		L6		L9		STP1		Total	
	No.	%	No.	%	No.	%	No.	%	No.	%	No.	%	No.	%
Organic tubular	5	12.8	18	46.2	1	2.6	2	5.1	3	7.7	2	5.1	31	79.5
Organic flat	6	15.4	2	5.1	0	0.0	0	0.0	0	0.0	0	0.0	8	20.5
Total	11	28.2	20	51.3	1	2.6	2	5.1	3	7.7	2	5.1	39	100.0

Table 3.18 Stratigraphic distribution of types of inclusions from site 8-B-76.

8-B-76	S		L1		L4		L6		L9		STP1		Total	
	No.	%	No.	%	No.	%	No.	%	No.	%	No.	%	No.	%
Walls	9	23.1	14	35.9	1	2.6	2	5.1	3	7.7	2	5.1	31	79.5
Rims	2	5.1	6	15.4	0	0.0	0	0.0	0	0.0	0	0.0	8	20.5
Total	11	28.2	20	51.3	1	2.6	2	5.1	3	7.7	2	5.1	39	100.0

Table 3.19 Stratigraphic distribution of body parts of the vessel from site 8-B-76.

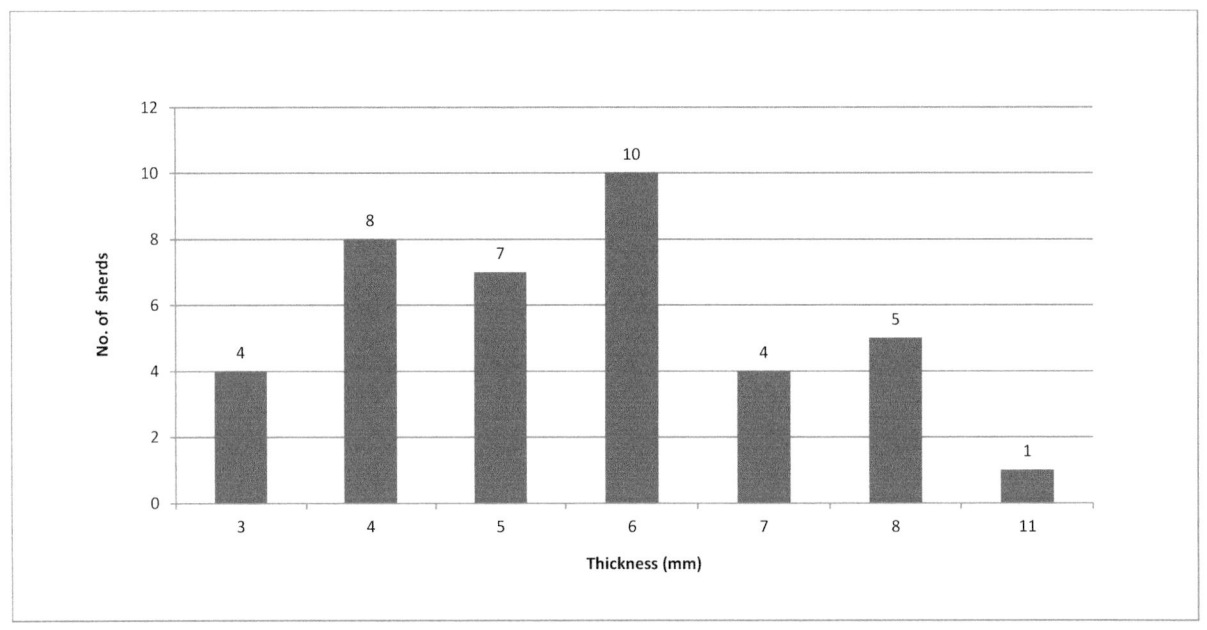

Figure 3.12 Sherd thicknesses from site 8-B-76.

Production: decoration techniques, implements, elements, motifs and structure of the decoration and surface treatment

Most of the sherds are not decorated **(Plate 3.13: 7)**. Seven sherds (17.9%) are decorated and all of them are rims **(Table 3.20)**. They are made of fine pastes and can show single incised decorations **(Plate 3.13: 8)**. Some of the sherds exhibit ripples marks on the surface. One piece shows milled impressions on the lip.

Burnishing is found in only a limited number of sherds. Of the burnished pieces, 23.1% (9 fr.) are burnished on both surfaces, while 10.3% (4 fr.) are burnished only on the inside **(Figure 3.13)**. The inner surfaces can be brown, or more often black. The outer surfaces are always brown **(Table 3.21)**. None of the ceramics from this site are polished.

ANALYSIS OF THE MACROSCOPIC DATA

8-B-76	S		L1		L4		L6		L9		STP1		Total	
	No.	%	No.	%	No.	%	No.	%	No.	%	No.	%	No.	%
Decorated	1	2.6	6	15.4	0	0.0	0	0.0	0	0.0	0	0.0	7	17.9
Undecorated	10	25.6	14	35.9	1	2.6	2	5.1	3	7.7	2	5.1	32	82.1
Total	11	28.2	20	51.3	1	2.6	2	5.1	3	7.7	2	5.1	39	100.0

Table 3.20 Stratigraphic distribution of decorated and undecorated sherds from site 8-B-76.

8-B-76		S		L1		L4		L6		L9		STP1		Total	
Burnishing in	Burnishing out	No.	%	No.	%	No.	%	No.	%	No.	%	No.	%	No.	%
N.p.	N.p.	11	28.2	13	33.3	0	0.0	0	0.0	1	2.6	1	2.6	26	66.7
Black	N.p.	0	0.0	3	7.7	0	0.0	0	0.0	0	0.0	0	0.0	3	7.7
Black	Brown	0	0.0	3	7.7	1	2.6	0	0.0	0	0.0	0	0.0	4	10.3
Brown	N.p.	0	0.0	0	0.0	0	0.0	0	0.0	0	0.0	1	2.6	1	2.6
Brown	Brown	0	0.0	1	2.6	0	0.0	2	5.1	2	5.1	0	0.0	5	12.8
Total		11	28.2	20	51.3	1	2.6	2	5.1	3	7.7	2	5.1	39	100.0

Table 3.21 Stratigraphic distribution of inside and outside burnishing from site 8-B-76.

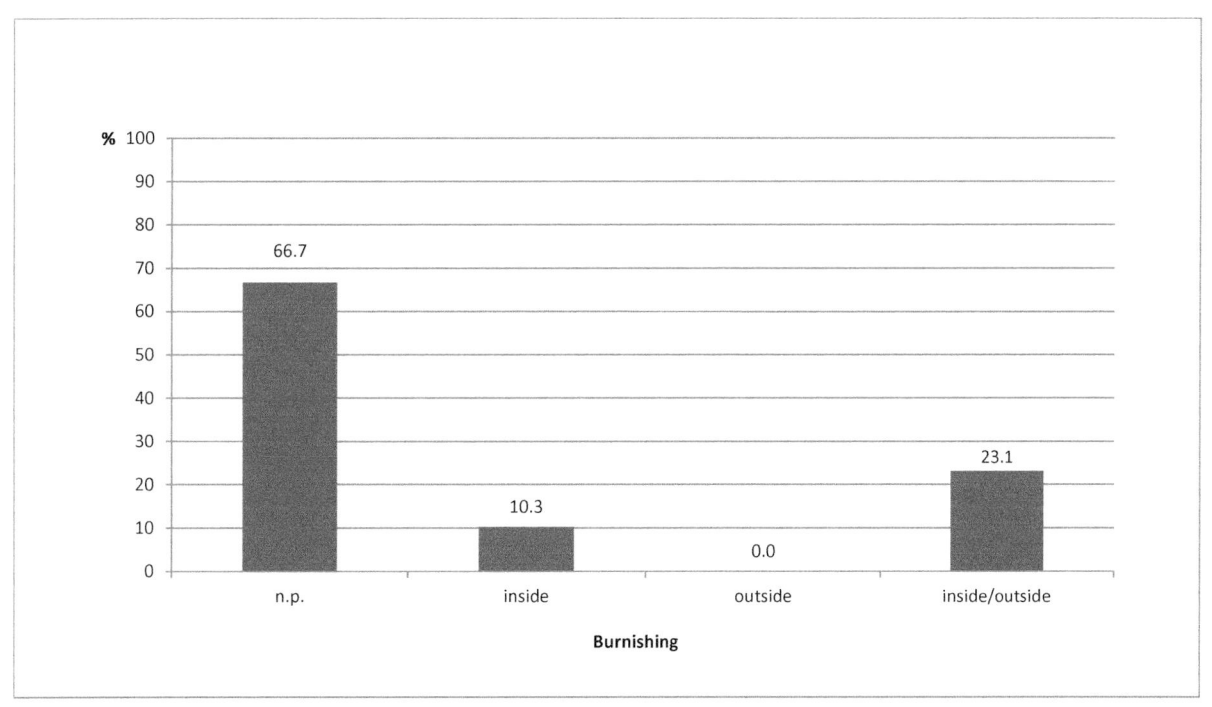

Figure 3.13 Percentages of burnishing from site 8-B-76.

Site 8-B-52A

State of preservation

The sample from site 8-B-52A is composed of a total of 448 sherds, of which 441 (98.4%) could be classified **(Table 3.22)**. 57 pieces come from silo 7; 271 from silo 39; and 120 from silo 44 **(Table 3.22)**. The index of conservation is high for each of the silos.

8-B-52A	Silo 7		Silo 39		Silo 44		Total	
	No.	%	No.	%	No.	%	No.	%
Classifiable	54	12.1	268	59.8	119	26.6	441	98.4
Unclassifiable	3	0.7	3	0.7	1	0.2	7	1.6
Total	57	12.7	271	60.5	120	26.8	448	100.0

Table 3.22 Classifiable and unclassifiable sherds from site 8-B-52A.

Each silo was considered as a single unit. They may have been in use over long periods, with episodes of being intentionally filled and emptied, and later, after the site was abandoned, subject to natural agents such as wind. Consequently, the original position of the artefacts inside the pits cannot be determined with any certainty. Similarly, material from outside could have fallen into the pits after the site was abandoned, getting mixed up with the original contents (cf., Hildebrand and Schilling 2016; see also Chapter 2).

Preparation: clay processing and addition of non-plastic inclusions

The pottery of 8-B-52A, assessed macroscopically, is well-fired with red or brown surfaces and fractures which are for the most part dark grey or black, because of the high level of organic content in the pastes. 88.4% of all the pottery from the three pits contains tubular organic fibres, probably of animal origin (dung from herbivores); while 51 pieces out of 441 (11.6%) contain larger organic inclusions, flat and fibrous in appearance, seemingly vegetable remains, grass leaves or chaff (cf., Livingstone Smith 2001) **(Table 3.23) (Plate 3.23: 4)**.

Mineral inclusions are more often rounded (50.1%) than angular (36.3%) **(Figure 3.14)**, and are always found in these pastes with the addition of organic substances as well. Their degree of sphericity is normally medium to high: about 40% (187 fr.) of the artefacts from the 3 pits have a high proportion of inclusions, with a medium degree of sphericity, and of these, 20.6% (91 fr.) are rounded and the remaining half (96 fr.) are angular; while 16.8% of the samples (74 fr.) have a high proportion of inclusions, with rounded morphology, and a high degree of sphericity **(Table 3.24) (Plate 3.23: 2, 3)**. In all three pits, rounded inclusions are more frequent than angular **(Table 3.24)**.

Taking all three silos together, 76.9% of the pottery (339 fr.) has a medium texture; 20.2% (89 fr.) of the samples have fine pastes; and just 2.7% (12 fr.) are made from coarse pastes **(Table 3.25)**.

Production: shaping

Description of morphological aspects is conditioned by the fact that only a limited number of potsherds offering useful information have been found. 90.5% of the classified fragments from the three pits (399 fr.) are walls, while rims and bottoms make up only 7.7% (34 fr.) and 1.8% (8 fr.) respectively **(Table 3.26)**. Peculiar on the site is the presence of some re-used sherds **(Plate 3.15)**.

8-B-52A	Silo 7		Silo 39		Silo 44		Total	
	No.	%	No.	%	No.	%	No.	%
Organic tubular	50	11.3	234	53.1	106	24.0	390	88.4
Organic flat	4	0.9	34	7.7	13	2.9	51	11.6
Total	54	12.2	268	60.8	119	27.0	441	100.0

Table 3.23 Types of inclusions from site 8-B-52A.

ANALYSIS OF THE MACROSCOPIC DATA

8-B-52A			Silo 7		Silo 39		Silo 44		Total	
Inclusions	Sphericity	Angularity	No.	%	No.	%	No.	%	No.	%
N.c.	N.c.	N.c.	4	0.9	31	7.0	25	5.7	60	13.6
Common	High	Angular	0	0.0	2	0.5	3	0.7	5	1.1
Common	High	Rounded	2	0.5	9	2.0	3	0.7	14	3.2
Common	Low	Angular	0	0.0	0	0.0	1	0.2	1	0.2
Common	Low	Rounded	1	0.2	5	1.1	0	0.0	6	1.4
Common	Medium	Angular	8	1.8	26	5.9	11	2.5	45	10.2
Common	Medium	Rounded	2	0.5	12	2.7	4	0.9	18	4.1
Frequent	N.c.	Rounded	0	0.0	0	0.0	1	0.2	1	0.2
Frequent	High	Angular	1	0.2	4	0.9	1	0.2	6	1.4
Frequent	High	Rounded	10	2.3	44	10.0	20	4.5	74	16.8
Frequent	Low	Angular	0	0.0	7	1.6	0	0.0	7	1.6
Frequent	Low	Rounded	2	0.5	9	2.0	5	1.1	16	3.6
Frequent	Medium	Angular	6	1.4	69	15.6	21	4.8	96	21.8
Frequent	Medium	Rounded	18	4.1	49	11.1	24	5.4	91	20.6
Rare	Medium	Rounded	0	0.0	1	0.2	0	0.0	1	0.2
Total			54	12.2	268	60.8	119	27.0	441	100.0

Table 3.24 Frequency, sphericity and angularity of inclusions from site 8-B-52A.

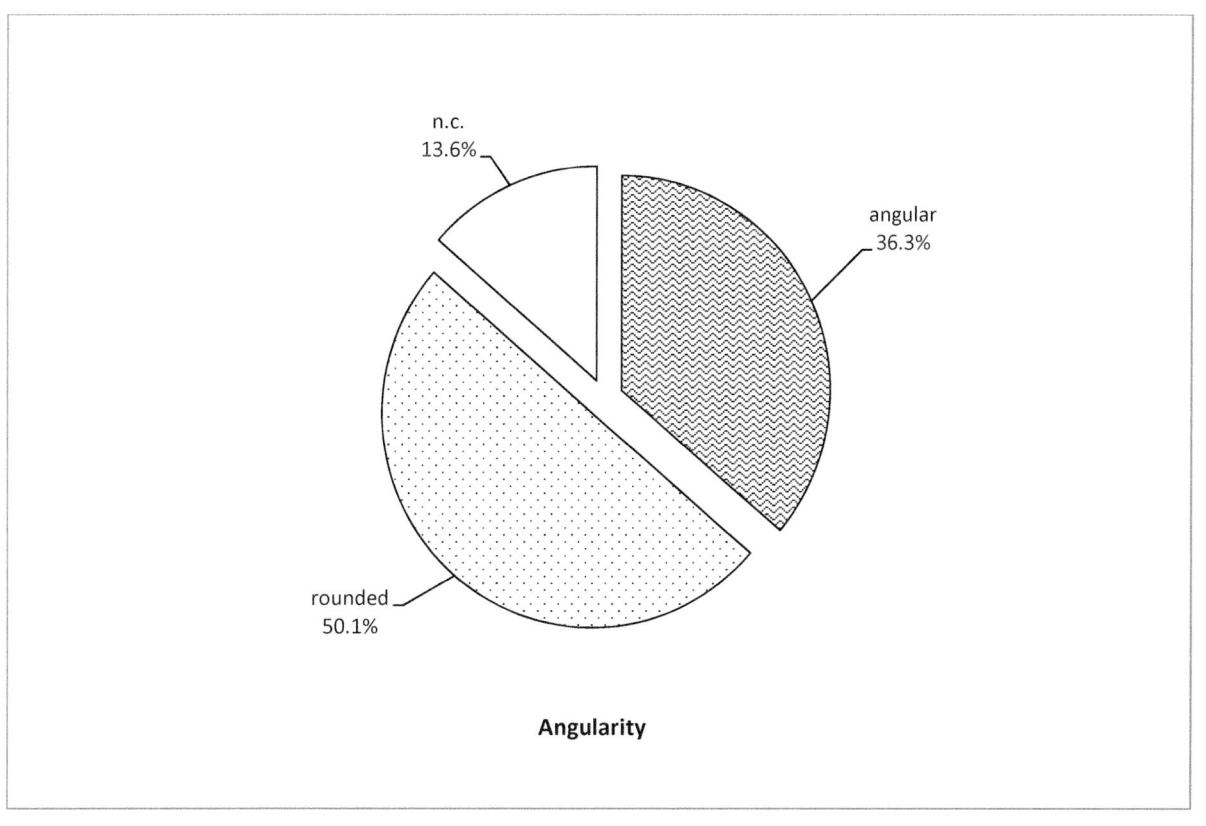

Figure 3.14 Angularity of mineral inclusions from site 8-B-52A.

8-B-52A	Silo 7		Silo 39		Silo 44		Total	
	No.	%	No.	%	No.	%	No.	%
Fine	12	2.7	47	10.7	30	6.8	89	20.2
Medium	37	8.4	218	49.4	84	19.0	339	76.9
Coarse	5	1.1	3	0.7	4	0.9	12	2.7
n.c.	0	0.0	0	0.0	1	0.2	1	0.2
Total	54	12.2	268	60.8	119	27.0	441	100.0

Table 3.25 Types of textures from site 8-B-52A.

8-B-52A	Silo 7		Silo 39		Silo 44		Total	
	No.	%	No.	%	No.	%	No.	%
Walls	44	10.0	241	54.6	114	25.9	399	90.5
Bottoms	1	0.2	5	1.1	2	0.5	8	1.8
Rims	9	2.0	22	5.0	3	0.7	34	7.7
Total	54	12.2	268	60.8	119	27.0	441	100.0

Table 3.26 Body parts of the vessel from site 8-B-52A.

The different vessel shapes include open forms such as bowls, of varying depth and diameter, straight walled **(Plate 3.2: 1-4)** or slightly convex **(Plate 3.2: 5-7)** and smaller bowls that are either saucer-shaped **(Plate 3.2: 8)** or almost hemispherical in shape **(Plate 3.2: 9-10)**. There are also closed forms, such as various ovoid jars **(Plate 3.3: 1-5)** or straight walled jars **(Plate 3.3: 6-9)**.

The open forms vary in the shape of the rim. Some have a simple rim with a rounded lip **(Plate 3.2: 1-2, 4)** while others have an inverted rim, with a lip which may be flat **(Plate 3.2: 5)**, rounded **(Plate 3.2: 6,**

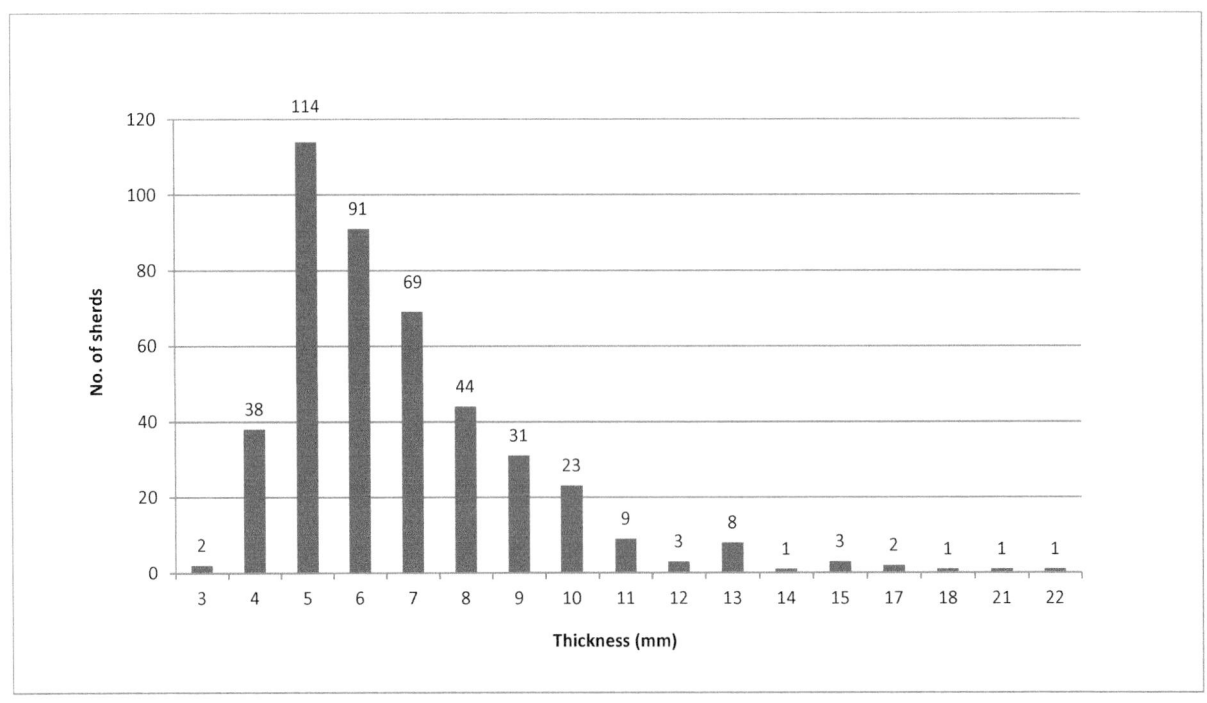

Figure 3.15 Sherd thicknesses from site 8-B-52A.

9) or thinned **(Plate 3.2: 10)**. Some of the bowls, on the other hand, have slightly everted rims that are rounded **(Plate 3.2: 3, 8)** or flat **(Plate 3.2: 7)** at the lip.

The jars generally have simple or very slightly inverted rims, with rounded **(Plate 3.3: 1)**, flat **(Plate 3.3: 2)** or thinned lips **(Plate 3.3: 4-5)**.

The straight walled jars have slightly inverted **(Plate 3.3: 6, 9)**, everted **(Plate 3.3: 7)** or simple **(Plate 3.3: 8)** rims.

The walls are between 4 and 8mm thick, with a peak at 5mm. A more limited but still significant percentage of sherds are thicker, between 9 and 13mm. Just a few samples, presumably coming from heavy-duty storage jars, are thicker than 13mm **(Figure 3.15)**.

Production: decoration techniques, implements, elements, motifs and structure of the decoration

Only 13.8% (61 fr.) of samples were decorated **(Plates 3.14 – 3.18)** and the highest number of decorated pieces, in absolute terms, comes from silo 39 (45 fr.) **(Table 3.27)**.

The rocker stamp is the most used decorative technique **(Plates 3.14 and 3.16)**. It is used on more than half the decorated pieces (52.5%). 23% of the samples have incised decoration **(Plate 3.17)**; 18% use an alternately pivoting stamp **(Plate 3.18)**; and 6.6% have motifs produced by simple impression **(Table 3.28)**.

If the results from silos 7 and 39 are compared (only 2 fragments come from silo 44, both decorated with incisions), it emerges that the principal technique in silo 7 is incision, whereas in silo 39, it is the rocker stamp. The proportions of sherds decorated with impressions or with alternately pivoting stamp do not vary significantly between the two pits **(Table 3.28)**. The ceramics decorated with the alternately pivoting stamp technique are distinguished for the most part by their fine texture. The other methods of decoration appear on fine and medium wares indifferently. At this site, coarse wares are never decorated.

8-B-52A	Silo 7		Silo 39		Silo 44		Total	
	No.	%	No.	%	No.	%	No.	%
Decorated	14	3.2	45	10.2	2	0.5	61	13.8
Undecorated	40	9.1	223	50.6	117	26.5	380	86.2
Total	54	12.2	268	60.8	119	27.0	441	100.0

Table 3.27 Decorated and undecorated sherds from site 8-B-52A.

8-B-52A	Silo 7		Silo 39		Silo 44		Total	
	No.	%	No.	%	No.	%	No.	%
APS	3	4.9	8	13.1	0	0.0	11	18.0
Incision	6	9.8	6	9.8	2	3.3	14	23.0
Rocker stamp	4	6.6	28	45.9	0	0.0	32	52.5
Simple impression	1	1.6	3	4.9	0	0.0	4	6.6
Total	14	23.0	45	73.7	2	3.3	61	100.0

Table 3.28 Types of decorative techniques from site 8-B-52A.

8-B-52A	Silo 7		Silo 39		Total	
	No.	%	No.	%	No.	%
Evenly serrated edge	2	6.3	9	28.1	11	34.4
Plain edge	1	3.1	17	53.1	18	56.3
Unevenly serrated edge	1	3.1	2	6.3	3	9.4
Total	4	12.5	28	87.5	32	100.0

Table 3.29 Tools used for rocker stamping from site 8-B-52A.

In silo 39, the rocker stamp technique is often applied using a plain edged tool (53.1%) or, less often, using a comb with an evenly serrated edge (28.1%). In only two cases has a tool with an unevenly serrated edge been used **(Table 3.29)**. In silo 7, on two samples, a comb with an evenly serrated edge has been used, and in the remaining two cases, one piece has been decorated using a comb with an unevenly serrated edge, and the other with a plain edged implement **(Table 3.29)**. In both silos 39 and 7, plain edged tools are the most common instruments used.

Among the decorative motifs in silo 39, the most common are straight zigzags made with a plain edged tool using the rocker stamp technique (19.7%) **(Plates 3.14 and 3.16: 1-3)**, followed by paired lines, both incised and impressed (13.1%), and single lines (9.8%). The paired lines were produced by a double pronged implement, using the alternately pivoting stamp technique, to make parallel lines of dots, or more often, of triangles **(Table 3.30) (Plate 3.18: 2-6)**. Occasionally, there are zigzags composed of dots or of horizontal or vertical dashes, which are packed, spaced or curved (4.9% of cases) **(Plate 3.16: 4-5)**; while a single sample has zigzags made of v's and dots (1.6%) produced using a comb with an unevenly serrated edge **(Table 3.30) (Plate 3.16: 6)**. In silo 7, the most common motif is of incised single lines (9.8%), while a lower percentage of sherds are decorated with packed zigzags (3.3%) **(Table 3.30)**.

8-B-52A		Silo 7		Silo 39		Silo 44		Total	
Tools	Motifs	No.	%	No.	%	No.	%	No.	%
Double pronged	Paired lines	2	3.3	8	13.1			10	16.4
Double pronged	Smocking	1	1.6					1	1.6
Stylus	Single lines	6	9.8	6	9.8	2	3.3	14	23.0
Evenly serrated edge	N.c.			1	1.6			1	1.6
Evenly serrated edge	Packed zig-zags	2	3.3	3	4.9			5	8.2
Evenly serrated edge	Ripple ware			2	3.3			2	3.3
Evenly serrated edge	Spaced zig-zags			3	4.9			3	4.9
Plain edge	N.c.			1	1.6			1	1.6
Plain edge	Curved zig-zags			3	4.9			3	4.9
Plain edge	Ripple ware			1	1.6			1	1.6
Plain edge	Straight zig-zags	1	1.6	12	19.7			13	21.3
Unevenly serrated edge	Packed zig-zags of "v"and dots	1	1.6	1	1.6			2	3.3
Uunevenly serrated edge	Spaced zig-zags			1	1.6			1	1.6
Unevenly serrated edge	Single lines	1	1.6					1	1.6
Stylus	Dots			3	4.9			3	4.9
Total		14	23.0	45	73.8	2	3.3	61	100.0

Table 3.30 Tools and decorative motifs from site 8-B-52A.

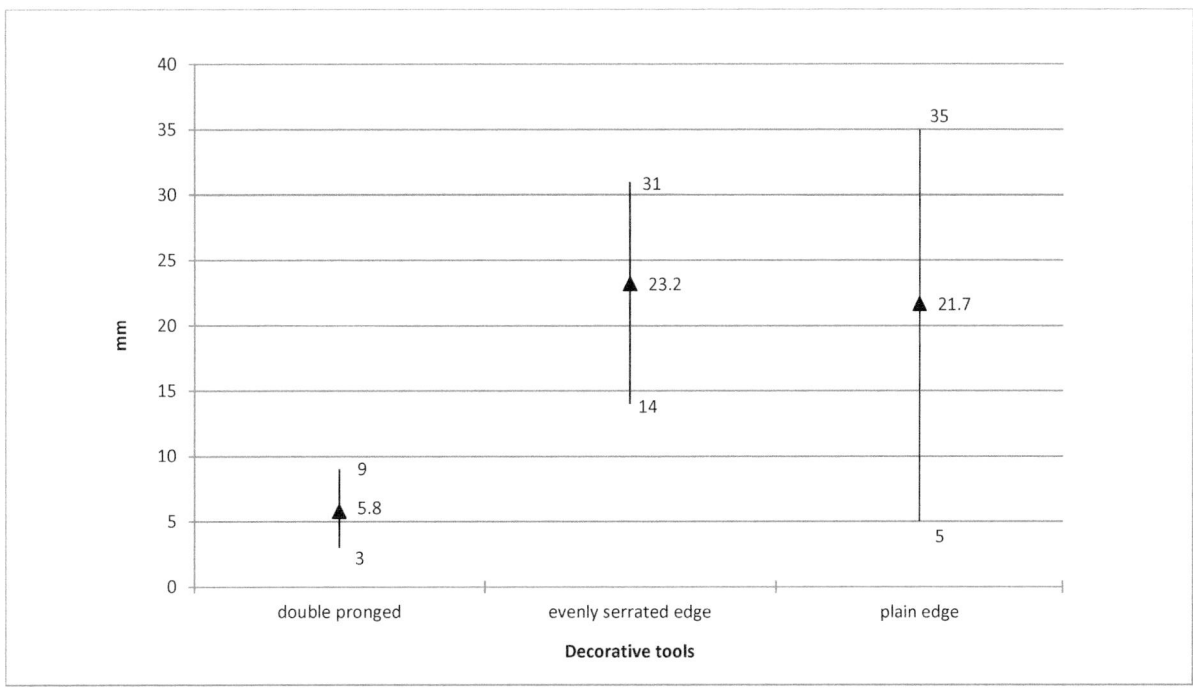

Figure 3.16 Ranges of the lengths of combs and double-pronged tools from site 8-B-52A (minimum, maximum and average).

Some of the ceramics, after being decorated with the rocker stamp technique, were then burnished. This treatment produced a characteristic rippled effect – for this reason the pottery is known as ripple ware – which is common to both Pre-Kerma (Honegger 2004b) and Middle-A Group pottery (Gatto 2006c; Nordström 1972) (see also Chapters 1 and 6) **(Table 3.30)**.

When the sizes of the tools used to decorate the samples from all three pits are compared, some sign of standardisation can be seen. The combs with evenly serrated edges measure between 14 and 31mm in width, with an average of 23.2mm. Similarly, the plain edged tools are an average of 21.7mm wide, but with a greater range of sizes, going from a minimum of 5mm to a maximum of 35mm **(Figure 3.16)**. The double pronged implements employed in the alternately pivoting stamp technique, are always smaller, between 3 and 9mm wide, with an average of 5.8mm **(Figure 3.16)**.

The structure of the decoration is usually continuous **(Plate 3.14, Plate 3.16 and Plate 3.18: 2–3, 5–6)**. Only on a few single fragments are motifs set out in bands **(Plate 3.18: 4)**, whereas panels appear to be relatively common **(Plate 3.17: 1)** (8.2% of the samples from silo 39, and 3.3% from silo 7) **(Table 3.31)**.

8-B-52A	Silo 7		Silo 39		Silo 44		Total	
	No.	%	No.	%	No.	%	No.	%
N.c.	7	11.5	19	31.1	0	0.0	26	42.6
Banded	1	1.6	2	3.3	0	0.0	3	4.9
Continuous	4	6.6	18	29.5	0	0.0	22	36.1
Geometric	0	0.0	1	1.6	0	0.0	1	1.6
Herring-bone	0	0.0	0	0.0	2	3.3	2	3.3
Panelled	2	3.3	5	8.2	0	0.0	7	11.5
Total	14	23.0	45	73.7	2	3.3	61	100.0

Table 3.31 Types of decorative structures from site 8-B-52A.

Geometrical or herring-bone structures are comparatively rare **(Table 3.31)**, and appear principally on rims **(Plate 3.17: 4)**. Rims can also be decorated with single incised lines or with simple impressions **(Plate 3.17: 2-3, 5-7)**.

The samples examined are almost always examples of black-topped ware: the outer surfaces of the pottery are red or brown, either burnished or polished, with inner walls black, just like the upper part of the vessels. As with ripple ware, black-topped ware is typical of Pre-Kerma sites, as it is of the Group-A culture, suggesting the possibility of close contacts, both stylistic and technological, between the two cultural groups (cf., Garcea and Hildebrand 2009).

Production: surface treatment

The ceramics analysed from the three silos have been burnished both internally and externally in 88.9% of cases. Just 1.6% of the samples have been burnished only on the inside, and 6.1% only on the outside **(Figure 3.17)**. Pottery with a black interior, but where the outside is red (14.3%) or, more often, brown (28.6%), is the most common in this assemblage. The production of vessels with walls that are only brown (22%) or only black (7.7%) is also common, but only a very limited number of samples have red interiors **(Table 3.32)**.

The completely black vessels were probably not manufactured by firing in a controlled reducing atmosphere, but through the use of organic combustibles (leaves, grass, dung from herbivores) added during the firing process, after the vessels had reached the correct degree of hardness, with the aim of blackening both surfaces. This process can also explain the black tops: after firing, but while still hot, vessels were placed upside down, with their tops in direct contact with the smoke coming from the burning organic matter (Nordström 1972: 45).

8-B-52A		Silo 7		Silo 39		Silo 44		Total	
Burnishing in	Burnishing out	No.	%	No.	%	No.	%	No.	%
N.p.	N.p.	1	0.2	13	2.9	1	0.2	15	3.4
N.p.	Black	0	0.0	0	0.0	1	0.2	1	0.2
N.p.	Brown	4	0.9	12	2.7	3	0.7	19	4.3
N.p.	Red	0	0.0	4	0.9	3	0.7	7	1.6
Black	N.p.	0	0.0	2	0.5	0	0.0	2	0.5
Black	Black	6	1.4	20	4.5	8	1.8	34	7.7
Black	Brown	13	2.9	74	16.8	39	8.8	126	28.6
Black	Red	5	1.1	39	8.8	19	4.3	63	14.3
Brown	N.p.	0	0.0	2	0.5	1	0.2	3	0.7
Brown	Black	4	0.9	17	3.9	13	2.9	34	7.7
Brown	Brown	18	4.1	60	13.6	19	4.3	97	22.0
Brown	Red	2	0.5	11	2.5	5	1.1	18	4.1
Red	N.p.	0	0.0	2	0.5	0	0.0	2	0.5
Red	Black	0	0.0	1	0.2	0	0.0	1	0.2
Red	Brown	1	0.2	0	0.0	2	0.5	3	0.7
Red	Red	0	0.0	11	2.5	5	1.1	16	3.6
Total		54	12.2	268	60.8	119	27.0	441	100.0

Table 3.32 Inside and outside burnishing from site 8-B-52A.

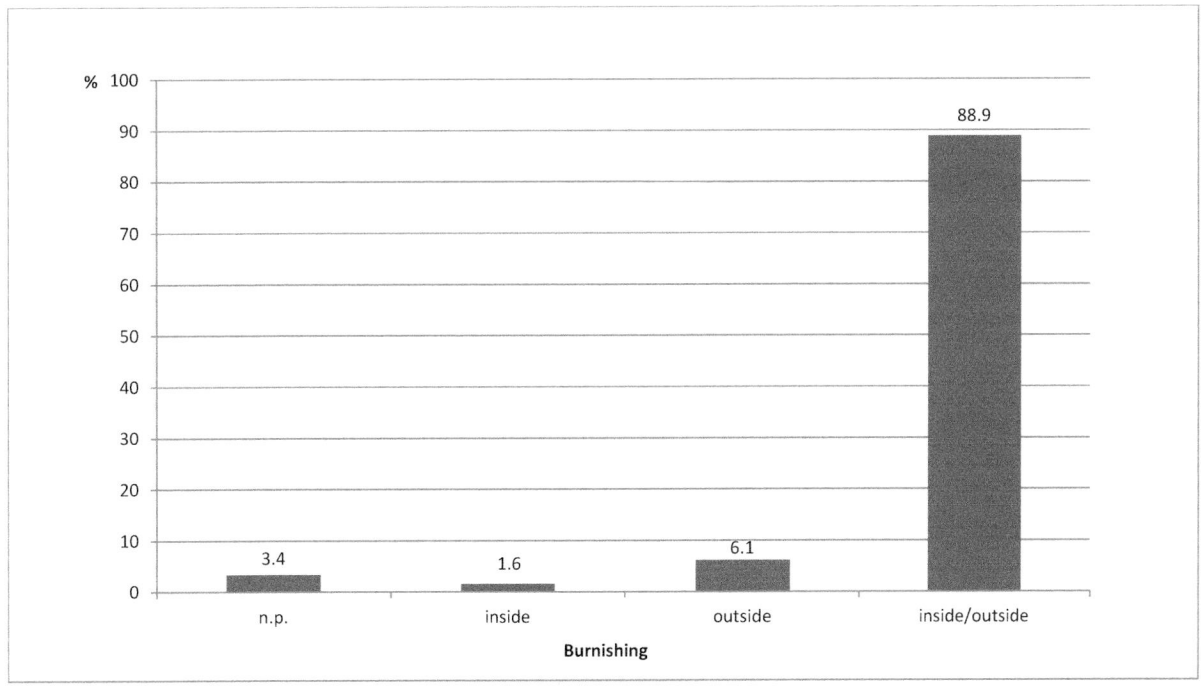

Figure 3.17 Percentages of burnishing from site 8-B-52A.

8-B-52A	Silo 7		Silo 39		Silo 44		Total	
	No.	%	No.	%	No.	%	No.	%
N.p.	15	3.4	108	24.5	19	4.3	142	32.2
Inside	23	5.2	86	19.5	62	14.1	171	38.8
Outside	2	0.5	14	3.2	6	1.4	22	5.0
Inside/Outside	14	3.2	60	13.6	32	7.3	106	24.0
Total	54	12.2	268	60.8	119	27.0	441	100.0

Table 3.33 Types of polishing from site 8-B-52A.

At this site, there is a marked contrast between burnishing and polishing practices. The commonest procedure is for vessels to be polished only on the inside (38.8%). Polishing both inside and outside is less usual (24%) and polishing only the outside is even rarer (5%) **(Table 3.33)**.

It is possible to explain this differing tendency – almost always burnishing inside and outside, but usually polishing only the inside – by attributing a different function to the two treatments. Burnishing seems to have been carried out for aesthetic reasons, possibly aiming to enhance the effect of the decoration (in the case of ripple ware, for example, the burnishing is superimposed on the decoration, almost erasing it, in order to produce the characteristic ripple effect on the outer surface); while polishing was treated as a more functional procedure. The inner surfaces were polished in order to make them more impermeable, and to make the pots more suitable for their function as storage vessels (see Chapter 5). When the data on polishing and decoration are compared, it turns out that inside polishing is mainly associated with undecorated sherds (49.5%). For decorated artefacts, however, there are about the same number of pieces that are polished just on the inside as are polished both inside and outside **(Table 3.34)**. This latter case could be explained as a possible response to an aesthetic purpose.

8-B-52A	Decorated		Undecorated		Total	
	No.	%	No.	%	No.	%
Inside	23	7.7	148	49.5	171	57.2
Outside	3	1.0	19	6.4	22	7.4
Inside/Outside	21	7.0	85	28.4	106	35.5
Total	47	15.7	252	84.3	299	100.0

Table 3.34 Distribution of polishing in relation to decorated and undecorated sherds from site 8-B-52A.

Site 8-B-10A

State of preservation

The sample examined from site 8-B-10A is composed of a total of 2495 ceramic fragments, of which 1309 come from the eastern sector (squares N20/E48-49) and 1186 from the western sector (squares N20/E18-19, N20/E20-21, N21/E18-19, N21/E20-21, N22/E18-19 and N23/E18-19) (see Chapter 2). Altogether, 41.8% of the material (1044 fr.) has been classified **(Table 3.35)**. The proportion of non-classifiable samples is particularly high (38.2%) in the western sector. In contrast, only 20% of the samples from the eastern sector are unclassifiable **(Table 3.35)**.

The spatial distribution of the samples from the western sector seems to be the result of a process of deflation. The material comes almost exclusively from the surface, while the other stratigraphic levels (from 1 to 11) are almost sterile, for the most part devoid of ceramics **(Table 3.36)**.

8-B-10A	East		West		Total	
	No.	%	No.	%	No.	%
Classifiable	811	32.5	233	9.3	1044	41.8
Unclassifiable	498	20.0	953	38.2	1451	58.2
Total	1309	52.5	1186	47.5	2495	100.0

Table 3.35 Classifiable and unclassifiable sherds from site 8-B-10A.

West	Classifiable		Unclassifiable		Total	
	No.	%	No.	%	No.	%
S	208	17.5	912	76.9	1120	94.4
L 1	9	0.8	32	2.7	41	3.5
L 2	6	0.5	2	0.2	8	0.7
L 3	1	0.1	6	0.5	7	0.6
L 5	2	0.2	0	0.0	2	0.2
L 6	2	0.2	1	0.1	3	0.3
L 7	2	0.2	0	0.0	2	0.2
L 8	1	0.1	0	0.0	1	0.1
L 9	1	0.1	0	0.0	1	0.1
L 11	1	0.1	0	0.0	1	0.1
Total	233	19.6	953	80.4	1186	100.0

Table 3.36 Stratigraphic distribution of classifiable and unclassifiable sherds from sector west of site 8-B-10A.

East	Classifiable		Unclassifiable		Total	
	No.	%	No.	%	No.	%
S	84	6.4	271	20.7	355	27.1
L1	9	0.7	16	1.2	25	1.9
L2	20	1.5	30	2.3	50	3.8
L3	58	4.4	57	4.4	115	8.8
L4	12	0.9	12	0.9	24	1.8
L5	9	0.7	4	0.3	13	1.0
L6	15	1.1	2	0.2	17	1.3
L7	85	6.5	14	1.1	99	7.6
L8	21	1.6	1	0.1	22	1.7
L9	129	9.9	20	1.5	149	11.4
L10	333	25.4	62	4.7	395	30.2
L11	36	2.8	9	0.7	45	3.4
Total	811	62.0	498	38.0	1309	100.0

Table 3.37 Stratigraphic distribution of classifiable and unclassifiable sherds from sector east of site 8-B-10A.

In the eastern sector, the stratigraphic distribution of the samples seems on the whole to be more uniform. Not just the surface but also the deeper levels have yielded a significant number of potsherds **(Table 3.37)**. On the lower levels, from level 5 downwards, there is a higher proportion of classifiable fragments than non-classifiable. On the surface and in levels 1 and 2, there is a higher proportion of unclassifiable fragments, and in levels 3 and 4, the proportions are evenly divided **(Table 3.37)**.

The analysis of this data suggests that the ceramics from this site – unlike the samples from site 8-B-52A, probably naturally preserved by the storage pits they were found in – have suffered significant damage over the years from a combination of human, environmental and taphonomic factors. This damage is obviously greater in that part of the assemblage found on or near the surface, and so exposed to the elements and to all sorts of other possible disturbance after the abandonment of the site.

Preparation: clay processing and addition of non-plastic inclusions

In technological terms, the ceramics from site 8-B-10A seem to be very similar to the pottery from site 8-B-52A. The pastes are often dark grey or black, given the high proportion of organic matter in them. The outer surfaces are red or brown. Completely oxidised sherds are found less often, as are pieces where, in section, the colour varies, the core being dark and the inner and outer surfaces reddish brown, the width of the coloured zones varying with the intensity of the oxidisation process.

In 76.8% of cases (802 fr.), the organic content of the pottery is made up of thin tubular fibres, possibly herbivore dung. 18.4% of the samples (192 fr.) contain flat organic fibres. Just 4.8% (50 fr.) contain only mineral inclusions **(Table 3.38)**. Pastes with flat organic fibres are more numerous in the eastern sector (17.4%), but still, even in this sector, pastes with tubular fibres are the most common.

8-B-10A	East		West		Total	
	No.	%	No.	%	No.	%
Organic tubular	588	56.3	214	20.5	802	76.8
Organic flat	182	17.4	10	1.0	192	18.4
Only mineral	41	3.9	9	0.9	50	4.8
Total	811	77.7	233	22.3	1044	100.0

Table 3.38 Types of inclusions from site 8-B-10A.

Ceramic manufacturing techniques and cultural traditions in Nubia

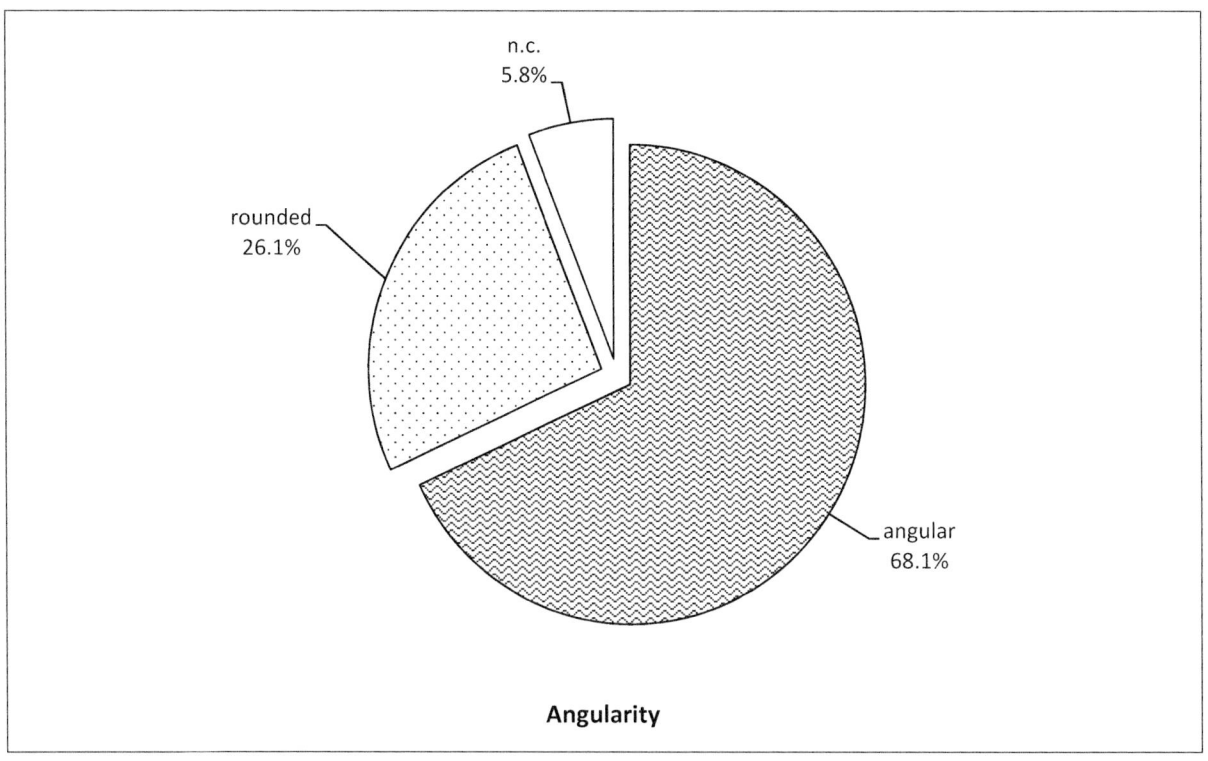

Figure 3.18 Angularity of mineral inclusions from site 8-B-10A.

8-B-10A			East		West		Total	
Inclusions	Sphericity	Angularity	No.	%	No.	%	No.	%
N.c.	N.c.	N.c.	54	5.2	1	0.1	55	5.3
N.c.	Low	N.c.	2	0.2	0	0.0	2	0.2
N.c.	Medium	N.c.	1	0.1	0	0.0	1	0.1
Common	Low	Angular	0	0.0	1	0.1	1	0.1
Common	N.c.	N.c.	2	0.2	0	0.0	2	0.2
Common	High	Angular	9	0.9	1	0.1	10	1.0
Common	High	Rounded	8	0.8	0	0.0	8	0.8
Common	Low	Angular	18	1.7	19	1.8	37	3.5
Common	Low	Rounded	9	0.9	1	0.1	10	1.0
Common	Medium	Angular	154	14.8	106	10.2	260	24.9
Common	Medium	Rounded	30	2.9	5	0.5	35	3.4
Frequent	N.c.	N.c.	1	0.1	0	0.0	1	0.1
Frequent	High	Angular	11	1.1	1	0.1	12	1.1
Frequent	High	Rounded	45	4.3	1	0.1	46	4.4
Frequent	Low	Angular	44	4.2	12	1.1	56	5.4
Frequent	Low	Rounded	39	3.7	5	0.5	44	4.2
Frequent	Medium	Angular	260	24.9	70	6.7	330	31.6
Frequent	Medium	Rounded	120	11.5	9	0.9	129	12.4
Rare	Medium	Angular	4	0.4	1	0.1	5	0.5
Total			811	77.7	233	22.3	1044	100.0

Table 3.39 Frequency, sphericity and angularity of inclusions from site 8-B-10A.

8-B-10A	East		West		Total	
	No.	%	No.	%	No.	%
Fine	168	16.1	32	3.1	200	19.2
Medium	639	61.2	199	19.1	838	80.3
Coarse	2	0.2	1	0.1	3	0.3
N.c.	2	0.2	1	0.1	3	0.3
Total	811	77.7	233	22.3	1044	100.0

Table 3.40 Types of textures from site 8-B-10A.

The mineral inclusions found in the Pre-Kerma pottery from site 8-B-10A, in contrast to what was found on the site 8-B-52A, are more often angular (68.1%) than rounded (26.1%) **(Figure 3.18)**. In the western sector, the predominant ceramic type contains common (10.2%) or frequent (6.7%) angular inclusions with a medium degree of sphericity. In the eastern sector, these two categories are again the most frequent, but pastes with frequent angular inclusions occur more often (24.9%) **(Table 3.39)**. Pastes with rounded inclusions are found only occasionally in the western sector, while in the eastern sector, 11.5% of the samples contain frequent, rounded inclusions with a medium level of sphericity **(Table 3.39)**.

80.3% of the pastes from the site have a medium texture (838 fr.). In 19.2% of the samples, the texture is fine (200 fr.), while it is coarse in only 0.3% of cases (three fr.). In both sectors ceramics with medium texture prevail but from the eastern sector comes an higher percentage of fine wares (16.1 % vs. 3.1% at the western sector) **(Table 3.40)**.

Production: shaping

Out of all the sherds classified, 8.8% (92 fr.) have been identified as rims, while over 90% of the assemblage (951 fr.) is composed of walls. One single fragment from the neck of a jar comes from the eastern sector **(Table 3.41)**. The state of preservation of the rims is such that no measurement of the diameter or size of the vessel is possible. From the inclination of the rims, however, it is possible to distinguish between open forms (bowls), vessels with straight walls, and closed forms (jars).

In the eastern sector, open forms predominate: hemispherical and straight walled bowls – the principal type identified on the site – of varying diameter and depth **(Plates 3.4–3.7)**. The open forms mainly have simple rims, or very slightly everted rims, which at the lip are rounded **(Plate 3.4: 1-4 and Plate 3.5: 1-2)**, thinned **(Plate 3.4: 5-10 and Plate 3.5: 3-4)**, or flat **(Plate 3.4: 11-12)**. More rarely, they have thickened rims, which are slightly inverted **(Plate 3.6: 1-2 and Plate 3.7: 3)** or simple **(Plate 3.6: 3-5)**. A few fragments, parts of small bowls, have a thinned rim **(Plate 3.6: 6-9 and Plate 3.7: 1-2)**, occasionally bevelled on the inside.

The straight walled jars **(Plate 3.8 and Plate 3.9: 1)** almost always have simple rims, which at the lip are flat **(Plate 3.8: 1-5)**, rounded **(Plate 3.8: 6-8)** or thinned **(Plate 3.8: 9-11)**. The flat lips were probably produced using a spatula or a smooth edged tool to remove the excess clay from the outer

8-B-10A	East		West		Total	
	No.	%	No.	%	No.	%
Walls	739	70.8	212	20.3	951	91.1
Necks	1	0.1	0	0.0	1	0.1
Rims	71	6.8	21	2.0	92	8.8
Total	811	77.7	233	22.3	1044	100.0

Table 3.41 Body parts of the vessel from site 8-B-10A.

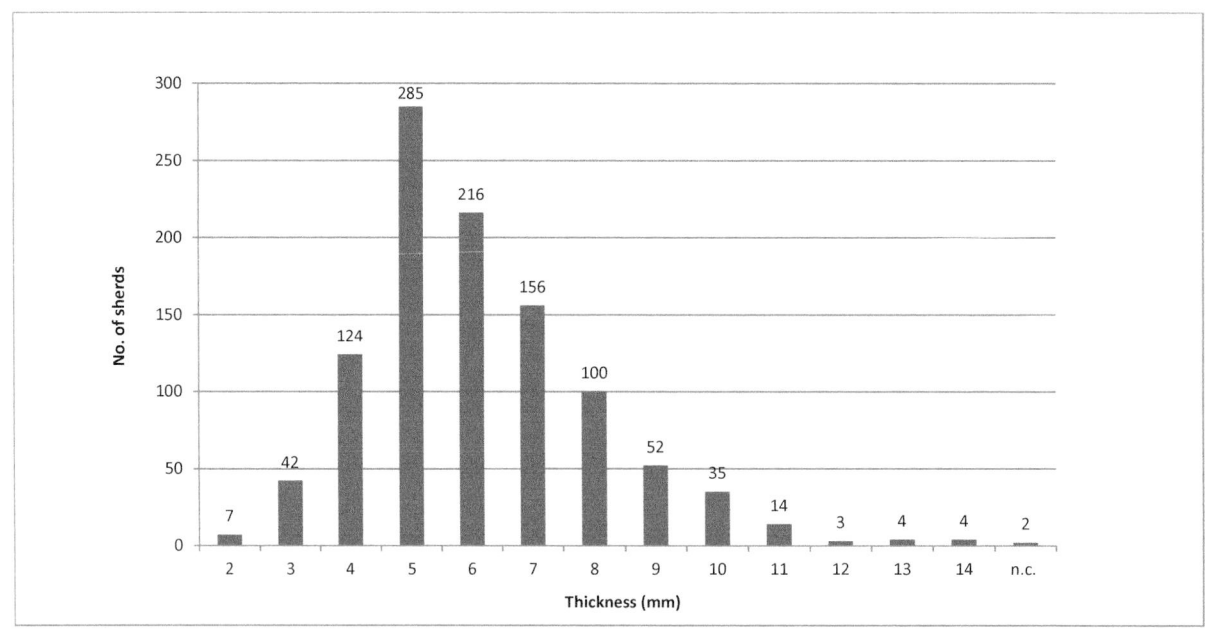

Figure 3.19 Sherd thicknesses from site 8-B-10A.

edge. Occasional samples from straight walled jars have a thickened rim with a flat lip **(Plate 3.9: 1)**.

The few rims of closed forms are almost always simple, with a rounded lip **(Plate 3.9: 2-7 and Plate 3.10: 1)**. Only occasionally is the lip flat **(Plate 3.10: 2-3)**, thickened and inverted **(Plate 3.10: 4)** or everted **(Plate 3.10: 5)**.

In the western sector, besides hemispherical bowls **(Plate 3.10: 11-12)**, slightly convex bowls **(Plate 3.11: 1-2)** and straight walled bowls **(Plate 3.11: 3)**, all with varying types of rim, different types of jars are found, ranging from globular to ovoid **(Plate 3.10: 6-10)** as well as straight walled **(Plate 3.11: 4-7)**. The straight walled jars have simple rims with thinned lips **(Plate 3.11: 4)** or else flat lips **(Plate 3.11: 5-7)**.

In both the sectors, the walls of the vessels tend to be between 4 and 8mm thick, with a peak at 5mm, just as they were on site 8-B-52A. Ceramics with very thin walls (less than 4mm) or with walls more than 8 mm thick are less common **(Figure 3.19)**.

Production: decoration techniques, implements, elements, motifs and structure of decoration

Only 8.1% of the pottery (85 fr., of which 60 from the eastern and 25 from the western sectors) is decorated **(Table 3.42) (Plates 3.19–3.22)**.

The most frequent decorative technique is incision (56.6%) **(Plate 3.19)**. Simple impression is found on 14.1% of the samples **(Plate 3.20: 9-11)**. Only 3.5% of the samples were decorated by rocker-stamping **(Plate 3.20: 7-8)**, while 5.9% have decorations made with the alternately pivoting stamp technique **(Table 3.43)**.

8-B-10A	East		West		Total	
	No.	%	No.	%	No.	%
Decorated	60	5.7	25	2.4	85	8.1
Undecorated	751	71.9	208	19.9	959	91.9
Total	811	77.7	233	22.3	1044	100.0

Table 3.42 Decorated and undecorated sherds from site 8-B-10A.

8-B-10A	East		West		Total	
	No.	%	No.	%	No.	%
APS	5	5.9	0	0.0	5	5.9
Incision	28	32.9	20	23.5	48	56.5
Rocker stamp	1	1.2	2	2.4	3	3.5
Rouletting	17	20.0	0	0.0	17	20.0
Simple impression	9	10.6	3	3.5	12	14.1
Total	60	70.6	25	29.4	85	100.0

Table 3.43 Types of decorative techniques from site 8-B-10A.

Some few samples (20%) showed a decoration which has been possibly obtained by the rouletting technique, using a cord-wrapped roulette (cf., Haour and Manning 2010) **(Table 3.43)**. At site 8-B-10C, the roulette was pressed against the surface of a vessel – like a comb – and alternating ends were moved forward while the opposite end was held in place (rocker stamp technique). At this site, the technique of rouletting was possibly employed, that is, the roulette was 'rolled over the surface of wet clay to leave a continuous band of impressions that repeat themselves at each revolution' (Soper 1985: 30). The individual impressions are of differing shapes and sizes (they are often oval or elongated) depending on the design of the implement used: the type of core could be either rigid or flexible, or else the type of cord or the way the cord was wrapped around the core could vary. At the same time, the depth of the impressions and their orientation could vary with the pressure exerted on the roulette and with the angle it was held at **(Plate 3.21 and Plate 3.22: 1)**.

Rouletting and the alternating pivoting stamp technique have been found only on samples taken from the eastern sector while incisions and simple impressions are common to both sectors **(Table 3.43)**.

An examination of the stratigraphic distribution of the different techniques in the eastern sector reveals an apparent change of practice over time. Decorations produced by impression appear only in the lower levels (10% on level 10). The same is true for incised decorations, which again appear principally in level 10 (25%) and then in levels 9 and 11. On the other hand, alternately pivoting stamp designs come mainly

East	APS		Incision		Rocker stamp		Simple impression		Rouletting		Total	
	No.	%	No.	%	No.	%	No.	%	No.	%	No.	%
S	4	6.7	3	5.0	0	0.0	0	0.0	2	3.3	9	15.0
L1	0	0.0	0	0.0	0	0.0	0	0.0	0	0.0	0	0.0
L2	0	0.0	0	0.0	0	0.0	0	0.0	0	0.0	0	0.0
L3	0	0.0	1	1.7	0	0.0	0	0.0	3	5.0	4	6.7
L4	0	0.0	1	1.7	0	0.0	0	0.0	1	1.7	2	3.3
L5	0	0.0	0	0.0	0	0.0	0	0.0	0	0.0	0	0.0
L6	0	0.0	1	1.7	0	0.0	0	0.0	1	1.7	2	3.3
L7	0	0.0	1	1.7	0	0.0	1	1.7	9	15.0	11	18.3
L8	0	0.0	0	0.0	0	0.0	0	0.0	0	0.0	0	0.0
L9	0	0.0	3	5.0	1	1.7	1	1.7	1	1.7	6	10.0
L10	1	1.7	15	25.0	0	0.0	6	10.0	0	0.0	22	36.7
L11	0	0.0	3	5.0	0	0.0	1	1.7	0	0.0	4	6.7
Total	5	8.3	28	46.7	1	1.7	9	15.0	17	28.3	60	100.0

Table 3.44 Stratigraphic distribution of decorative techniques from sector east of site 8-B-10A.

8-B-10A	East		West		Total	
	No.	%	No.	%	No.	%
Continuous	5	5.9	0	0.0	5	5.9
Criss-cross	2	2.4	2	2.4	4	4.7
Geometric	1	1.2	1	1.2	2	2.4
Herring-bone	1	1.2	7	8.2	8	9.4
Panelled	12	14.1	5	5.9	17	20.0
N.c.	39	45.9	10	11.8	49	57.6
Total	60	70.6	25	29.4	85	100.0

Table 3.45 Types of decorative structures from site 8-B-10A.

from the surface (6.7%). The incidence of rouletting is at its highest in level 7 (15%), reaches 5% in level 3, and is 3.3% on the surface, but is almost nil in the lower levels, from 8 to 11 **(Table 3.44)**.

The decorative motifs predominantly impressed or incised single lines are often organized in panels (20%). Similarly common are herring-bone (9.4%), criss-cross (4.7%) and geometrical designs (2.4%), which are typical of the decoration of rims, and are found sometimes on black-topped wares **(Table 3.45) (Plate 3.19 and Plate 3.20)**. None of the decorations are banded. 5.9% of the motifs are continuous **(Table 3.45)**. Some of the motifs combine different techniques: there are, for example, panels combining incised triangles with impressed dots inside them **(Plate 3.20: 10)**.

The dimensions of the tools can be deduced from only two of the motifs: the motif consisting of single lines of dots made by simple impression (from 7 to 21mm, with an average width of 11.7mm) and the paired lines made by alternately pivoting stamp (from 2 to 15mm, with an average of 8.4mm) **(Figure 3.20)**. Excluding the two pronged implements, the rest of the combs used for rocker-stamping or to make simple impressions had a minimum of six and a maximum of eight teeth.

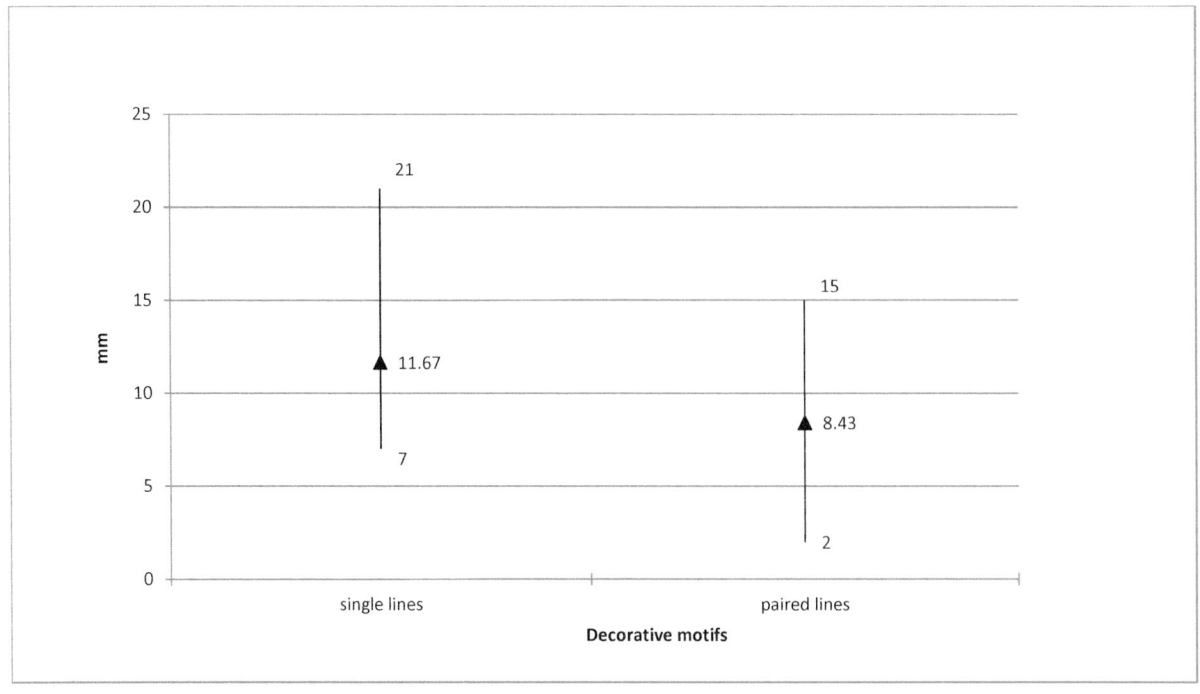

Figure 3.20 Ranges of the lengths of decorative motifs from site 8-B-10A (minimum, maximum and average).

Production: surface treatment

Most of the pottery from site 8-B-10A is burnished both inside and out (76.4%), just like the pottery from site 8-B-52A **(Figure 3.21)**. 17% of the samples are burnished just on the outside, but it is much rarer for them to be burnished only inside (2.6%) **(Figure 3.21)**.

In both sectors, the most common type of ceramic is black on the inside, with outer walls brown (41.5% in the eastern, 7.0% in the western sector) or, in second place, red (6.3% in the eastern, 5.1% in the western sector) **(Table 3.46)**. Pottery which is unburnished inside and brown outside makes up about 13% of the production from the two sectors. Less common are totally black vessels (5.7% in the eastern, 2.1% in the western sector) or totally brown (5.7% in the eastern, 2.1% in the western sector) **(Table 3.46)**.

Only a minority of the samples have been polished. Of these, 11.1% are polished inside, 8% outside, and 3.5% both inside and outside **(Table 3.47)**.

Unpolished pottery is the most common type in both sectors (60.8% in the east, 16.5% in the west). There are a higher percentage of vessels with external polishing in the western sector (4.9%) and a higher percentage in the eastern sector of vessels polished internally (10.6%) **(Table 3.47)**.

In the two sectors together, the ceramics with a medium texture are usually not polished (67.2%). For the fine pastes, similar proportions are polished and not polished (9.4% polished, 9.8% unpolished) **(Table 3.48)**.

When the data on polishing and on decoration are compared, it can be seen that polishing is reserved almost exclusively for the non-decorated wares **(Table 3.49)**.

8-B-10A		East		West		Total	
Burnishing in	**Burnishing out**	No.	%	No.	%	No.	%
N.p.	N.p.	36	3.4	5	0.5	41	3.9
N.p.	Black	3	0.3	4	0.4	7	0.7
N.p.	Brown	107	10.2	31	3.0	138	13.2
N.p.	Red	14	1.3	19	1.8	33	3.2
Black	N.p.	22	2.1	2	0.2	24	2.3
Black	Black	60	5.7	12	1.1	72	6.9
Black	Brown	433	41.5	73	7.0	506	48.5
Black	Red	66	6.3	53	5.1	119	11.4
Brown	N.p.	1	0.1	2	0.2	3	0.3
Brown	Black	1	0.1	0	0.0	1	0.1
Brown	Brown	60	5.7	22	2.1	82	7.9
Brown	Red	6	0.6	4	0.4	10	1.0
Red	Black	0	0.0	1	0.1	1	0.1
Red	Brown	1	0.1	1	0.1	2	0.2
Red	Red	1	0.1	4	0.4	5	0.5
Total		811	77.7	233	22.3	1044	100.0

Table 3.46 Inside and outside burnishing from site 8-B-10A.

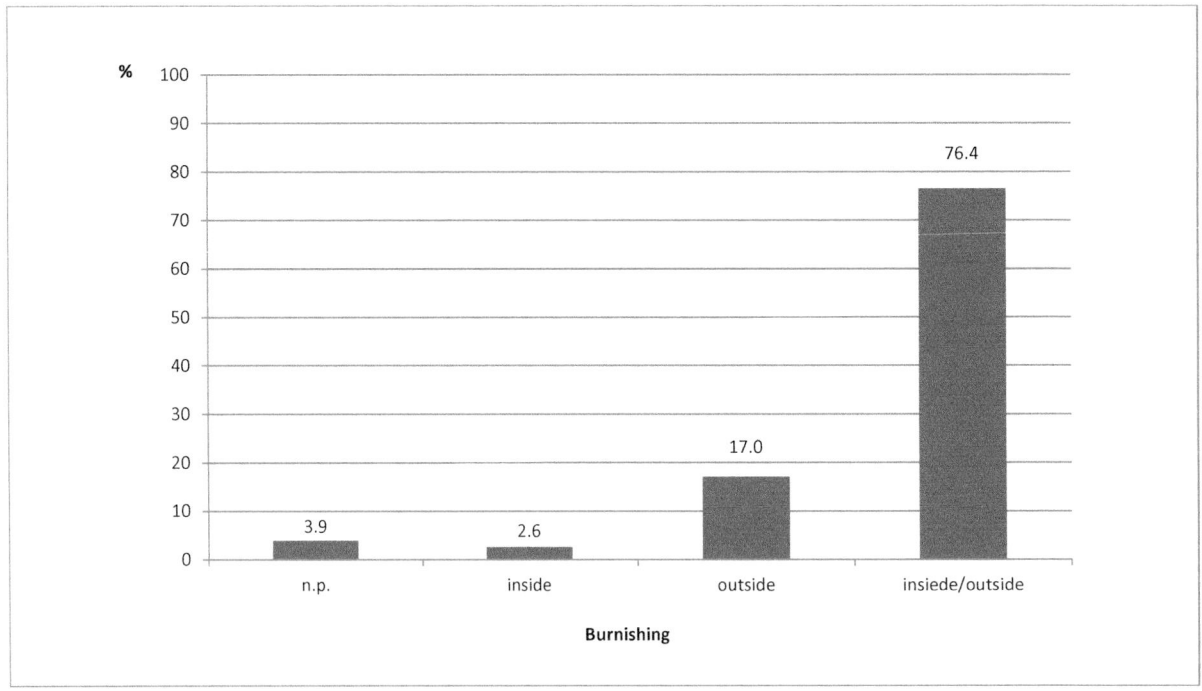

Figure 3.21 Percentages of burnishing from site 8-B-10A.

8-B-10A	East		West		Total	
	No.	%	No.	%	No.	%
N.p.	635	60.8	172	16.5	807	77.3
Inside	111	10.6	5	0.5	116	11.1
Outside	33	3.2	51	4.9	84	8.0
Inside/outside	32	3.1	5	0.5	37	3.5
Total	811	77.7	233	22.3	1044	100.0

Table 3.47 Types of polishing from site 8-B-10A.

A rather different picture emerges here from what was seen on site 8-B-52A. There, polishing primarily met a functional requirement, and only secondarily responded to aesthetic purposes (mainly the interiors of vessels were polished); but here polishing was much rarer, and is associated more often with the site's fine and undecorated wares.

8-B-10A	Present		Not present		Total	
	No.	%	No.	%	No.	%
Fine	98	9.4	102	9.8	200	19.2
Medium	136	13.0	702	67.2	838	80.3
Coarse	1	0.1	2	0.2	3	0.3
N.c.	2	0.2	1	0.1	3	0.3
Total	237	22.7	807	77.3	1044	100.0

Table 3.48 Distribution of textures in relation to the presence /absence of polishing from site 8-B-10A.

8-B-10A	Decorated		Undecorated		Total	
	No.	%	No.	%	No.	%
Inside	6	2.5	110	46.4	116	48.9
Outside	1	0.4	83	35.0	84	35.4
Inside/outside	0	0.0	37	15.6	37	15.6
Total	7	3.0	230	97.0	237	100.0

Table 3.49 Distribution of polishing in relation to decorated and undecorated sherds from site 8-B-10A.

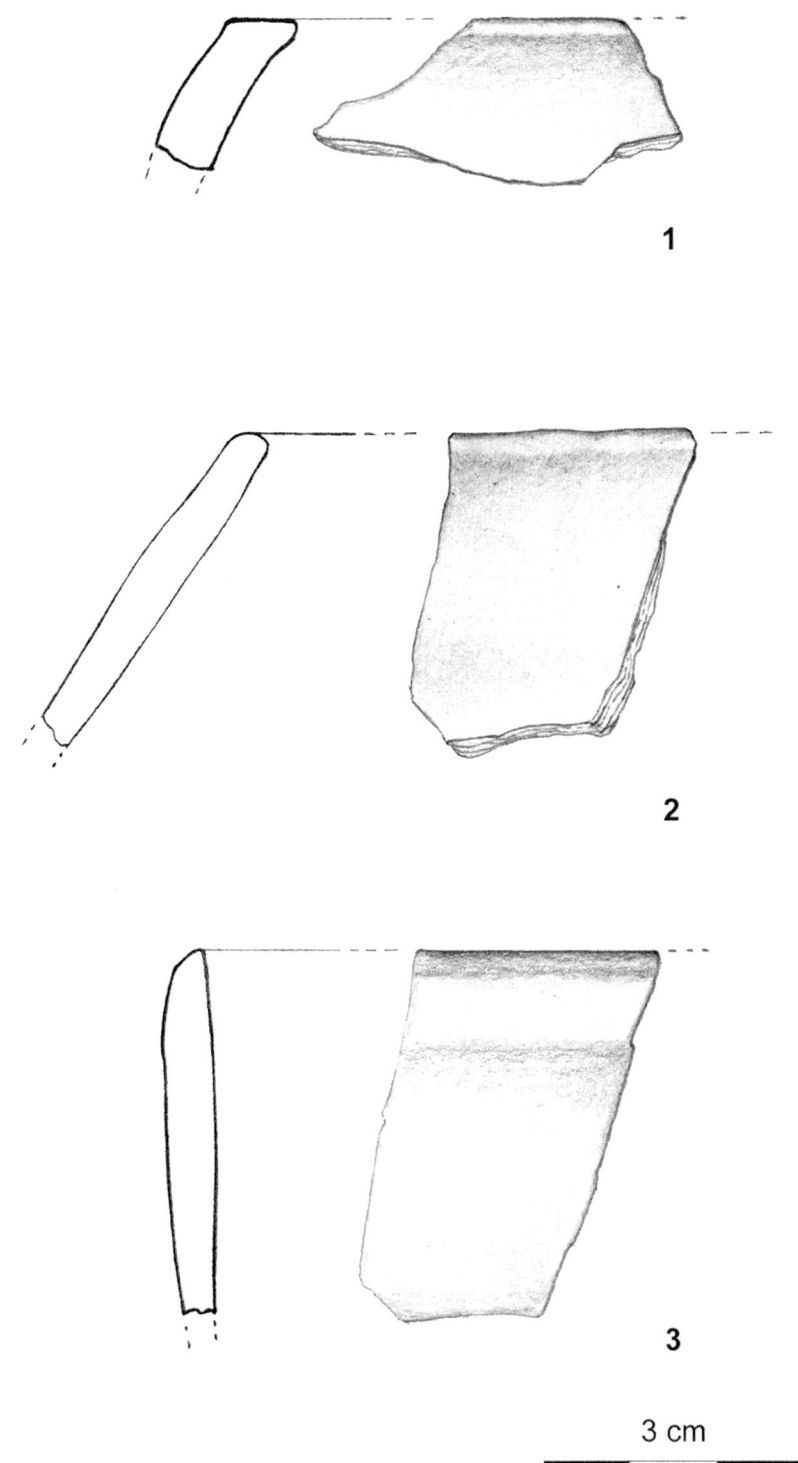

Plate 3.1 Abkan pottery from site 8-B-76. (1-3) Globular and straight walled jars. Drawings by T. D'Este.

ANALYSIS OF THE MACROSCOPIC DATA

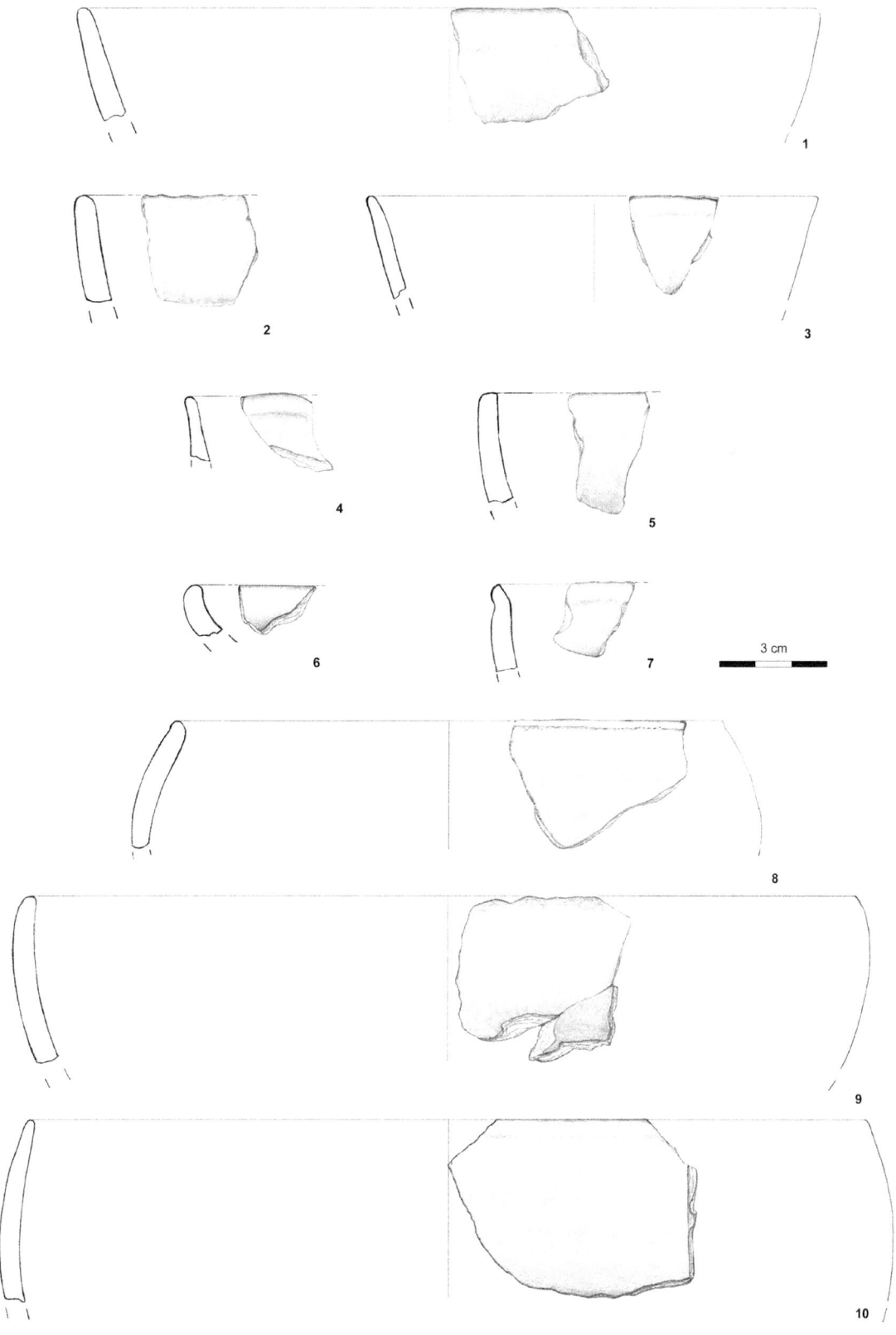

Plate 3.2 Pre-Kerma pottery from site 8-B-52A. (1-4) Bowls with straight walls; (5-7) slightly convex bowls; (8) saucer-shaped bowl; (9-10) hemispherical bowls. Drawings by T. D'Este.

Ceramic manufacturing techniques and cultural traditions in Nubia

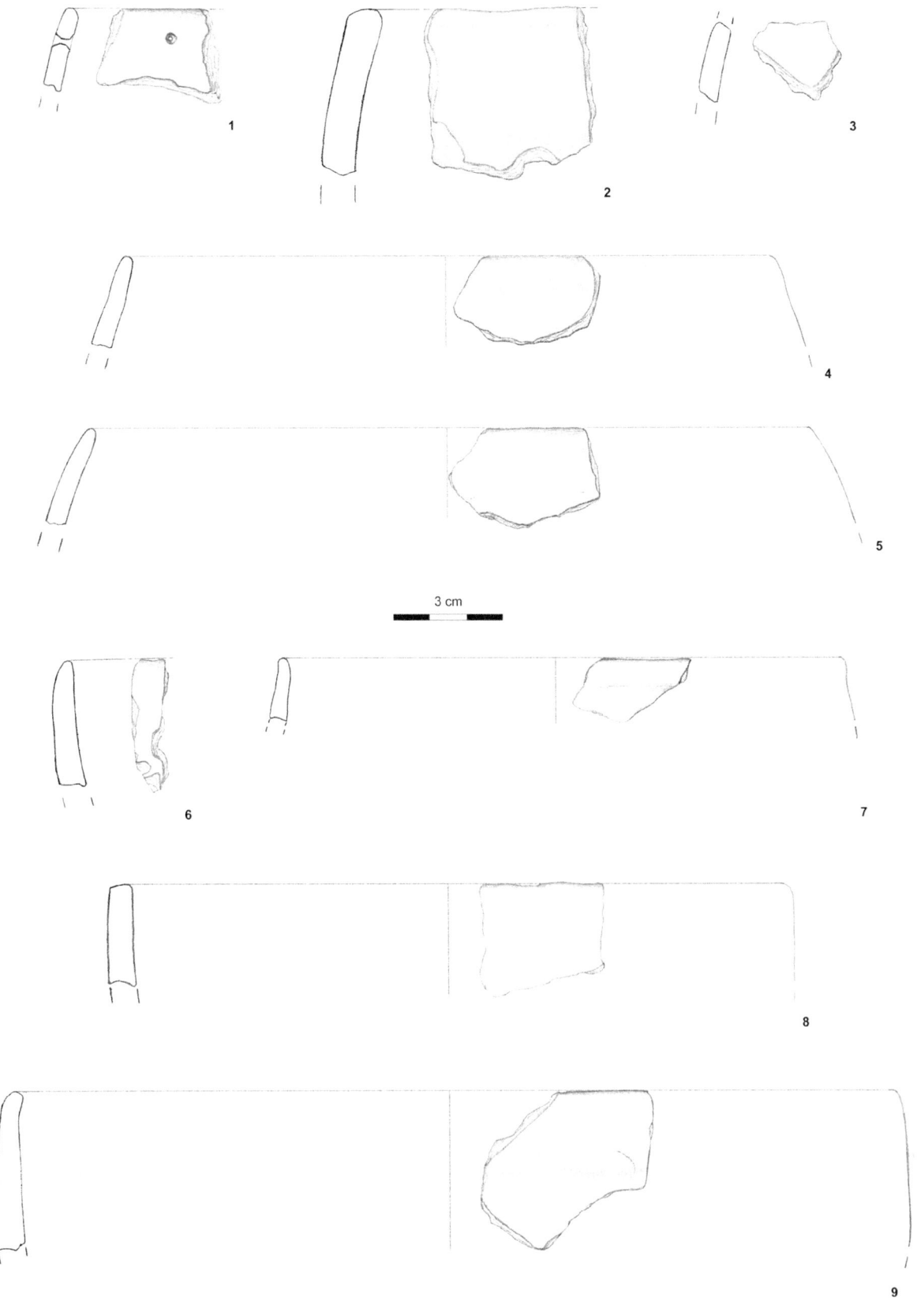

Plate 3.3 Pre-Kerma pottery from site 8-B-52A. (1-5) Ovoid jars; (6-9) jars with straight walls. Drawings by T. D'Este.

ANALYSIS OF THE MACROSCOPIC DATA

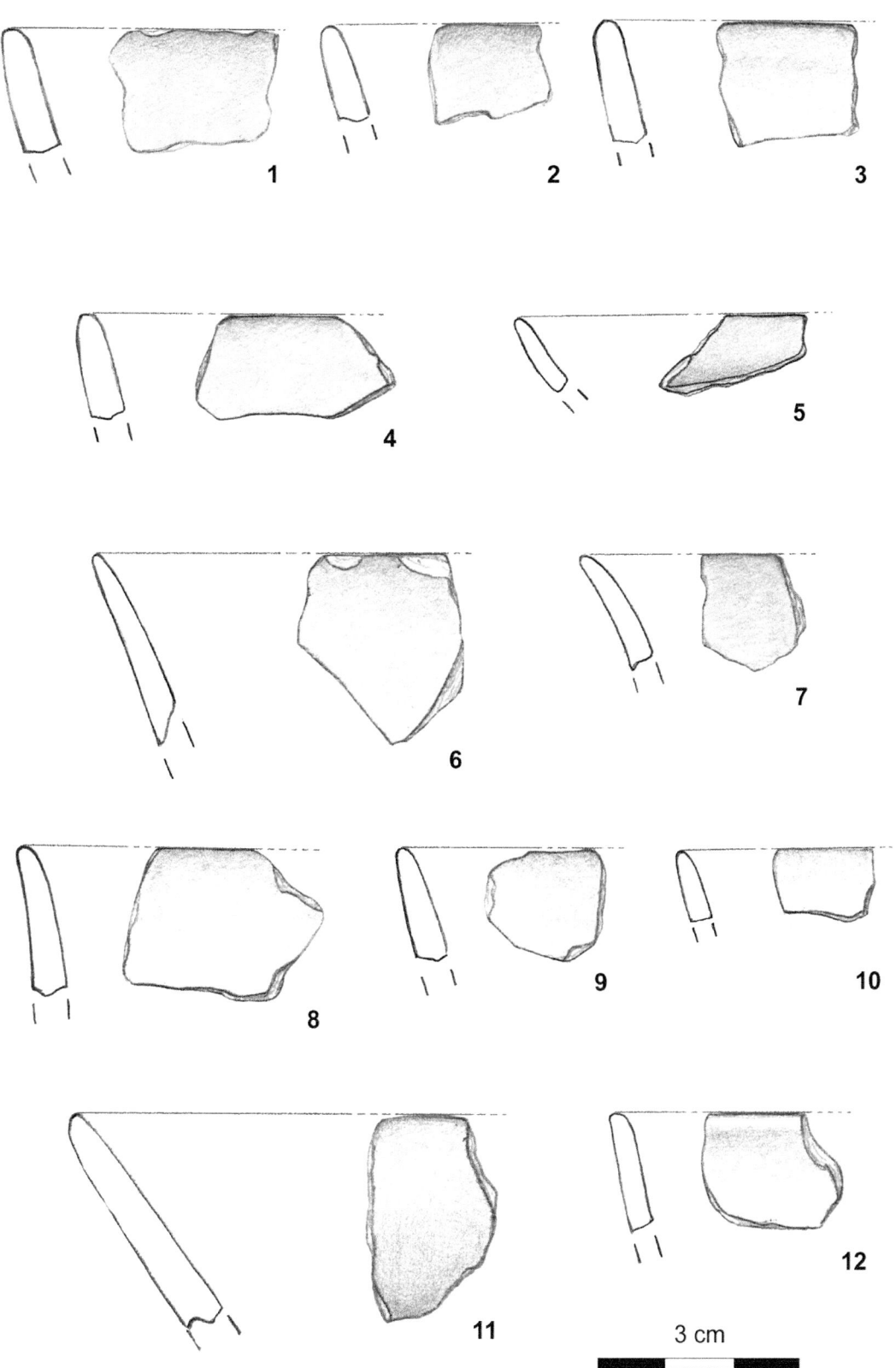

Plate 3.4 Pre-Kerma pottery from site 8-B-10A. Sector East. (1-12) Slightly convex bowls and bowls with straight walls. Drawings by T. D'Este.

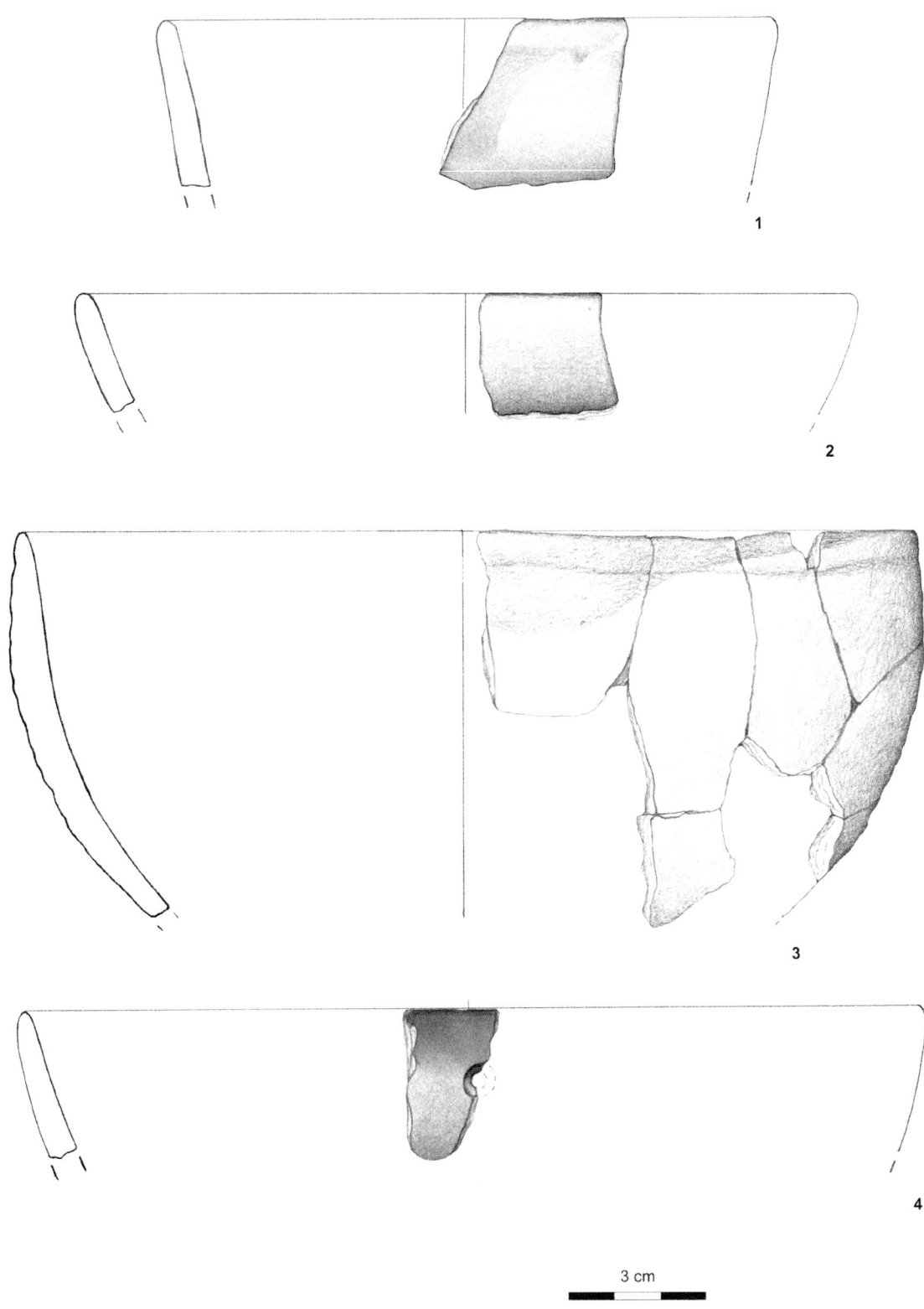

Plate 3.5 Pre-Kerma pottery from site 8-B-10A. Sector East. (1-4) Hemispherical and straight walled bowls. Drawings by T. D'Este.

ANALYSIS OF THE MACROSCOPIC DATA

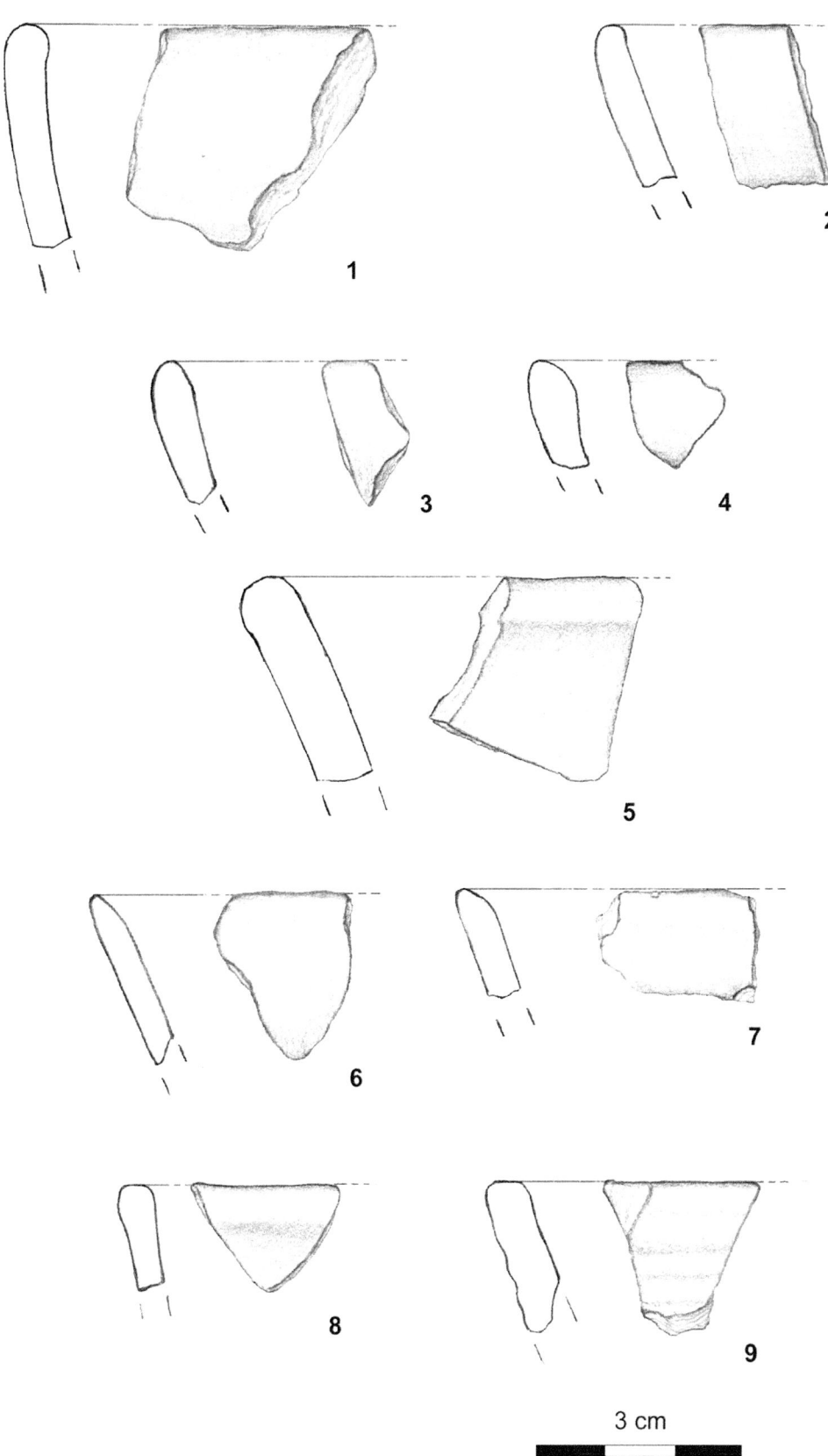

Plate 3.6 Pre-Kerma pottery from site 8-B-10A. Sector East. (1–9) Hemispherical and straight walled bowls. Drawings by T. D'Este.

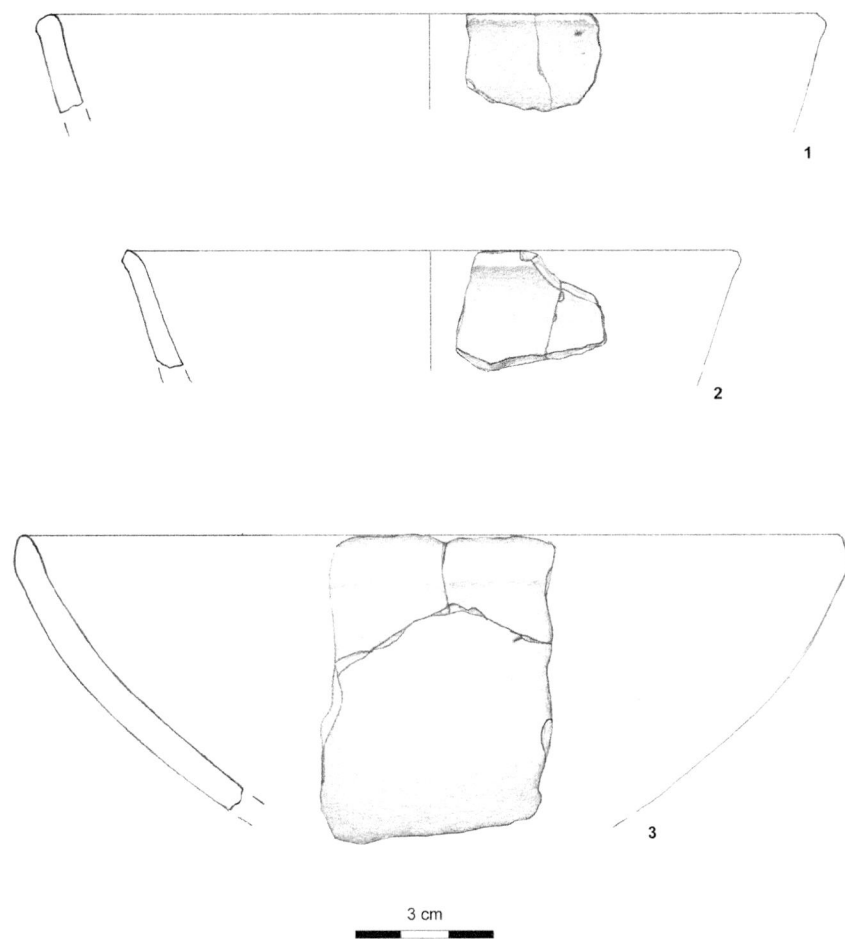

Plate 3.7 Pre-Kerma pottery from site 8-B-10A. Sector East. (1-3) Hemispherical and straight walled bowls. Drawings by T. D'Este.

ANALYSIS OF THE MACROSCOPIC DATA

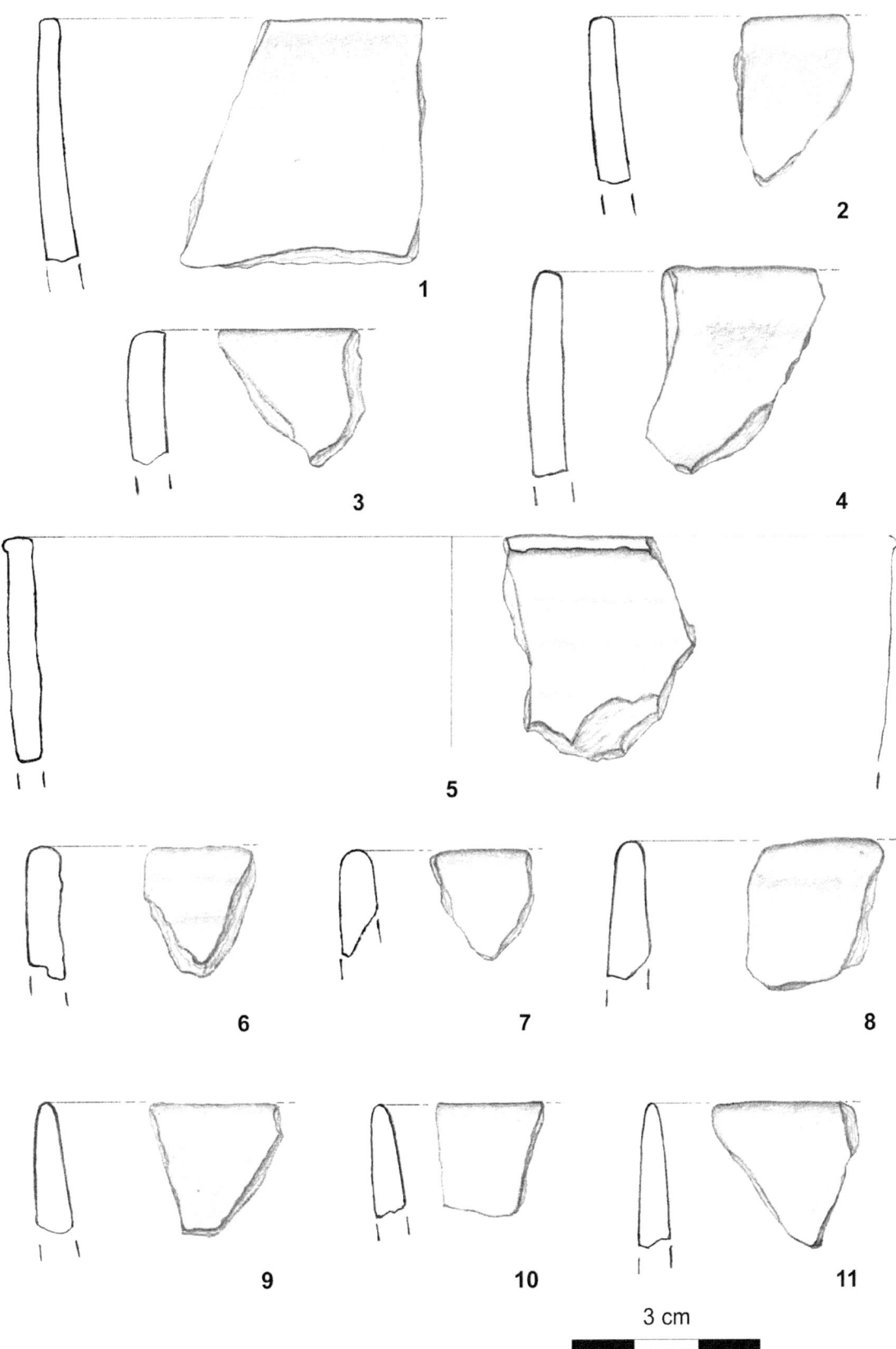

Plate 3.8 Pre-Kerma pottery from site 8-B-10A. Sector East. (1-11) Jars with straight walls. Drawings by T. D'Este.

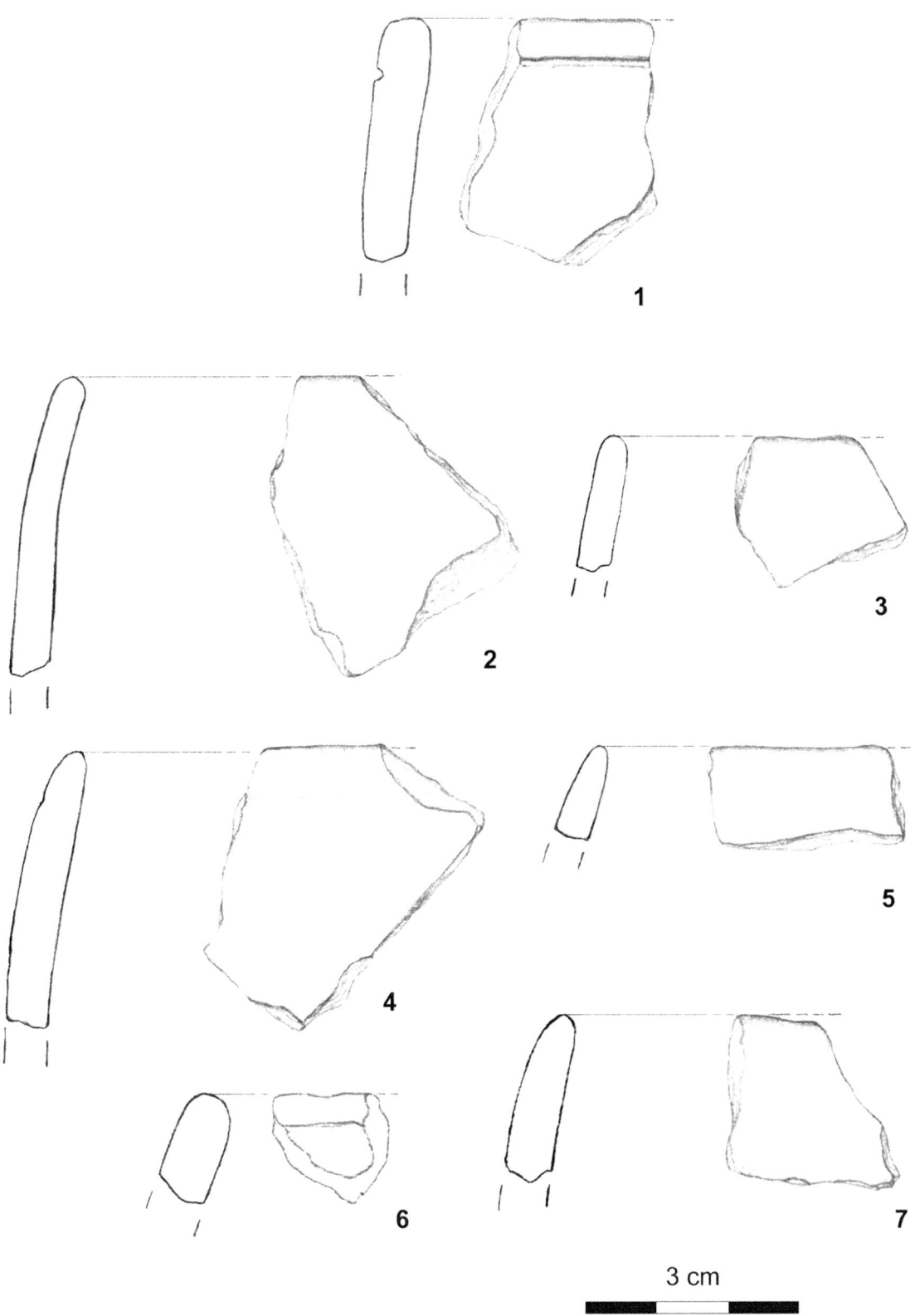

Plate 3.9 Pre-Kerma pottery from site 8-B-10A. Sector East. (1-7) Closed forms. Drawings by T. D'Este.

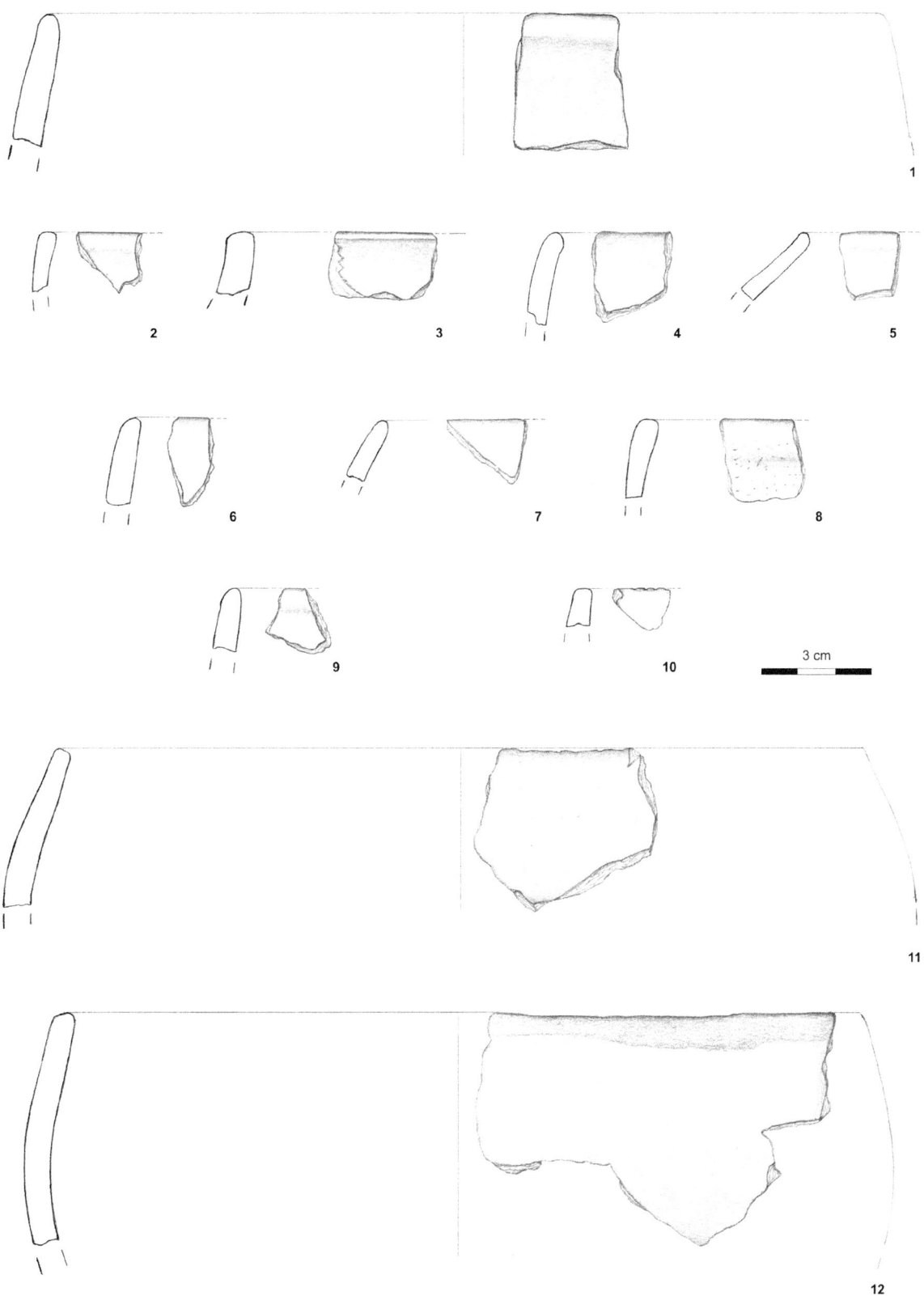

Plate 3.10 Pre-Kerma pottery from site 8-B-10A. (1-5) Sector East. Closed forms; (6-12) Sector West. Globular or ovoid jars and hemispherical bowls. Drawings by T. D'Este.

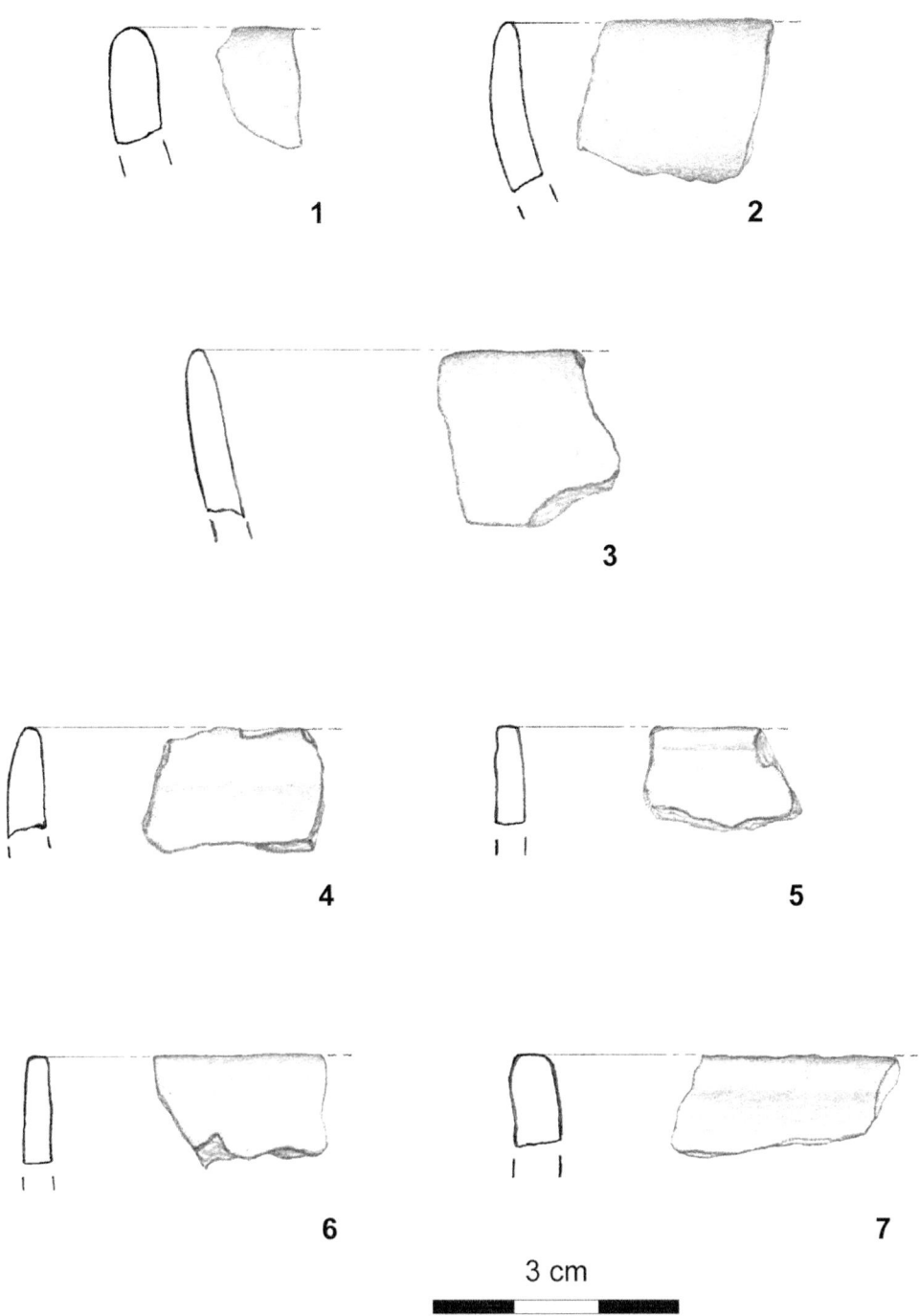

Plate 3.11 Pre-Kerma pottery from site 8-B-10A. Sector West. (1-2) Slightly convex bowls; (3) bowl with straight walls; (4-7) jars with straight walls. Drawings by T. D'Este.

ANALYSIS OF THE MACROSCOPIC DATA

Plate 3.12 Khartoum Variant pottery from site 8-B-10C. (1) Rim decorated with milled impressions; (2) ceramic sherd possibily used as comb; (3-7) dotted wavy line decorations. Photos by R. Ceccacci.

Plate 3.13 (1-6) Khartoum Variant pottery from site 8-B-10C. Zigzag patterns produced using the rocker stamp technique with a roulette; (7-8) Abkan pottery from site 8-B-76. Photos by R. Ceccacci.

ANALYSIS OF THE MACROSCOPIC DATA

Plate 3.14 Pre-Kerma pottery from site 8-B-52A. Large sherd decorated using the rocker stamp technique with a plain edge tool. Photos by R. Ceccacci.

Plate 3.15 (1) External and (2) internal surface of a ceramic sherd possibly re-used as 'spoon'. Photos by R. Ceccacci.

ANALYSIS OF THE MACROSCOPIC DATA

Plate 3.16 Pre-Kerma pottery from site 8-B-52A. Sherds decorated using the rocker stamp technique with (1-3) a plain edge tool; (4-5) a comb with an evenly serrated edge; (6) a comb with an unevenly serrated edge. Photos by R. Ceccacci.

Plate 3.17 Pre-Kerma pottery from site 8-B-52A. (1-7) Rimsherds decorated with simple incisions or impressions. Photos by R. Ceccacci.

ANALYSIS OF THE MACROSCOPIC DATA

Plate 3.18 Pre-Kerma pottery from site 8-B-52A. (1) Undecorated sherd with a little hole close to the rim; (2-6) sherds decorated using the alternately pivoting stamp technique (APS). Photos by R. Ceccacci.

Plate 3.19 Pre-Kerma pottery from site 8-B-10A. (1-13) Sherds decorated with incisions with (1-8) simple; (9) criss-cross; (10-13) herring-bone patterns. Photos by R. Ceccacci.

ANALYSIS OF THE MACROSCOPIC DATA

Plate 3.20 Pre-Kerma pottery from site 8-B-10A. (1-6) Black-topped rims; (7-8) sherds decorated using the rocker stamp technique; (9-11) sherds decorated with simple impressions. Photos by R. Ceccacci.

Plate 3.21 Pre-Kerma pottery from site 8-B-10A. (1-14) Sherds decorated using the rouletting technique. Photos by R. Ceccacci.

ANALYSIS OF THE MACROSCOPIC DATA

Plate 3.22 Pre-Kerma pottery from site 8-B-10A. (On top) Sherd decorated using the rouletting technique; (below) sherd showing marks of burnishing on the internal surface. Photos by R. Ceccacci.

Plate 3.23 Photos of fractures showing (1) angular inclusions; (2) rounded inclusions; (3) calcareous inclusions; (4) organic inclusions. Photos by R. Ceccacci.

4. Archaeometric analysis

Introduction

The macroscopic examination of ceramic specimens taken from the four sites described above (see Chapter 3) was followed by an archaeometric analysis of 92 samples (71 potsherds, five clay lumps and 16 local sediments) from the same four sites **(Table 4.1)**.

The principal aim of the archaeometric analysis was to characterize (petrographically, mineralogically and chemically) the ceramics. In Chapter 5, the results from the different sites will be then compared and discussed with the aim of identifying where variation may occur, either in the selection of raw materials and tempers or in the manufacturing process, over the course of the successive horizons or indeed over the lifetime of each separate cultural complex. Through a comparison of the ceramics and of the samples of sediment, the relationship between the ceramic petrofacies and the geological substrate of the island can be assessed.

All the archaeometric analyses were performed at the laboratories of the Department of Earth and Geoenvironmental Sciences, University of Bari 'Aldo Moro', Italy.

The archaeometric investigation consisted of the following phases:

- selection and preparation of the samples;
- laboratory analyses;
- processing of the data (results)

Selection and preparation of the samples

The ceramic samples

The ceramic samples were systematically selected from the four sites and excavated structures. An equivalent number of specimens were selected from each site, trying to represent each cultural and chronological phase equally, and bearing in mind the stylistic and technological variation within each horizon, as well as the differing stratigraphic origins of the various artefacts. For each site, the sample was chosen from different stratigraphic levels and from the different macroscopic categories identified previously (see Chapter 3). Out of a total of 71 samples, 22 came from site 8-B-10C, 19 from site 8-B-76, 17 from site 8-B-52A, and 13 from site 8-B-10A. Since these analyses involve the partial destruction of the samples, mostly body sherds were chosen, large enough to provide material for the analysis (about six grams altogether). Samples from the stratigraphic levels were preferred to material from the surface **(Table 4.2)**.

Site	Description	Cultural horizon	# ceramic samples	# clay lumps	# local sediment samples
8-B-10C	settlement	KV	22		5
8-B-76	settlement	ABK	19		6
8-B-52A	grain storage site	PK	17		
8-B-10A	settlement	PK	13	5	5

Table 4.1 Total number of analysed samples (Acronyms ABK; KV; PK stand for: Abkan; Khartoum Variant; Pre-Kerma).

Ceramic manufacturing techniques and cultural traditions in Nubia

# Sample	Cultural horizon	Site	Location/Sector	Layer	Vessel part	Thickness (mm)	Texture	Organics	Surface internal/external	Decorative technique
SAI 01	PK	8-B-10A	N21/E19	5	wall	6	medium	X	polished / polished	
SAI 02	PK	8-B-10A	N20/E48	2	wall	8	medium	X	polished / burnished	
SAI 03	PK	8-B-10A	N20/E49	3	wall	7	fine	X	burnished / polished	
SAI 04	PK	8-B-10A	N20/E49	3	wall	7	medium	X	burnished / burnished	
SAI 05	PK	8-B-10A	N20/E48	5	wall	8	medium	X	burnished / burnished	
SAI 06	PK	8-B-10A	N20/E49	7	wall	8	medium	X	polished / burnished	
SAI 07	PK	8-B-10A	N20/E49	7	wall	12	medium	X	burnished / burnished	
SAI 08	PK	8-B-10A	N20/E48	10	wall	9	medium	X	burnished / burnished	
SAI 09	PK	8-B-10A	N20/E48	10	wall	9	medium	X	burnished / burnished	
SAI 10	PK	8-B-10A	N20/E48	10	wall	8	medium	X	burnished / polished	
SAI 11	PK	8-B-52A	silo 7		wall	5	fine	X	polished / polished	Rocker
SAI 12	PK	8-B-52A	silo 7		wall	4	medium	X	polished / polished	
SAI 13	PK	8-B-52A	silo 7		wall	5	medium	X	polished / burnished	Rocker
SAI 14	PK	8-B-52A	silo 39		wall	7	fine	X	polished / polished	APS
SAI 15	PK	8-B-52A	silo 39		wall	7	medium	X	burnished / burnished	
SAI 16	PK	8-B-52A	silo 39		wall	9	coarse	X	burnished / burnished	

Table 4.2a Provenance and macroscopic features of the analysed samples (Acronyms ABK; APS; BT; DWL; KV; PK; S stand for: Abkan; Alternately Pivoting Stamp; Black Topped; Dotted Wavy Line; Khartoum Variant; Pre-Kerma; Surface).

4. Archaeometric analysis

SAI 17	PK	8-B-52A	silo 39		wall	5	fine	x	polished polished
SAI 18	PK	8-B-52A	silo 44		wall	7	medium	x	polished burnished
SAI 19	PK	8-B-52A	silo 44		wall	15	coarse	x	
SAI 20	PK	8-B-52A	silo 44		wall	6	fine	x	polished polished
SAI 21	PK	8-B-52A	silo 44		wall	9	medium		polished polished
SAI 22	ABK	8-B-76		1	wall	5	medium	x	
SAI 23	ABK	8-B-76		1	wall	5	fine	x	
SAI 24	ABK	8-B-76		1	rim	6	fine	x	burnished burnished
SAI 25	ABK	8-B-76		1	wall	3	fine	x	burnished
SAI 26	KV	8-B-10C	92N/106E	2	wall	14	coarse		
SAI 27	KV	8-B-10C	98N/106E	2	wall	6	medium		burnished
SAI 28	KV	8-B-10C	101N/106E	2	wall	7	medium		burnished
SAI 29	KV	8-B-10C	101N/106E	2	wall	7	medium		burnished
SAI 30	KV	8-B-10C	103N/106E	2	wall	9	medium		burnished
SAI 31	KV	8-B-10C	103N/103E	1	wall	5	fine	x	burnished
SAI 32	KV	8-B-10C	101N/102E	1	wall	5	fine	x	
SAI 33	KV	8-B-10C	103N/106E	1	wall	9	medium		
SAI 34	KV	8-B-10C	103N/106E	1	rim	6	medium		
SAI 35	KV	8-B-10C	103N/105E	1	wall	8	coarse		
SAI 36	KV	8-B-10C	103N/106E	1	rim	7	coarse		
SAI 37	KV	8-B-10C	102N/106E	1	wall	9	coarse		DWL
SAI 38	KV	8-B-10C	102N/106E	1	wall	8	coarse		DWL
SAI 39	KV	8-B-10C	102N/105E	1	wall	8	fine		
SAI 40	PK	8-B-10A	N20/E48	S	rim	6	medium	x	burnished burnished	Incision
SAI 41	PK	8-B-10A	N20/E48	7	rim	7	medium	x	burnished burnished	Incision

Table 4.2b *Provenance and macroscopic features of the analysed samples (Acronyms ABK; APS; BT; DWL; KV; PK; S stand for: Abkan; Alternately Pivoting Stamp; Black Topped; Dotted Wavy Line; Khartoum Variant; Pre-Kerma; Surface).*

Sample	Culture	Site	Location	Unit	Part	#	Fabric	X	Surface	Decoration
SAI 42	PK	8-B-10A	N20/E48	9	wall	6	medium	X	burnished / burnished	Rocker
SAI 43	PK	8-B-52A	silo 39		wall	5	fine	X	polished / polished	APS
SAI 44	PK	8-B-52A	silo 39		wall	7	medium	X	burnished / burnished	Rocker
SAI 45	PK	8-B-52A	silo 39		rim	5	medium	X	burnished / burnished	Incision
SAI 46	PK	8-B-52A	silo 39		wall	7	fine	X	polished / polished	APS
SAI 47	PK	8-B-52A	square 15		rim	6	fine	X	polished / polished	BT
SAI 48	PK	8-B-52A	square 19		rim	5	fine	X	polished / polished	Rocker (BT)
SAI 49	ABK	8-B-76		4	wall	11	medium	X	burnished / burnished	
SAI 50	ABK	8-B-76		1	wall	7	medium	X		
SAI 51	ABK	8-B-76		1	rim	6	fine	X		Incision
SAI 52	PK	8-B-10A		5	clay lump					
SAI 53	PK	8-B-10A	N20/E48	6	clay lump					
SAI 54	PK	8-B-10A	N20/E49	6	clay lump					
SAI 55	PK	8-B-10A	N20/E48	7	clay lump					
SAI 56	PK	8-B-10A	N20/E49	8	clay lump					
SAI 58	ABK	8-B-76	STP1		wall	6	medium	X	burnished	
SAI 59	ABK	8-B-76	STP1		wall	8	medium	X	burnished	
SAI 60	ABK	8-B-76		1	wall	5	fine	X	burnished	
SAI 61	ABK	8-B-76		1	wall	5	fine	X		
SAI 62	ABK	8-B-76		1	rim	6	fine	X	burnished / burnished	Milled rim
SAI 63	ABK	8-B-76		1	wall	5	medium	X		
SAI 64	ABK	8-B-76		1	wall	4	fine	X	burnished	
SAI 65	ABK	8-B-76		S	wall	7	medium	X		

Table 4.2c Provenance and macroscopic features of the analysed samples (Acronyms ABK; APS; BT; DWL; KV; PK; S stand for: Abkan; Alternately Pivoting Stamp; Black Topped; Dotted Wavy Line; Khartoum Variant; Pre-Kerma; Surface).

4. Archaeometric analysis

SAI 66	ABK	8-B-76			S	6	wall	medium	X
SAI 67	ABK	8-B-76			1	7	rim	medium	X
SAI 68	ABK	8-B-76			1	7	wall	medium	X
SAI 69	ABK	8-B-76			S	5	wall	fine	X
SAI 70	KV	8-B-10C			3	10	wall	fine	burnished
SAI 71	KV	8-B-10C			3	8	wall	medium	burnished
SAI 72	KV	8-B-10C			3	9	wall	coarse	
SAI 73	KV	8-B-10C	Simple impression		1	7	rim	medium	polished
SAI 74	KV	8-B-10C			1	6	rim	fine	polished	X
SAI 75	KV	8-B-10C			1	6	wall	fine	polished	X
SAI 76	KV	8-B-10C			1	8	wall	fine	
SAI 77	KV	8-B-10C	Rocker		S	9	wall	fine	

Table 4.2d Provenance and macroscopic features of the analysed samples (Acronyms ABK; APS; BT; DWL; KV; PK; S stay for: Abkan; Alternately Pivoting Stamp; Black Topped; Dotted Wavy Line; Khartoum Variant; Pre-Kerma; Surface).

Every sample was first entered on the database used for the macroscopic classification described in Chapter 3, given an identification label (the letters SAI followed by numbers starting from 01), and then photographed.

A standard procedure has been followed for the preparation. Once the thin section was made, the remaining part was crushed and mechanically ground in an agate mortar to a fine powder for the succeeding mineralogical and chemical analyses (D'Ercole *et al.* 2015).

The sediment samples

Sixteen samples of sediment from the three settlements were added to the ceramics samples: five samples from site 8-B-10C, six from 8-B-76, and five from 8-B-10A. The local sediments, like the ceramics, were chosen at each site from different depths, taking into account the internal variability of the stratigraphic sequence **(Table 4.3 and Figure 4.1)**.

They were prepared in the following way: first, each sample was divided into homogeneous portions by means of quartering; the part of the resulting portion (about 6g) intended for mineralogical and chemical analysis was then ground up, like the ceramic samples, into fine powder. The remaining portions of sediment were placed in glass beakers for a week, with distilled water added at intervals, until a firm but still malleable paste was obtained. Once all the water had evaporated, this compound was placed into 2 X 3 X 1cm briquettes and then heated in an electric kiln at 400°C for 1h, in order to harden the samples without significantly altering the original mineralogical content for the preparation of thin sections. The resulting briquettes were impregnated with epoxy resin for thin sectioning. Only 12 of the 16 sediments were prepared for petrographic analysis since their mineralogical and chemical bulk compositions were compatible with those of the potsherds analysed (i.e., Ca-poor) (D'Ercole *et al.* 2015).

It was notable how, after the drying, each of the samples reacted in different ways, according to the composition of the different sediments. The sandy sediments were clearly less malleable, and became crumbly. The clay sediments instead remained firmer but developed shrinkage cracks.

# Sample	Cultural horizon	Site	Depth (cm)
SAI 78	PK	8-B-10A	0-20
SAI 79	PK	8-B-10A	30-40
SAI 80	PK	8-B-10A	57-64
SAI 81	PK	8-B-10A	90-100
SAI 82	PK	8-B-10A	120-130
SAI 83	ABK	8-B-76	10
SAI 84	ABK	8-B-76	20
SAI 85	ABK	8-B-76	30
SAI 86	ABK	8-B-76	40
SAI 87	ABK	8-B-76	60
SAI 88	ABK	8-B-76	60
SAI 89	KV	8-B-10C	74-90
SAI 90	KV	8-B-10C	62-74
SAI 91	KV	8-B-10C	46-62
SAI 92	KV	8-B-10C	36-46
SAI 93	KV	8-B-10C	0-36

Table 4.3 Provenance of the analysed local sediment samples
(Acronyms ABK; KV; PK stand for: Abkan; Khartoum Variant; Pre-Kerma).

Figure 4.1 Profile of site 8-B-76. Arrows indicate the depths from which the sediments samples were taken (figure by G. D'Ercole).

Laboratory analyses

An integrated approach was adopted for each of the samples, carrying out optical microscopy (OM), X-ray powder diffraction analysis (XRPD) and X-ray fluorescence analysis (XRF) of major oxides (SiO_2, TiO_2, Al_2O_3, Fe_2O_3, MnO, MgO, CaO, Na_2O, K_2O and P_2O_5) and trace elements (Rb, Sr, Y, Zr and Nb). In addition, certain samples were also analysed using a scanning electron microscope/energy dispersive spectrometry (SEM/EDS).

Optical microscopy (OM)

Microscopic observation on thin sections aims to observe the optical properties of minerals, which take on characteristic colourings when a beam of polarized light passes through them. This analysis allows a qualitative determination of the mineralogical phases of the pottery (the non-plastic inclusions), as well as a classification of their morphology and dimensions. This method also provides information on matrix and porosity (iso-orientation, size and distribution of pores). It does not however allow the identification of the clay minerals as they are smaller than 2μm and cannot be detected with an optical microscope (cf., Barclay 2001; Cuomo di Caprio 2007; Orton *et al.* 1993; Quinn 2013; Rice 1987; Velde and Druc 1999). In this study, this analysis has been directed at characterizing the different fabrics and examining in greater detail particular technological aspects of the pottery manufacturing sequence. These include the preparation of the pastes, the addition of mineral and/or organic tempers, the manufacturing methods and lastly the atmosphere used in and the length of the firing process.

The petrographic analysis required the removal of a portion of the sample (an area of circa 2cm²), cut transversally to the surface, and then ground down so as to obtain a section so thin (30μm thick) that the polarized light of the petrographic microscope could pass through it.

Petrographic observations on thin sections were carried out with a Carl Zeiss 'Axioskop 40 pol' polarized light microscope (D'Ercole *et al.* 2015). The abundance of the non-plastic inclusions (NPIs) and macro-porosity was visually estimated using comparison charts (Matthew *et al.* 1991).

X-ray powder diffraction analysis (XRPD)

X-ray powder diffraction analysis (XRPD) was used, in combination with optical microscopy, to determine the mineral phases present in the pastes, including the clay phases not visible with a petrographic microscope. It also provided information on the firing processes the pottery went through. Certain minerals (in particular the clay minerals among the silicates) can degrade, disappear or be altered at given temperatures, because the crystalline structure collapses through the process of dehydroxylation (Maggetti 1982; Rice 1987). This analysis is based on the phenomenon of diffraction of electromagnetic radiation, and works by exploiting the fact that X-rays falling on crystalline planes in minerals are reflected at varying known angles (cf., Barclay 2001; Cuomo di Caprio 2007; Orton *et al.* 1993; Quinn 2013; Rice 1987; Velde and Druc 1999).

The procedure requires that a tiny portion of the sample be ground up (about 1g of powder). The powdered sample is then placed in the diffractometer's sample well and pressed flat. Once it is inside the instrument, level and at a constant distance from the X-ray source, a detector can capture the diffraction effects produced by the X-rays, and convert them into a clear and readable signal (Laviano 2002).

At the end of the procedure, a diffraction spectrum is digitally produced. Each mineral has peaks placed at determined angular distances, expressed in degrees 2θ, of heights varying according to the intensity of the diffraction phenomenon. The identification of these peaks permits a qualitative and semi-quantitative estimate of the minerals present in a sample. The proportion of each mineral within a sample was quantified on a table, giving each mineralogical phase a number x (from 0 to 5, according to quantity), assessed by comparing the heights of the main peaks (Laviano 2002).

The powdered samples were analysed with a PANalytical X'Pert pro MDS powder diffractometer using CuKα radiation (40 kV; 40 mA) in step scan mode (0.02° 2θ), with each step measured for 2s. The X-ray data were collected in a Bragg–Brentano (θ/2θ) vertical geometry (flat reflection mode) between 2° and 65° (2θ). The diffraction peaks of the XRPD spectra were compared to a JCPDS–ICDD diffraction chart and the crystalline phases thereby identified (D'Ercole *et al.* 2015).

X-ray fluorescence analysis (XRF)

The identification of the major and trace elements in the samples was carried out using wavelength dispersive X-ray fluorescence (WDXRF). This methodology allows both qualitative and quantitative analysis of the chemical composition of the ceramic material and sediment.

In this procedure, the chemicals in the sample are irradiated with (primary) X-rays, absorb that energy and then emit it in the form of fluorescence, that is, in the form of (secondary) X-rays whose wavelength (λ) identifies the element. The intensity of these secondary X-rays allows an estimate of the quantity of each element since the intensity is in proportion to the concentrations in the sample (cf., Barclay 2001; Cuomo di Caprio 2007; Maggetti 1990; Rice 1987; Tite 2008; Velde and Druc 1999).

For the analysis, about 4g of the sample powder was bound with Elvacite® dissolved in acetone and then dried for pill preparation. Pills were obtained by pressing the dry powder in small aluminium cups under a hydraulic press. For some of the smaller potsherds, the analysis was carried out with less than four grams of powder, preparing the pills with a larger boric acid base, concentrating the sample in the centre where the irradiation was the greatest.

Determinations of major oxide (SiO_2, TiO_2, Al_2O_3, Fe_2O_3, MnO, MgO, CaO, Na_2O, K_2O and P_2O_5) and trace element (Rb, Sr, Y, Zr and Nb) concentrations were performed using a Philips PW 1480/10 spectrometer, using the analytical techniques outlined by Franzini *et al.* (1972, 1975) and Leoni and Saitta (1976). The detection limit for major element oxides was 0.01 wt% and for trace elements 1–5 ppm. Loss on ignition (LOI) was determined on previously dried samples by heating at 1000°C for 12h (D'Ercole *et al.* 2015).

Finally, five samples (SAI 09, 63, 64, 65, and 71) were examined through a scanning electron microscope/energy dispersive spectrometry (SEM/EDS) in order to determine the structure and chemical composition of particular mineral phases identified during the petrographic or XRPD analyses. A LEO EVO-50XVP microscope was used, fitted with a PentaFET Si (Li) detector for energy dispersive microanalysis, which can give very accurate qualitative and quantitative analyses of the surface of samples of sizes varying from 15cm to a couple of micrometres. The samples were prepared following the standard protocol: thin sections were attached to aluminium mounts with colloidal graphite in order to ensure electrical conductivity over the entire surface of the slide, and then metallized.

Results: the ceramic samples

X-ray powder diffraction analysis

The mineralogical analysis by XRPD showed quartz (Qtz) to be the principal constituent of all of the samples, accompanied by varying amounts of feldspar, either potassium/alkali feldspar (Kfs) or plagioclase feldspar (Pl), and of micas (Mca). There are small amounts or traces of clay minerals (C.M.), calcite (Cal), pyroxene (Px), amphibole (Amp) and zircon (Zrn) in a more limited number of the samples.

4. Archaeometric analysis

# Sample	XRPD Group	C.M.	Micas	Qtz	Kfs	Pl	Cal	Px	Amp	Zrn
SAI 01	QPl			XXXXX	tr	X	tr			
SAI 02	QPl			XXXXX	tr	X	tr	tr		
SAI 03	QPl			XXXXX	tr	X	tr	tr	tr	
SAI 04	QPl			XXXXX	tr	X	X			
SAI 05	QPl			XXXXX	tr	X	tr	tr		
SAI 06	QPl		X	XXXX	X	XX	tr	tr	tr	
SAI 07	QPl		tr	XXXXX	X	X	tr	tr		
SAI 08	QPl		X	XXXXX	tr	X	tr	tr	tr	
SAI 09	QPl		X	XXXXX		X	tr	tr	tr	
SAI 10	QPl		tr	XXXXX	tr	X	tr	tr		
SAI 11	QPl	tr	X	XXXX	tr	XX	tr	tr	tr	
SAI 12	QPl		X	XXXXX	tr	X	X		tr	
SAI 13	QPl		X	XXXXX	tr	X	tr	X		
SAI 14	QPl		X	XXXXX	tr	X	tr	tr	tr	
SAI 15	QPl	tr	X	XXXXX	tr	X	X	tr		
SAI 16	QPl		tr	XXXXX	tr	X	tr		tr	
SAI 17	QPl	tr	X	XXXX	tr	XX	tr	tr	tr	
SAI 18	QPl			XXXXX	tr	X	tr	tr	tr	
SAI 19	QPl		X	XXXXX	tr	X	tr		tr	
SAI 20	QPl			XXXXX	tr	X	tr	tr		
SAI 21	QPl		tr	XXXXX	tr	X	X			
SAI 22	QPl			XXXXX	tr	tr	tr			
SAI 23	QPl			XXXXX	tr	tr				X
SAI 24	QPl			XXXXX	tr	X	tr	tr		
SAI 25	QPl			XXXXX	tr	X	tr	tr	tr	
SAI 40	QPl			XXXXX	tr	X	tr	tr	tr	
SAI 41	QPl		X	XXXX	X	XXX	tr	tr	tr	
SAI 42	QPl	tr	tr	XXXXX	tr	X	tr	tr	tr	
SAI 43	QPl		X	XXXXX	tr	X	tr	tr	tr	
SAI 44	QPl		X	XXXXX	tr	X	X			
SAI 45	QPl			XXXXX	tr	tr	tr			
SAI 46	QPl		X	XXXXX	tr	XX	tr	tr		
SAI 47	QPl		tr	XXXXX	tr	tr	tr			
SAI 48	QPl		tr	XXXXX		X	X			
SAI 49	QPl	X	X	XXXXX	X	X	tr			
SAI 51	QPl			XXXXX	tr	tr				X
SAI 52	QPl	X	X	XXXX	X	XX	tr		tr	
SAI 53	QPl	X	tr	XXXXX	X	X	tr		tr	
SAI 54	QPl	X	X	XXXX	X	XX	tr		tr	

Table 4.4a Mineral composition of the analysed ceramic samples and clay lumps (C.M.: Clay Minerals; Qtz: Quartz; Kfs: K-feldspar; Pl: Plagioclase; Cal: Calcite; Px: Pyroxene; Amp: Amphibole; Zrn: Zircon). Quantities: XXXXX - predominant; XXXX - abundant; XXX - good; XX - moderate; X - scarce; tr – traces.

Sample	Group	C.M.	Qtz	Kfs	Pl	Cal	Px	Amp	Zrn
SAI 55	QPl	X	X	XXXX	X	XXX	tr		tr
SAI 56	QPl		X	XXXX	X	XX	tr		tr
SAI 59	QPl			XXXXX	tr	tr			
SAI 60	QPl	X	tr	XXXXX	X	X	tr		
SAI 61	QPl		tr	XXXXX	tr	tr	X		
SAI 62	QPl			XXXXX	tr	tr	tr		
SAI 63	QPl			XXXXX		tr	tr	tr	X
SAI 64	QPl		tr	XXXXX		X	tr	tr	tr
SAI 65	QPl		X	XXXXX	tr	X		tr	tr
SAI 66	QPl	X	X	XXXXX	tr	X	tr	tr	tr
SAI 67	QPl			XXXXX	tr	tr			
SAI 69	QPl			XXXXX	tr	X	tr		
SAI 26	QKfs		X	XXX	XXXX				
SAI 28	QKfs		tr	XXXX	XXXX				
SAI 29	QKfs	XXX	XXXX	XXXX	XX	X	X	tr	
SAI 30	QKfs		X	XXX	XXXX				
SAI 32	QKfs			XXXXX	X		tr		
SAI 33	QKfs		XX	XXXX	XXX				
SAI 34	QKfs		XX	XXXX	XXXX				
SAI 35	QKfs		XX	XXXX	XXXX				
SAI 36	QKfs		X	XXXX	XXXX				
SAI 37	QKfs		XX	XXXX	XXXX			tr	
SAI 38	QKfs		XXX	XXXX	XXXX		tr	tr	
SAI 39	QKfs		XX	XXXX	XXX			tr	
SAI 50	QKfs		tr	XXXXX	X	tr			tr
SAI 58	QKfs		X	XXXX	XX	XX	tr	tr	tr
SAI 68	QKfs		XX	XXXX	XX	XX		tr	tr
SAI 70	QKfs		X	XXXX	XX	X	tr	tr	tr
SAI 71	QKfs		X	XXXX	XXXX	XX	X	tr	
SAI 72	QKfs		XX	XXXX	XXX	tr	tr	tr	
SAI 73	QKfs		X	XXXX	XXX	X	tr	tr	
SAI 74	QKfs		tr	XXXX	XX				
SAI 75	QKfs			XXXX	XX				
SAI 76	QKfs		tr	XXXXX	XX	X	tr	tr	tr
SAI 77	QKfs		XX	XXXX	XXXX	tr	tr	tr	
SAI 27	Q		tr	XXXXX					
SAI 31	Q		tr	XXXXX					

Table 4.4b Mineral composition of the analysed ceramic samples and clay lumps (C.M.: Clay Minerals; Qtz: Quartz; Kfs: K-feldspar; Pl: Plagioclase; Cal: Calcite; Px: Pyroxene; Amp: Amphibole; Zrn: Zircon). Quantities: XXXXX – predominant; XXXX – abundant; XXX – good; XX – moderate; X – scarce; tr – traces.

On the basis of the different phase associations and the varying quantities of individual minerals observed, three main petro-mineralogical groups were distinguished: QPl (quartz-plagioclase), QKfs (quartz-K-feldspar) and Q (quartz) (D'Ercole *et al*. 2015) **(Table 4.4 and Figure 4.2)**.

In the first group, QPl (SAI 01-25, 40-49, 51-56, 59-67 and 69), samples are composed mainly of quartz (Qtz), together with a variable proportion of plagioclase (Pl) and by scarce amounts or traces of K-feldspar (Kfs), pyroxene (Px), and calcite (Cal). Micas, when they occur, appear only as traces or are at most scarce. Some samples from this group show reflexes of clay minerals (C.M.), while others have traces of amphibole (Amp) or, more rarely, zircon (Zrn) (SAI 23, 51 and 63). This group is the largest and includes all samples from site 8-B-10A (SAI 01-10, 40-42 and the clay lumps SAI 52-56) and from 8-B-52A (SAI 11-21 and 43-48) and most of the ceramics from 8-B-76 (SAI 22-25, 49, 51, 59-67 and 69).

In the second group, QKfs (SAI 26, 28-30, 32-39, 50, 58, 68 and 70-77), the diffraction spectrum is characterised above all by the presence of quartz (Qtz), K-feldspar (Kfs) and micas. K-feldspar (microcline and/or anorthoclase) occur most of the time (SAI 26, 28-30, 33-39, 71-73 and 77) in very high proportions (good to abundant) and in certain samples (SAI 26 and 30) is more common than quartz. Micas too, unlike in the QPl group, are almost always present in significant amounts, while plagioclase, when occur, is normally less common than K-feldspar. Only occasionally are traces of calcite (Cal), pyroxene (Px), or amphibole (Amp) recorded. The samples SAI 32, 50, 58, 68, 70 and 74-76, though part of this group, contain however a decidedly lower proportion of K-feldspar. In these samples, micas often appear only as traces, or at least in very limited quantities. Additionally, in comparison with the rest of the group, the balance between K-feldspar and plagioclase is more even. Of the twenty-three pottery samples in the QKfs group, 20 come from site 8-B-10C, and 3 (SAI 50, 58 and 68) from site 8-B-76.

Finally, two samples (SAI 27 and 31) showed only the typical peak of quartz in the diffraction spectra. These samples were assigned to group Q (quartz) and come from site 8-B-10C.

Optical microscopy

Microscopic observations on thin sections were consistent with the groupings that came out of the diffraction analysis, and confirm the use of non-calcareous clays as the raw material for all of these ceramics. Nevertheless, some groups, particularly the first group (QPl) though substantially uniform from the point of view of their mineralogical composition, were shown through petrographic examination to have notable differences within them – differences which can for the most part be explained by technological factors – which permit a further level of classification (D'Ercole *et al.* 2015).

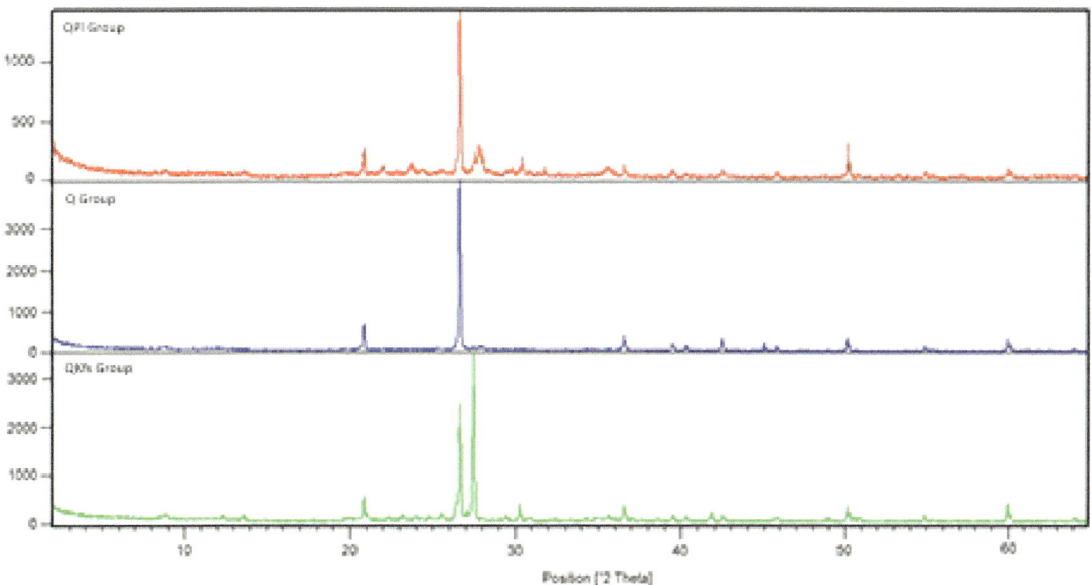

Figure 4.2 Diffractograms representative of the three identified XRD groups of ceramics: QPl (quartz-plagioclase); Q (quartz) and QKfs (quartz-K-feldspar).

1- QPl group (quartz-plagioclase) **(Plates 4.1, 4.2 and 4.3)**

Most of the samples from the QPl group (SAI 01-25, 40-49, 51-56, 59-67 and 69) display a unimodal texture, with grain size mainly ranging from the class of very fine (~ 0.062-0.125mm) to fine grained sand (~ 0.125-0.25mm).

The framework is composed mainly of mono-crystalline (predominant) and poly-crystalline quartz, with straight and undulose extinction, sub-angular to sub-rounded, with grain size regularly less than 0.5mm. Quartz grains larger than 0.5mm usually have well rounded edges. Among other mineral phases present were: plagioclase, K-feldspar (often altered and partially replaced by crystals of sericite mica), biotite and muscovite micas, iron oxides, as well as traces of clinopyroxene (i.e., augite), epidote, of varying type and relief, and amphibole, including green hornblende. Reasonably large aggregates of micrite calcite (between 0.5 and 2mm or even larger) were also common, distributed randomly through the paste, often tending to cement other clasts **(Figure 4.3: b)**; as well as rock fragments of metamorphic and, occasionally, volcanic origin (tephra) **(Plate 4.3: a)**.

In the examination by SEM, certain accessory minerals such as ilmenite (titanium-iron oxide) and zircon were also identified **(Figures 4.3: a-a1 and 4.4)**.

Organic inclusions (charcoal and/or vegetable matter), collophane (microcrystalline hydrated fluoroapatite) and clay pellets, in varying proportions, are characteristic of most of the samples in the QPl group.

Birefringence of the matrix is often masked by carbonaceous matter in the pastes, but is normally medium to medium-low. The porosity is usually very high (estimated at 20% to 35% of the volume), consisting in large part of voids left by burnt up organic matter. Shrinkage cracks, varying in frequency from sample to sample, often parallel to the surface, follow the morphology of the primary pores, sometimes enlarging them and/or cutting across them. In some of the samples these cracks are filled with secondary calcite.

In this group, four of different fabrics can be distinguished, based on the type and quantity of organic inclusions in the pastes (cf., D'Ercole *et al.* 2015).

Fabric 1 (QPl-Veg) - samples rich in organic inclusions (SAI 11-21 and 43-48) **(Plate 4.1)**: the common characteristic of this fabric is the significant presence of organic matter, most probably added intentionally to the paste as a temper. The organic inclusions can occur simply in the form of voids/pores, if during firing they have been completely burnt up, or else may have preserved, at least in part, the structure of the original vegetable fibre **(Plate 4.3: b-d)**. Most of them show an elongated tubular shape and seem to consist either of plant remains, grains and glumes, or of particles of animal dung in the form of crumbled plant remains resulting from herbivore digestion (cf., D'Ercole *et al.* 2015, Livingstone Smith 2001). Varying from sample to sample, the orientation of the organic fibres can be completely random, or else lie parallel or sub-parallel to the surface. The fibres are between approximately 0.5 and 3mm long, though in some samples the fibres are considerably longer.

The QPl-Veg fabric samples come exclusively from site 8-B-52A and can be divided into three further sub-groups, with similar composition, but with differing grain sizes. In samples SAI 12, 13, 15, 18 and 44 (Fabric 1a) the grain size is between 0.125 and 0.25mm **(Plate 4.1: a-b)**. Samples SAI 11, 14, 17, 20, 21, 43, 46 and 48 (Fabric 1b) have a finer texture (~ 0.062-0.125mm) and smaller organic inclusions **(Plate 4.1: c-d)**. These organics are iso-oriented and parallel to the surface of the samples in SAI 17 and 48, but in SAI 11, 14, 20, 21, 43 and 46 their distribution seems to be completely random. In addition, in samples

SAI 14, 21 and 46, as well as vegetable inclusions, sub-millimetric particles of charcoal can be seen. Samples SAI 16, 19, 45 and 47 display further differences in texture and form the third sub-group, Fabric 1c. They have bimodal grain size and highly developed primary porosity. They are also distinguished by the presence of large well-rounded grains (about 0.5 to 2mm) of mono- and polycrystalline quartz, perhaps added to the paste as a temper, and of a greater amount of micrite calcite, this too with rather large grains **(Plate 4.1: e-f)**.

Fabric 2 (QPl) - samples with few organic inclusions (SAI 01, 02, 04, 05, 40 and 42) **(Plate 4.2: a-b)**: these samples, all from site 8-B-10A, contain organic inclusions, but regularly in lower proportions than in the previous fabric. Consequently, estimates of the overall porosity give a lower percentage too.

Fabric 3 (QPl-Col) - samples containing collophane and clay pellets and with few organic inclusions (SAI 03, 06-10, 41, 65 and clay lumps SAI 52-56) **(Plate 4.2: c-d)**: these samples are distinguished by a particular texture, with a very fine grain size (~ 0.062-0.125mm), low shrinkage porosity and a higher proportion of plagioclase and clinopyroxene, corresponding to a more limited quantity of amphibole and metamorphic rock fragments. Finally, they contain a significant amount of clay pellets, collophane and carbon inclusions **(Figures 4.5 and 4.6)**. The samples made from this fabric come from site 8-B-10A, or else, in one case, sample SAI 65, from site 8-B-76.

Fabric 4 (QPl-Cha) - samples with charcoal inclusions (SAI 22-25, 49, 51, 59-64, 66-77 and 69) **(Plate 4.2: e-f)**: in these samples, the organic content consists almost entirely of sub-millimetric angular particles of charcoal, which were most likely deliberately added to the paste **(Figure 4.7: a)**. These inclusions are distributed through the matrix in varying quantities – more in SAI 24-25, 60-62, 64 and 66 (Fabric 4a), and less in SAI 22-23, 49, 51, 59, 63, 67 and 69 (Fabric 4b). Fabric 4b in most cases has bimodal grain size, and always has highly developed primary porosity. Additionally, partially burned vegetable fibres can be seen together with the charcoal. The QPl-Cha fabric samples are all from site 8-B-76. They share a medium or high level of shrinkage porosity, and quartz clasts with rounded edges. In the composition of the pastes, there are varying proportions of plagioclase, of K-feldspar, rock fragments, micrite calcite aggregates and micas. Certain accessory minerals, such as pyroxene, amphibole and epidote – identified with the SEM – are always present, even if as traces, as well as iron oxide and titanium (ilmenite) and zircon **(Figures 4.3: a-a1 and 4.4)**.

2- QKfs group (quartz-K-feldspar) **(Plates 4.3 and 4.4)**

OM analysis of the samples from the QKfs group (SAI 26, 28-30, 32-39, 50, 58, 68 and 70-77) led to the identification of certain differences in composition, already indicated in part by the diffraction data (i.e., the variation in quantities of K-feldspar and micas, noted on the mineralogical table 4.4). Based on these differences, two distinct fabrics can be defined (D'Ercole *et al.* 2015).

Fabric 1 (QKfs-Bt) - samples with biotite (Bt) and significant amounts of K-feldspar (SAI 26, 29-30, 33-39, 70-72 and 76) **(Plate 4.4: a-b)**: these samples are unimodal with a seriate range of grain sizes, where coarse sand (~ 0.5-1mm) is the predominant grain type. Among the non-plastic inclusions, the most important are grains of mono- and polycrystalline quartz, of medium or large dimensions, which are angular or sub-angular, and K-feldspar, micas (principally altered biotite laths), and rock fragments of exclusively metamorphic origin. Iron oxide, iron ores, titanium and zircon are found among the accessory phases **(Figure 4.7: b)**. In this fabric, calcite is rare or completely absent, and even the proportion of minerals of volcanic origin (pyroxene, amphibole and tephra) seems in general very low. Plagioclase, characteristic of the preceding group (QPl), is seen only in limited quantities. A distinctive characteristic is the absence of organic inclusions, collophane and clay pellets. The matrix always displays a medium-high or high level of birefringence. Additionally, as the mineral component increases (percentage of volume, 40-50) the porosity, including secondary/shrinkage porosity, decreases (percentage of volume, 10-15, rarely 20). Samples SAI 71 and 76, despite the lower amount of K-feldspar and finer texture (~ 0.25-0.5mm) are members of this fabric group because of the significant presence in their pastes of micas and in particular of biotite.

Fabric 2 (QKfs-Ms) - samples with muscovite (Ms) and smaller amounts of K-feldspar (SAI 28, 32, 73-75 and 77) **(Plate 4.4: c-d)**: these samples have a finer texture (~ 0.125-0.25 or 0.25-0.5mm) than the previous fabric, a medium rather than high birefringence of the matrix, and often more highly developed shrinkage porosity. The framework is formed predominantly from mono- and polycrystalline quartz, along with varying quantities of metamorphic rock fragments, K-feldspar (often altered and less than in the preceding fabric), and small amounts of plagioclase, clinopyroxene and epidote. Among the micas, muscovite and sericite, formed through the alteration of K-feldspar and visible where the feldspar geminate **(Plate 4.3: e)**, is more common than biotite. Traces of secondary calcite are common to all the samples of this fabric **(Plate 4.3: f)**. Samples SAI 32, 74 and 75 contain, in addition, few completely burnt up organic inclusions of vegetable origin. Samples SAI 28, 73 and 77, though they contain rather elevated levels of K-feldspar, belong in this fabric group because they contain only small amounts of micas (muscovite) and have a finer texture (~ 0.25-0.5mm) than the ceramics of the QKfs-Bt fabric.

Both the QKfs-Bt fabric samples and the QKfs-Ms samples come from site 8-B-10C.

Finally, in the QKfs group, the three samples SAI 50, 58 and 68 have certain particular characteristics: bimodal grain size, particularly significant shrinkage porosity, presence in the pastes – even if scarce

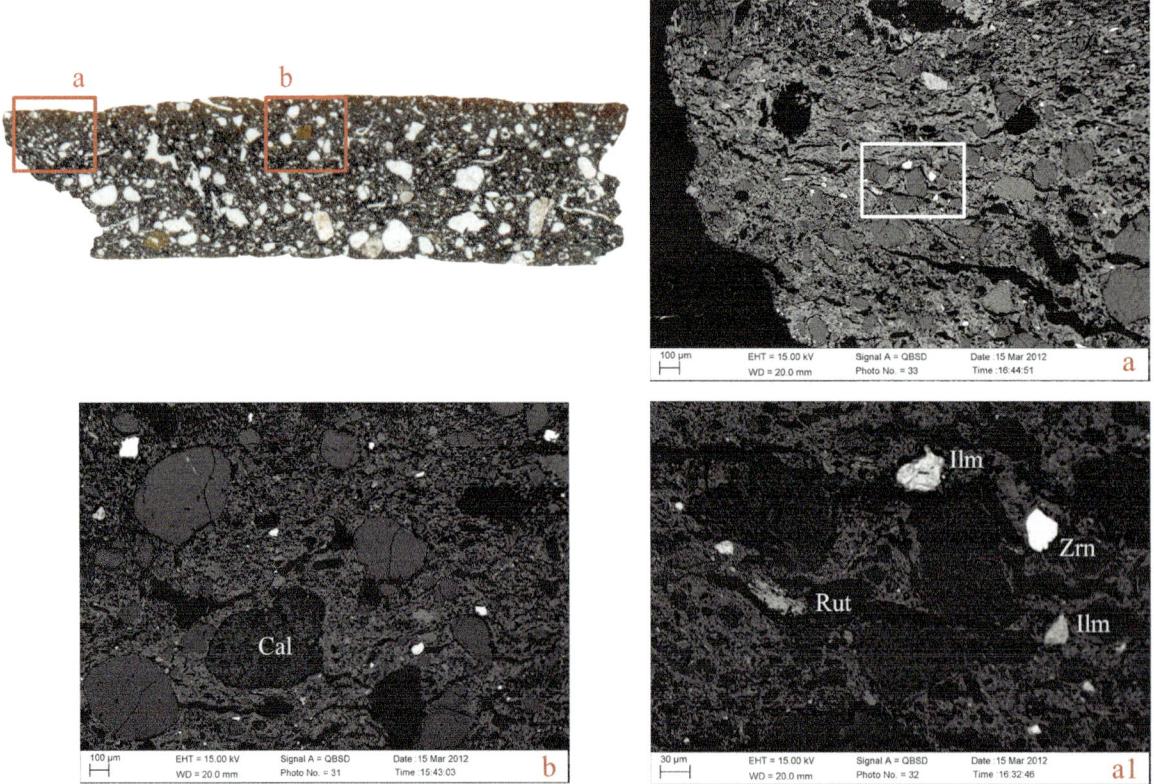

Figure 4.3 Thin section and SEM images showing an Abkan sample from site 8-B-76 with ilmenite (Ilm), zircon (Zrn) and rutile (Rut) inclusions (Images a-a1) and a micrite calcite aggregate (Image b) (modified after D'Ercole et al. 2017).

– of carbonized organic matter, altered K-feldspar and biotite mica. These three samples are from site 8-B-76.

3- Q Group (quartz) **(Plate 4.4: e-f)**

The remaining two samples, SAI 27 and 31, belonging to group Q **(Plate 4.4: e-f)**, have the same type of texture (~ 0.25-0.5mm) and a similar composition: a predominance of clasts of monocrystalline quartz with undulose extinctions and very rounded edges, traces of muscovite and heavy minerals such as clinopyroxene, amphibole and epidote. Samples SAI 27 and 31 come both from site 8-B-10C.

X-ray fluorescence analysis

The chemical data obtained by XRF analysis allow a more detailed appreciation of the differences between the three principal groups of ceramics QPl, QKfs and Q, as defined above **(Table 4.5)**.

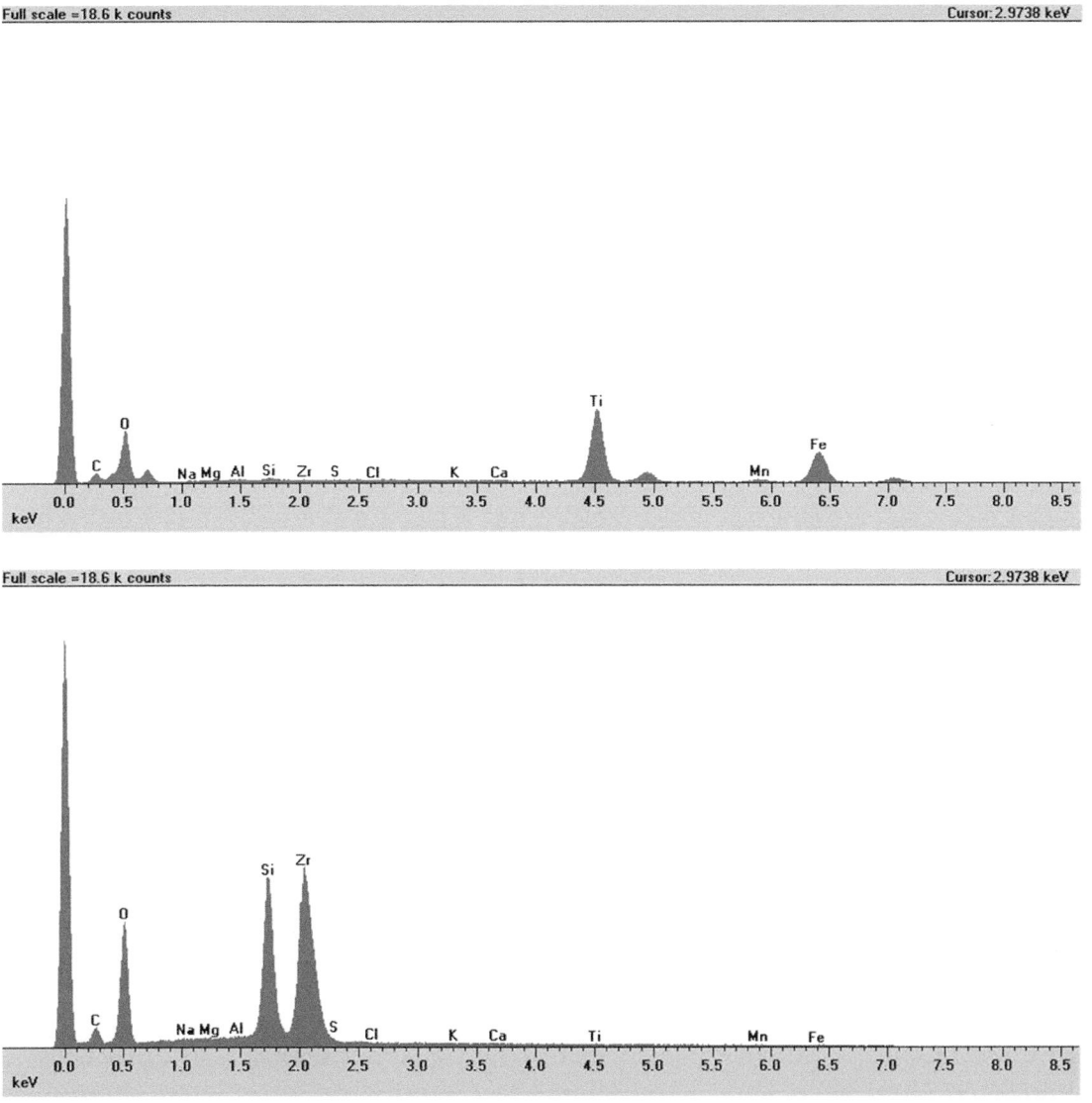

Figure 4.4 EDS spectrum showing the presence of ilmenite (FeTiO3) (on top) and zircon (ZrSiO4) (below).

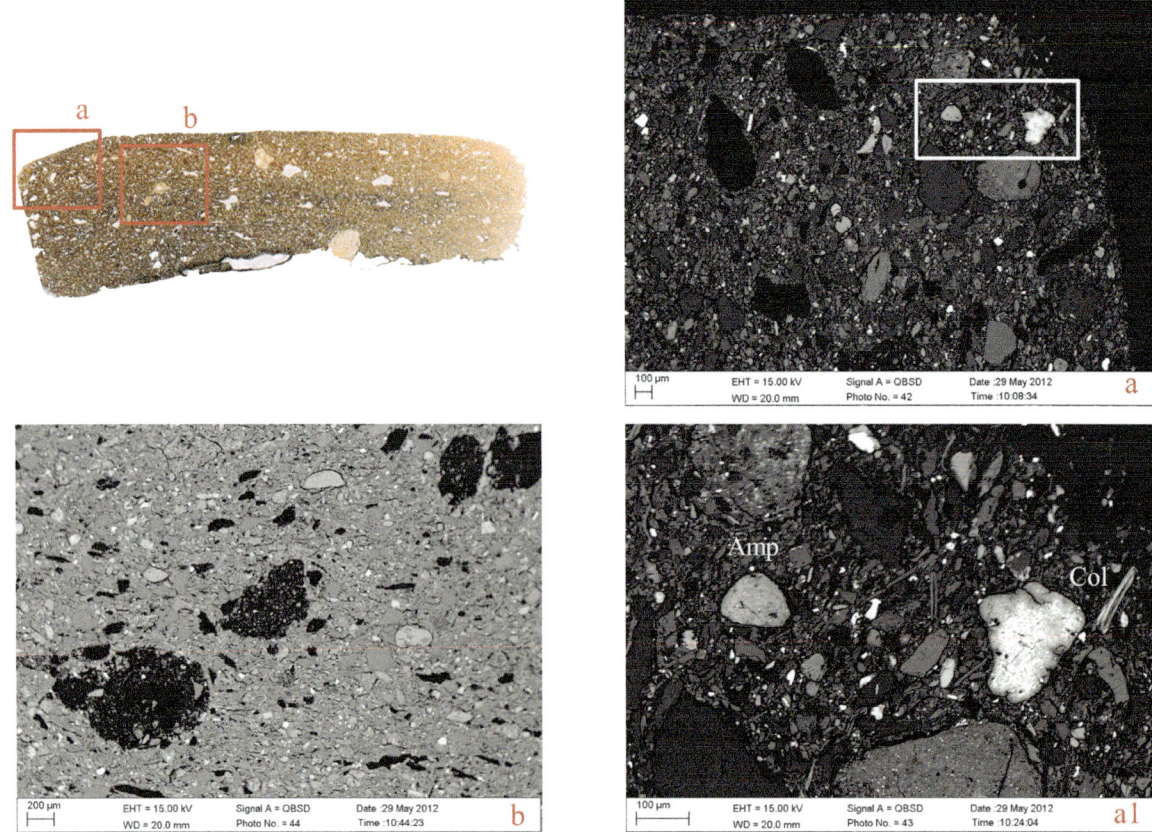

Figure 4.5 Thin section and SEM images showing a Pre-Kerma sample from site 8-B-10A with amphibole (Amp) and collophane (Col) inclusions (Images a-a1) and carbon inclusions (Image b) (modified after D'Ercole et al. 2017).

The oxides of silicon (SiO_2) and aluminium (Al_2O_3) are the most common in all three groups. The proportion they make up of the fabric however, deriving from the presence of silicate minerals, varies from case to case.

The highest levels of SiO_2 are found in the Q (quartz) group (x=64.53; σ=4.17) along with quite large quantities of Al_2O_3: x=15.12; σ=0.15. The other two groups, QPl (quartz-plagioclase) and QKfs (quartz-K-feldspar) contain levels of SiO_2 that are only slightly lower: x=56.32; σ=3.89 in the QPl group, and x=57.06; σ=3.57 in the QKfs group; while the amount of Al_2O_3, similar for the QPl group to the level reported above (x=15.65; σ=0.92), is greater for the QKfs group (x=18.74; σ=1.95).

The QPl and QKfs groups vary from each other additionally in the amounts of Fe_2O_3 and TiO_2 they contain, as well as in the levels of CaO and MgO. There are always higher amounts of iron oxide (Fe_2O_3) and titanium dioxide (TiO_2) in the samples from the QPl group than in the QKfs group: respectively, x=10.46; σ=1.23 as against x=6.66; σ=0.98 for Fe_2O_3, and x=1.80; σ=0.28 against x=0.78; σ=0.19 for TiO_2. This is also true for the oxides of calcium (CaO) and magnesium (MgO) since the samples from the QPl group contain double the quantities found in the QKfs group: x=4.06; σ=0.81 against x=2.09; σ=0.33 for CaO, and x=3.40; σ=0.60 against x=1.37; σ=0.65 for MgO. The percentages for CaO are due to the presence in the QPl group of plagioclase (and in particular, calcium plagioclase such as anorthite, $CaAl_2Si_2O_8$). The levels of iron and titanium oxides however are more probably related to the composition of the matrix itself (the clay-size fraction), and also to heavy mineral accessory phases (i.e. pyroxenes, ilmenite). The quartz group (Q) contain similar levels of CaO (x=2.07; σ=0.21) and MgO (x=1.85; σ=0.22) to the QKfs group,

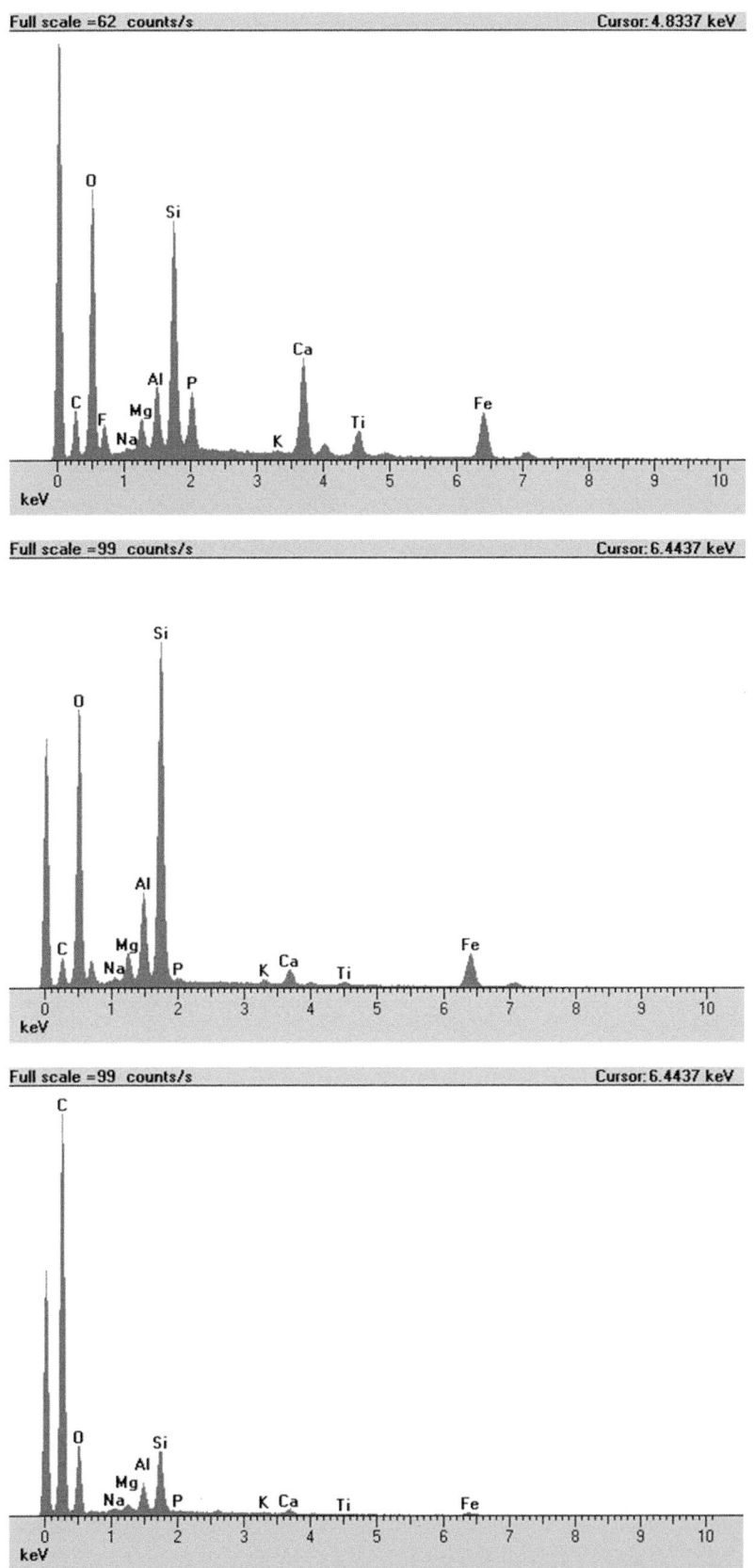

Figure 4.6 EDS spectrum showing the presence of collophane (microcrystalline hydrated fluoroapatite) (on top), amphibole (in the middle) and carbon inclusions (below).

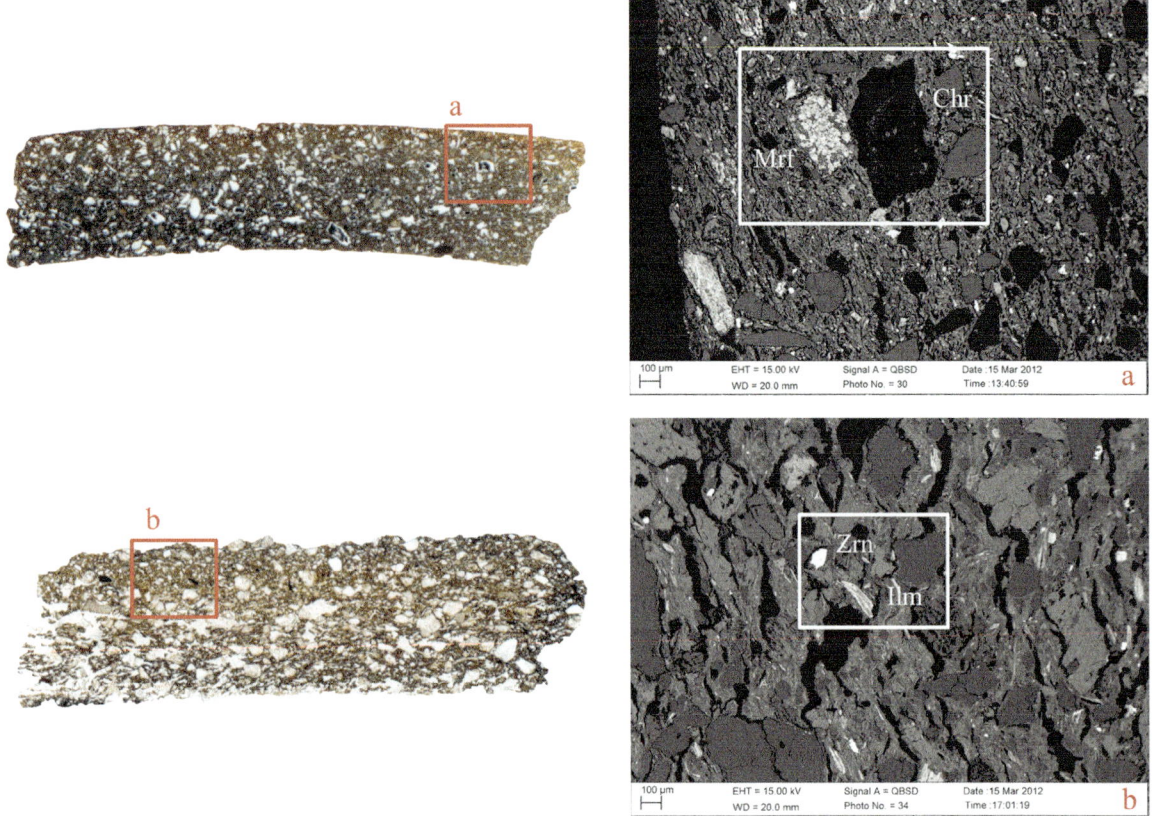

Figure 4.7 Thin sections and SEM images showing an Abkan sample (Image a) with a metamorphic rock fragment (Mrf) and a charcoal inclusion (Chr); and a Khartoum Variant sample from site 8-B-10C (Image b) with zircon (Zr) and ilmenite (Ilm) inclusions (modified after D'Ercole et al. 2017).

while levels of Fe2O3 (x=6.89; σ=0.54) and of TiO_2 (x=0.91; σ=0.09) are very slightly higher. The quantity of potassium oxide (K_2O) provides a similar means of differentiation: since it is related essentially to K-feldspar, it is not surprising that the highest amounts are found in the group QKfs (x=3.65; σ=0.95) and in particular in the samples SAI 26-30, 33-39, 70-73 and 77, which when analysed by XRPD are also distinguished by a higher potassium feldspar content. In the QPl and Q groups, distinctly lower levels of K_2O are found (x=1.12; σ=0.16 for QPl, and x=1.40; σ=0.18 for Q).

The chemical differences that have been found between the three groups of samples become clearer when the results for certain of the main oxides are plotted on scatter diagrams.

On the graph of CaO vs. K_2O **(Figure 4.8)**, the separation between the QPl samples (in blue) and the QKfs group (in orange) can be clearly seen. The values for the two oxides are inversely proportional: a higher content of K_2O corresponds to a lower content of CaO (QKfs samples) and vice versa (QPl samples). It should be noted how the samples from the QKfs group (orange) with a lower proportion of potassium feldspar occupy an intermediate position on the graph, detached from the rest of their group. SAI 27 and 31, from the Q group (in green), seem to fall into the QPl group because of having similar levels of K_2O, but are partially separated from it as a result of the lower levels of CaO.

The following graph, of MgO vs. K_2O **(Figure 4.9)** shows a similar pattern: the relationship between the two oxides, which again correlate negatively, clearly distinguishes the QPl group (blue) from the QKfs samples (orange), leaving separate in the centre the QKfs samples that are less rich in K_2O (SAI 32, 50, 58, 68, 74-76) and, to a lesser extent, the two samples, SAI 27 and 31 from the Q group (green).

4. Archaeometric analysis

#Sample	XRPD Group	SiO_2	TiO_2	Al_2O_3	Fe_2O_3	MnO	MgO	CaO	Na_2O	K_2O	P_2O_5	LOI
SAI 01	QPI	58.78	1.84	15.93	10.68	0.19	3.3	4.1	1.58	1.34	0.26	2.01
SAI 02	QPI	56.96	1.96	16.17	11.43	0.16	3.23	4.2	1.56	1.37	0.6	2.37
SAI 03	QPI	57.13	2.02	16.39	11.67	0.19	3.04	4.04	1.61	1.39	0.35	2.18
SAI 04	QPI	55.25	1.77	15.82	10.23	0.17	3.68	5.82	1.82	1.2	0.22	4.02
SAI 05	QPI	56.05	2.21	15.83	12	0.23	3.64	3.8	1.95	1.03	0.24	3.04
SAI 06	QPI	52.75	2.1	16.25	11.64	0.35	3.87	4.93	1.36	1.08	0.23	5.44
SAI 07	QPI	56.22	2.29	16.81	12.34	0.23	3.11	3.29	1.99	1.2	0.26	2.26
SAI 08	QPI	56.07	1.94	16.52	11.22	0.3	3.44	3.52	1.79	1.17	0.32	3.71
SAI 09	QPI	55.37	1.89	16.38	10.93	0.37	3.69	3.52	1.76	1.11	0.29	4.67
SAI 10	QPI	55.14	2.04	15.92	11.28	0.4	3.61	3.53	2	1.08	0.29	4.7
SAI 11	QPI	52.78	2.07	16.34	11.34	0.2	3.6	4.25	1.89	0.88	0.25	6.41
SAI 12	QPI	55.1	1.61	14.26	9.04	0.15	3.34	5.7	1.38	1.02	0.27	8.12
SAI 13	QPI	53.64	2.03	16.81	11.68	0.21	3.33	3.96	1.4	1.1	0.27	5.57
SAI 14	QPI	53.18	1.95	15.68	10.95	0.18	3.56	3.91	2.5	1.01	0.31	6.77
SAI 15	QPI	51.33	1.66	16.08	10.15	0.13	3.88	5.11	2.19	0.95	0.24	8.27
SAI 16	QPI	58.92	1.58	14.91	9.09	0.12	2.78	3.7	2.45	0.91	0.27	5.27
SAI 17	QPI	51.02	1.89	16.24	11.06	0.17	3.58	3.83	2.75	0.9	0.29	8.28
SAI 18	QPI	57.13	1.91	15.9	10.76	0.17	3.71	4.22	2.19	1.18	0.25	2.58
SAI 19	QPI	54.52	1.59	14.34	9.13	0.14	3.04	4.68	2.44	1.24	0.8	8.07
SAI 20	QPI	53.4	2.02	15.98	11.25	0.22	3.5	4.85	1.78	0.98	0.35	5.67
SAI 21	QPI	55.06	1.65	16.07	9.29	0.12	2.99	4.03	1.44	0.93	0.2	8.21
SAI 22	QPI	64.15	1.53	12.52	8.55	0.18	2.41	4.73	1.36	1.11	0.14	3.33
SAI 23	QPI	64.44	1.79	15.64	10.31	0.14	2.20	2.85	0.61	1.17	0.15	0.69
SAI 24	QPI	55.37	2.11	16.19	12.12	0.22	3.15	3.8	2.6	1.24	0.22	2.98
SAI 25	QPI	54.46	1.78	15.01	11.44	0.27	3.5	4.21	2.39	0.94	0.19	5.82
SAI 40	QPI	56.39	1.88	15.72	10.56	0.2	3.52	3.95	2.25	1.22	0.26	4.06
SAI 41	QPI	52.8	1.99	16.51	11.74	0.28	4.3	4.15	1.82	1.03	0.26	5.1
SAI 42	QPI	52.95	1.92	15	10.34	0.3	4.17	4.6	1.25	1	0.22	8.25
SAI 43	QPI	51.82	1.88	15.63	10.85	0.16	3.97	5.15	1.85	1.29	0.32	7.08
SAI 44	QPI	53.49	1.6	14.94	9.42	0.14	3.55	5.28	2.12	0.97	0.24	8.25
SAI 45	QPI	63.46	1.44	15.42	8.58	0.12	2.39	3.4	1.15	1.2	0.18	2.65
SAI 46	QPI	51.47	1.81	15.51	10.55	0.16	3.57	4.54	2.27	1.01	0.38	8.73
SAI 47	QPI	56.5	1.73	15.94	10.3	0.17	3.58	3.85	1.88	0.96	0.23	4.85
SAI 48	QPI	52.98	1.75	15.93	10.67	0.15	3.86	5.91	1.44	1.21	0.25	5.85
SAI 49	QPI	56.91	1.41	16.41	8.86	0.14	3.95	2.96	1.7	1.27	0.2	6.2
SAI 51	QPI	63.01	1.84	16.36	10.68	0.13	2.31	2.96	0.65	1.20	0.16	0.71
SAI 52	QPI	53.66	2.15	15.85	11.38	0.29	3.61	4.79	1.1	1.15	0.2	5.81
SAI 53	QPI	52.54	2.15	16.86	11.37	0.37	3.78	3.95	1.16	1.15	0.18	6.48
SAI 54	QPI	53.68	2.1	16.17	11.25	0.28	3.59	4.5	1.37	1.09	0.2	5.77
SAI 55	QPI	53.2	2.25	15.95	11.65	0.25	3.57	4.57	1.54	1.18	0.19	5.64
SAI 56	QPI	54.82	2.19	16.02	11.38	0.2	3.53	4.65	1.7	1.23	0.23	4.04

Table 4.5a Major elements composition of the analysed ceramic samples and clay lumps in relation to the identified mineralogical/petrographic groups (x: average composition, σ: standard deviation). Major elements and LOI (Loss on ignition) in weight percent (w%).

Sample	Group	SiO2	TiO2	Al2O3	Fe2O3	MnO	MgO	CaO	Na2O	K2O	P2O5	LOI
SAI 59	QPl	58.89	1.55	15.45	10.26	0.17	2.35	2.81	2.07	1.19	0.16	5.10
SAI 60	QPl	59.28	1.23	15.06	8.41	0.16	3.83	2.72	1.59	1.01	0.16	6.55
SAI 61	QPl	60.81	1.18	13.82	7.00	0.14	3.26	3.79	1.37	1.26	0.17	7.20
SAI 62	QPl	65.35	1.23	13.19	6.99	0.14	2.51	3.67	1.11	1.16	0.12	4.53
SAI 63	QPl	65.62	1.71	15.13	9.76	0.14	2.13	2.72	0.61	1.05	0.14	0.98
SAI 64	QPl	56.65	1.73	14.51	10.90	0.25	3.32	3.86	2.56	0.88	0.18	5.15
SAI 65	QPl	55.92	1.82	15.98	10.70	0.21	3.09	3.80	1.54	1.69	0.42	4.83
SAI 66	QPl	55.31	1.69	14.38	10.55	0.24	3.50	4.01	2.07	0.97	0.17	7.11
SAI 67	QPl	64.53	1.41	15.01	8.68	0.13	3.36	2.10	1.30	1.31	0.15	2.01
SAI 69	QPl	56.20	1.15	17.27	11.11	0.20	5.66	3.76	1.18	1.10	0.15	2.24
x		56.32	1.80	15.65	10.46	0.20	3.40	4.06	1.71	1.12	0.25	5.01
σ		3.89	0.28	0.92	1.23	0.07	0.60	0.81	0.51	0.16	0.11	2.20
SAI 26	QKfs	56.81	0.5	21.49	6.27	0.39	1.22	1.91	1.39	4.05	0.09	5.89
SAI 28	QKfs	57.44	0.61	22.52	5.43	1.21	0.91	1.62	1.71	4.22	0.09	4.23
SAI 29	QKfs	53.89	0.82	18.71	7.99	2.27	1.15	2.31	2.76	3.73	0.13	6.26
SAI 30	QKfs	56.84	0.73	17.77	6.66	2.16	1.29	1.93	2.45	4.73	0.15	5.27
SAI 32	QKfs	61.61	1	17.15	6.84	0.46	2.02	2.68	1.82	2.29	0.12	4
SAI 33	QKfs	57.89	0.65	18.45	6.1	1.58	0.73	1.91	2.03	4.73	0.13	5.81
SAI 34	QKfs	55.25	0.67	19.54	7.28	1.68	0.91	2.02	1.95	4.58	0.14	5.97
SAI 35	QKfs	56.02	0.66	19.19	6.72	1.42	0.7	1.82	2.68	4.52	0.13	6.15
SAI 36	QKfs	56.25	0.64	19.11	5.92	1.38	0.75	2.39	2.92	4.6	0.1	5.95
SAI 37	QKfs	54.11	0.74	20.54	7.47	1.6	0.89	2.19	1.7	4.53	0.1	6.12
SAI 38	QKfs	55.08	0.66	19.35	6.89	1.52	1	1.8	2.72	4.39	0.14	6.45
SAI 39	QKfs	47.83	0.68	18.8	6.8	1.19	1.1	2.15	1.19	4.26	0.18	15.81
SAI 50	QKfs	59.91	1.08	15.74	7.84	0.17	1.9	2.08	1.92	2.45	0.26	6.66
SAI 58	QKfs	61.21	0.95	16.37	7.65	0.16	2.78	2.71	2.65	2.55	0.14	2.82
SAI 68	QKfs	59.84	1.22	18.66	8.27	0.14	3.34	2.53	1.62	2.3	0.25	1.82
SAI 70	QKfs	52.6	0.86	17.93	8.44	0.86	1.49	2.55	1.93	3.11	0.09	10.14
SAI 71	QKfs	56.87	0.82	18.01	6.96	1.23	1.23	2.14	1.92	3.94	0.11	6.78
SAI 72	QKfs	56.23	0.71	17.99	6.31	1.65	1.37	2.03	1.32	4.66	0.11	7.61
SAI 73	QKfs	55	0.56	20.14	5.9	1.04	0.88	1.78	2.22	3.15	0.09	9.23
SAI 74	QKfs	63.38	1.04	15.66	5.42	0.34	1.66	1.89	2.07	2.43	0.08	6.03
SAI 75	QKfs	63.92	1.05	15.25	5.33	0.14	1.61	1.95	2.44	2.39	0.08	5.84
SAI 76	QKfs	57.86	0.8	21.14	5.65	0.45	1.6	2.24	2.76	2.39	0.07	5.03
SAI 77	QKfs	56.53	0.52	21.62	4.98	0.55	1.03	1.46	1.22	3.85	0.1	8.15
x		57.06	0.78	18.74	6.66	1.03	1.37	2.09	2.06	3.65	0.13	6.44
σ		3.57	0.19	1.95	0.99	0.67	0.65	0.33	0.53	0.95	0.05	2.73
SAI 27	Q	61.58	0.97	15.01	7.27	1.29	1.69	2.22	1.83	1.27	0.19	6.69
SAI 31	Q	67.48	0.84	15.22	6.51	0.25	2.00	1.92	1.12	1.52	0.28	2.86
x		64.53	0.91	15.12	6.89	0.77	1.85	2.07	1.48	1.40	0.24	4.78
σ		4.17	0.09	0.15	0.54	0.74	0.22	0.21	0.50	0.18	0.06	2.71

Table 4.5b Major elements composition of the analysed ceramic samples and clay lumps in relation to the identified mineralogical/petrographic groups (x: average composition, σ: standard deviation). Major elements and LOI (Loss on ignition) in weight percent (w%).

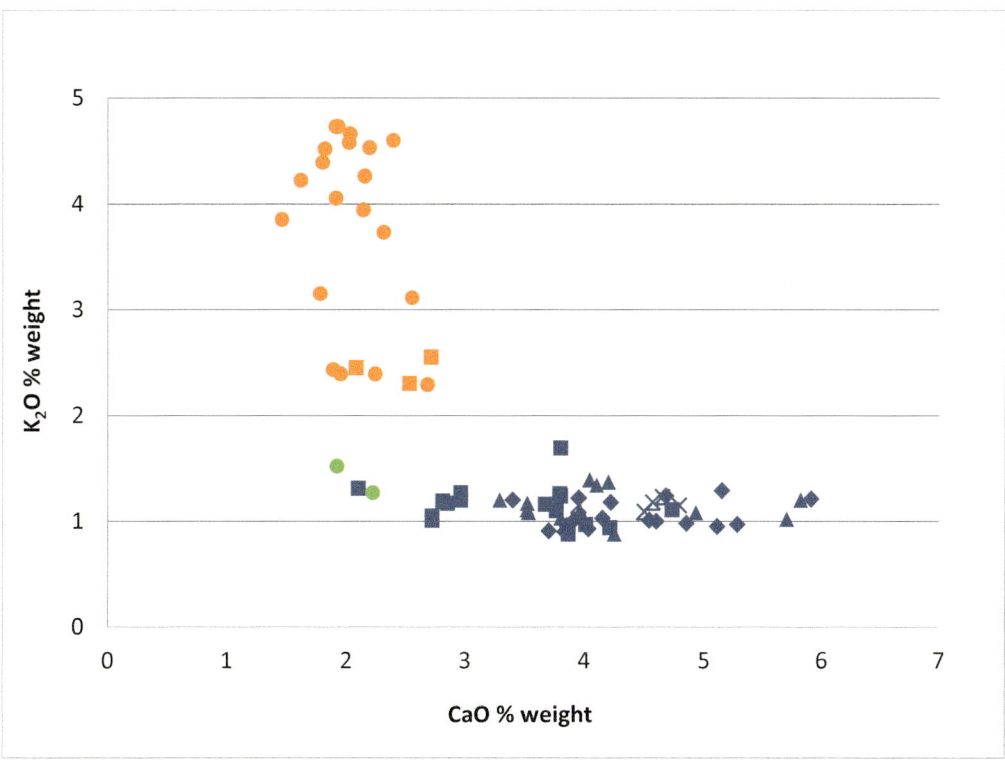

Figure 4.8 CaO vs. K_2O plot (wt %). The colour of symbols refers to the XRD groups: blue = QPl group; orange = QKfs group; green = Q group. The shape of the symbols refers to the sites: triangles are samples from site 8-B-10A; diamonds are samples from site 8-B-52-A; squares are samples from site 8-B-76; circles are samples from site 8-B-10C; crosses are the clay lumps from site 8-B-10A.

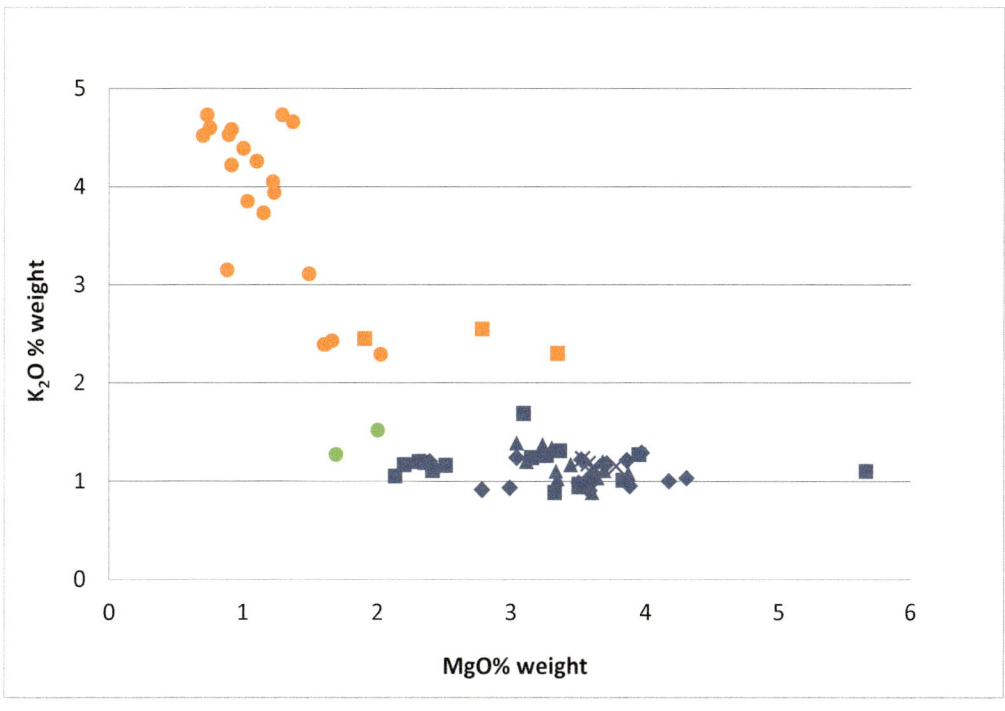

Figure 4.9 MgO vs. K_2O plot (wt %). The colour of symbols refers to the XRD groups: blue = QPl group; orange = QKfs group; green = Q group. The shape of the symbols refers to the sites: triangles are samples from site 8-B-10A; diamonds are samples from site 8-B-52-A; squares are samples from site 8-B-76; circles are samples from site 8-B-10C; crosses are the clay lumps from site 8-B-10A.

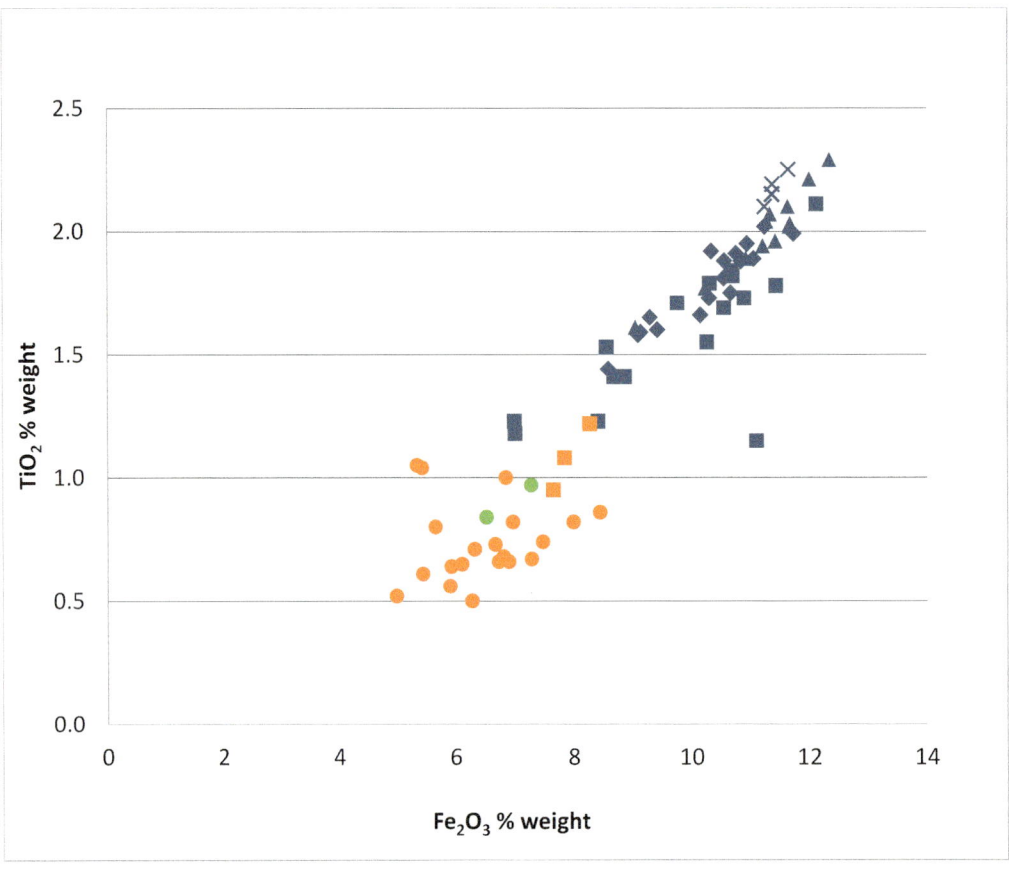

Figure 4.10 Fe₂O₃ vs. TiO₂ plot (wt %). The colour of symbols refers to the XRD groups: blue = QPl group; orange = QKfs group; green = Q group. The shape of the symbols refers to the sites: triangles are samples from site 8-B-10A; diamonds are samples from site 8-B-52A; squares are samples from site 8-B-76; circles are samples from site 8-B-10C; crosses are the clay lumps from site 8-B-10A.

The relationship between Fe_2O_3 and TiO_2 **(Figure 4.10)** shows a strong positive linear correlation. This means that the two oxides are geochemically related, and that as proportions of one of them increase so do proportions of the other (QPl samples, in blue, with higher levels of Fe_2O_3 and TiO_2) or vice versa (QKfs samples, in orange, and Q samples, in green, with distinctly lower levels of Fe_2O_3 and TiO_2). It is worth to note that the levels of Fe_2O_3 and TiO_2 clearly increase in the samples proportionally to their age.

The results for a group of trace elements – rubidium (Rb), strontium (Sr), ittrium (Y) zirconium (Zr) and niobium (Nb) – confirmed the distinctions between the three main groups made on the basis of petrographic and mineralogical analysis, in particular as indicated by the chemical variability observed for certain elements typical of resistate or absorbate phases of phyllosilicates, such as Rb or Zr, and, to a lesser extent, Y.

Values for both Rb and Zr are always higher for the QKfs group than for the QPl group: respectively, x=85; σ=20 as against x=45; σ=6 for Rb; and x=487; σ=168 against x=374; σ=87 for Zr **(Table 4.6)**. The Q group has the lowest levels of Zr (x=246; σ=81), while Rb is present in proportions similar to those for the QPl group (x=39; σ=2). Yttrium is found at very slightly higher levels in the QKfs group (x=46; σ=15) than in the QPl group (x=34; σ=3) or the Q group (x=36; σ=7). There do not seem however to be significant variations in the levels of Sr or Nb in the three groups.

# Sample	XRPD Group	Rb	Sr	Y	Zr	Nb
SAI 01	QPI	52	327	34	386	30
SAI 02	QPI	52	334	34	314	25
SAI 03	QPI	52	327	34	352	26
SAI 04	QPI	49	309	35	423	28
SAI 05	QPI	40	345	33	345	30
SAI 06	QPI	38	383	31	279	24
SAI 07	QPI	51	323	38	366	32
SAI 08	QPI	46	321	36	354	30
SAI 09	QPI	43	310	35	352	29
SAI 10	QPI	42	333	35	378	31
SAI 11	QPI	39	340	37	320	33
SAI 12	QPI	42	317	32	408	26
SAI 13	QPI	48	328	37	312	29
SAI 14	QPI	43	315	34	321	30
SAI 15	QPI	43	281	33	310	29
SAI 16	QPI	44	232	34	427	27
SAI 17	QPI	42	325	35	303	31
SAI 18	QPI	50	277	34	393	30
SAI 19	QPI	46	286	31	457	27
SAI 20	QPI	45	365	36	363	31
SAI 21	QPI	45	243	35	468	28
SAI 22	QPI	42	281	31	484	23
SAI 23	QPI	58	207	40	587	29
SAI 24	QPI	54	341	39	352	34
SAI 25	QPI	39	348	36	368	28
SAI 40	QPI	43	356	33	370	29
SAI 41	QPI	42	315	35	325	31
SAI 42	QPI	36	334	34	357	25
SAI 43	QPI	50	347	34	377	30
SAI 44	QPI	42	282	32	357	27
SAI 45	QPI	50	195	38	459	37
SAI 46	QPI	43	305	33	294	29
SAI 47	QPI	49	231	37	426	30
SAI 48	QPI	55	323	34	328	30
SAI 49	QPI	36	284	30	282	19
SAI 51	QPI	62	214	41	640	32
SAI 52	QPI	39	357	31	304	25
SAI 53	QPI	44	314	32	303	25
SAI 54	QPI	39	354	32	284	25
SAI 55	QPI	41	342	31	280	23
SAI 56	QPI	38	375	30	315	24

Table 4.6a Trace elements composition of the analysed ceramic samples and clay lumps in relation to the identified mineralogical/petrographic groups (x: average composition, σ: standard deviation). Trace elements in parts per million (ppm).

Sample	Group					
SAI 59	QPl	57	329	37	286	27
SAI 60	QPl	34	260	27	345	19
SAI 61	QPl	37	232	24	392	19
SAI 62	QPl	43	256	27	539	23
SAI 63	QPl	58	200	35	650	30
SAI 64	QPl	42	324	35	384	28
SAI 65	QPl	46	439	34	270	27
SAI 66	QPl	36	356	32	349	26
SAI 67	QPl	49	230	28	370	22
x		45	307	34	374	28
σ		6	52	3	87	4
SAI 26	QKfs	140	272	30	200	8
SAI 28	QKfs	116	242	47	239	10
SAI 29	QKfs	91	316	69	688	39
SAI 30	QKfs	95	322	72	622	32
SAI 32	QKfs	64	402	30	525	22
SAI 33	QKfs	87	320	54	508	27
SAI 34	QKfs	89	314	53	506	29
SAI 35	QKfs	77	315	56	528	29
SAI 36	QKfs	85	319	54	469	27
SAI 37	QKfs	85	422	51	588	33
SAI 38	QKfs	87	320	57	540	31
SAI 39	QKfs	89	343	56	565	33
SAI 50	QKfs	57	246	46	518	23
SAI 58	QKfs	76	272	28	337	35
SAI 68	QKfs	82	256	27	279	20
SAI 70	QKfs	75	296	57	753	39
SAI 71	QKfs	76	329	57	630	34
SAI 72	QKfs	92	311	65	645	31
SAI 73	QKfs	98	293	41	245	11
SAI 74	QKfs	63	271	26	638	22
SAI 75	QKfs	64	258	26	656	23
SAI 76	QKfs	59	283	29	277	14
SAI 77	QKfs	120	238	36	244	10
x		85	303	46	487	25
σ		20	46	15	168	9
SAI 27	Q	37	341	41	189	14
SAI 31	Q	41	214	32	303	16
x		39	277	36	246	15
σ		2	90	7	81	1

Table 4.6b Trace elements composition of the analysed ceramic samples and clay lumps in relation to the identified mineralogical/petrographic groups (x: average composition, σ: standard deviation). Trace elements in parts per million (ppm).

4. Archaeometric analysis

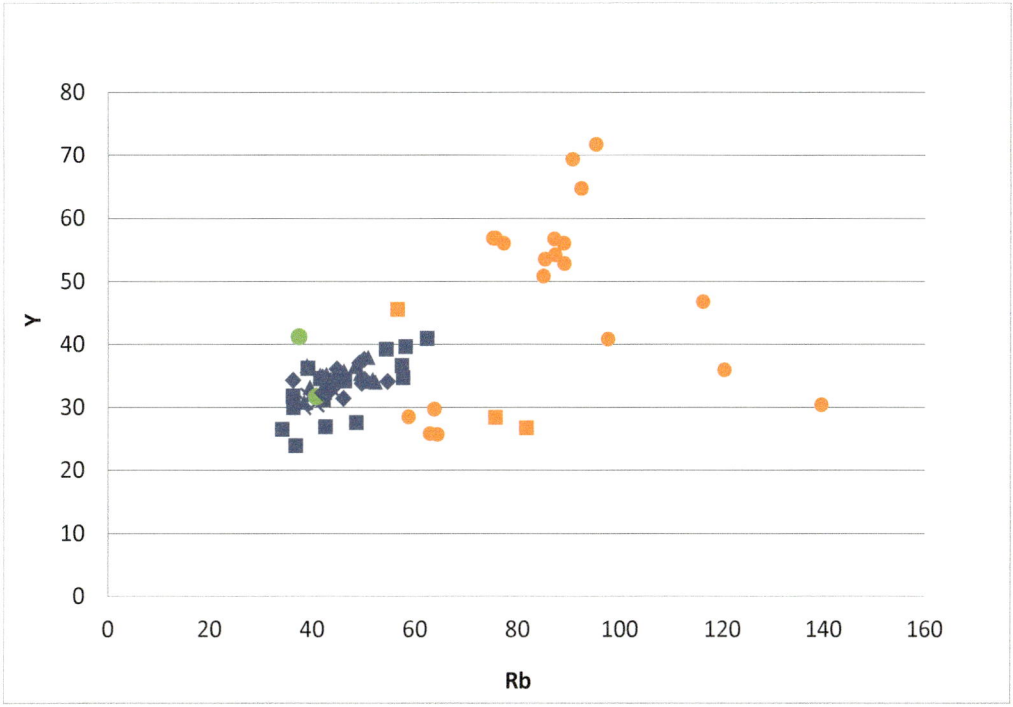

Figure 4.11 Rb vs. Y plot. The colour of symbols refers to the XRD groups: blue = QPl group; orange = QKfs group; green = Q group. The shape of the symbols refers to the sites: triangles are samples from site 8-B-10A; diamonds are samples from site 8-B-52-A; squares are samples from site 8-B-76; circles are samples from site 8-B-10C; crosses are the clay lumps from site 8-B-10A.

The scatter plot of Rb *vs.* Y **(Figure 4.11)** shows the positive correlation between these two elements; the higher the quantities of Rb, the higher the quantities of Y (most of the QKfs group samples, in orange) and vice versa (the QPl group, in blue, and the Q group, in green). SAI 26, 28, 73 and 77 from the QKfs group contain a higher level of rubidium; while SAI 32, 50, 58, 68 and 74-76 have levels of Rb and Y that place them on the graph among the ceramics of the QPl group. Since rubidium is found in K-feldspar as a cation, it is to be expected that lower levels should be found in those samples from the QKfs group with less K-feldspar in them (SAI 32, 50, 58, 68 and 74-76), and greater quantities where K-feldspar are particularly abundant (especially in SAI 26, 28, 73 and 77).

The relationship between Rb and Zr **(Figure 4.12)** shows again two elements with similar geochemical behaviour, and also allows a differentiation of the samples in the QKfs group (orange) from those in the QPl (blue) or Q (green) groups. The QKfs samples, in fact, as noted above, normally have higher levels of Zr, as well as of Rb. The Zr content is also particularly high in samples SAI 23, 51 and 63, where the presence of accessory mineral phases such as zircon was observed in the diffraction and the SEM analyses.

Results: the sediment samples

X-ray powder diffraction analysis

After X-ray powder diffraction analysis, the 16 samples of sediment (SAI 78-93) taken from the sites 8-B-10C, 8-B-76 and 8-B-10A could be divided into two distinct mineralogical groups: QPl (quartz-plagioclase) and QCal (quartz-calcite) (D'Ercole *et al.* 2015) **(Table 4.7 and Figure 4.13)**.

The first group, QPl, (samples SAI 78-88 and 93) has similar characteristics to the QPl group of ceramics described above: the dominant presence of quartz, associated with varying quantities of plagioclase

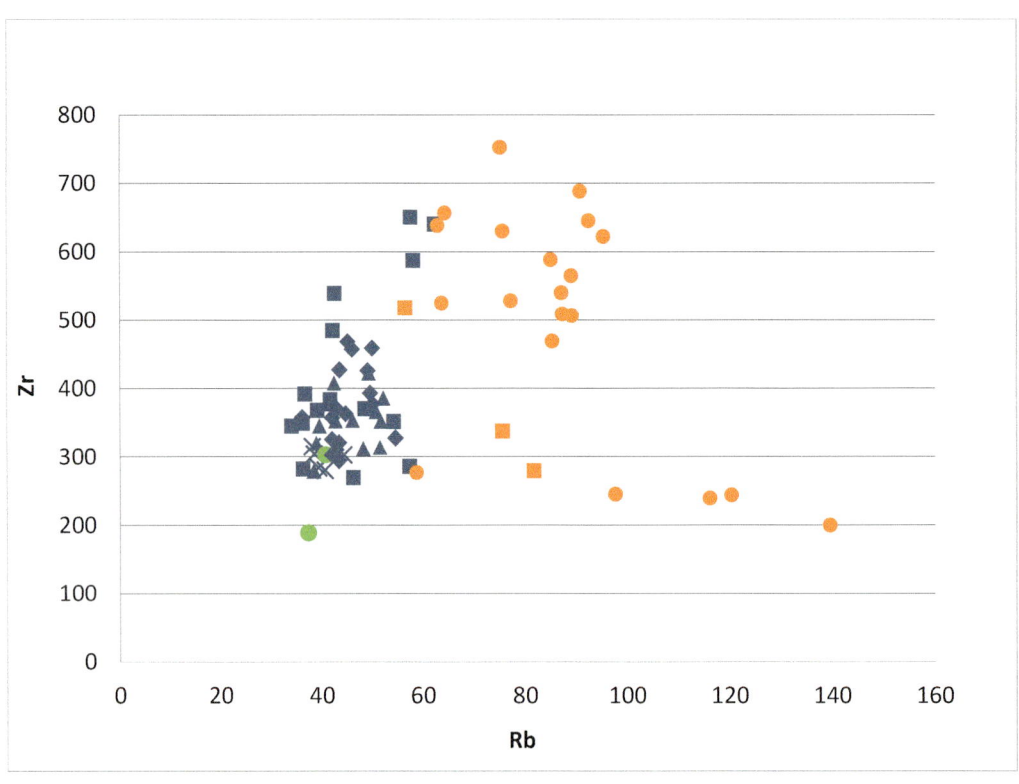

Figure 4.12 Rb vs. Zr plot. The colour of symbols refers to the XRD groups: blue = QPl group; orange = QKfs group; green = Q group. The shape of the symbols refers to the sites: triangles are samples from site 8-B-10A; diamonds are samples from site 8-B-52-A; squares are samples from site 8-B-76; circles are samples from site 8-B-10C; crosses are the clay lumps from site 8-B-10A.

# Sample	XRPD Group	C.M. (Chl)	Micas	Qtz	Kfs	Pl	Cal	Dol	Px
SAI 78	QPl			XXXXX	tr	X	tr		tr
SAI 79	QPl			XXXXX	tr	X	tr		tr
SAI 80	QPl	tr		XXXX	X	XX			tr
SAI 81	QPl	XX	tr	XXXXX		X	tr		tr
SAI 82	QPl	XX	X	XXXX		XXX			tr
SAI 83	QPl	tr	tr	XXXXX		X	tr		
SAI 84	QPl	X	tr	XXXXX	tr	X	tr		tr
SAI 85	QPl	tr	tr	XXXXX		X	tr		tr
SAI 86	QPl	X	tr	XXXXX	tr	X			tr
SAI 87	QPl	X	X	XXXX	tr	XX	tr		tr
SAI 88	QPl	XX	tr	XXXXX	tr	X	tr		tr
SAI 93	QPl	tr	tr	XXXXX	tr	X	X		tr
SAI 89	QCal	XXXX	tr	XX			XXX	tr	
SAI 90	QCal	XXXX		XX			XXXX	X	
SAI 91	QCal	XXXX	X	XXX		tr	XXX	X	
SAI 92	QCal	X	X	XXXX		X	XX	X	

Table 4.7 Mineral composition of the analysed local sediment samples (C.M.: Clay Minerals; Chl: Chlorite; Qtz: Quartz; Kfs: K-feldspar; Pl: Plagioclase; Cal: Calcite; Dol: Dolomite; Px: Pyroxene). Quantities: XXXXX - predominant; XXXX - abundant; XXX - good; XX - moderate; X - scarce; tr - traces.

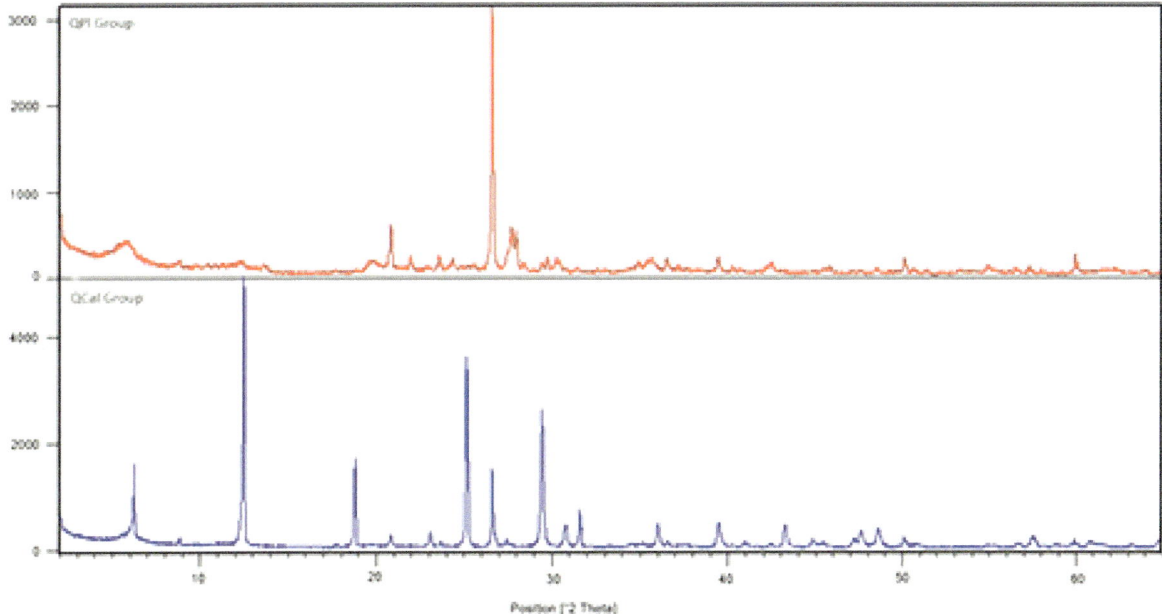

Figure 4.13 Diffractograms representative of the two identified XRD groups of local sediments: QPl (quartz-plagioclase) and QCal (quartz-calcite).

with traces of K-feldspar (Kfs) and micas. Traces of pyroxene (Px) and calcite (Cal) can also be observed. However, unlike the ceramics, most of the sediment samples in this group show the presence of clay minerals (C.M.). The amounts present are scarce in SAI 84, 86 and 87, moderate in SAI 81, 82 and 88, and there are only traces in SAI 80, 83, 85 and 93. Samples 78-82 come from site 8-B-10A, and SAI 83-88 from site 8-B-76. SAI 93 is the only sample in this group that comes from site 8-B-10C.

The other four samples (SAI 89-92) from site 8-B-10C turned out to have a completely different mineralogical composition. They were made of calcareous sediment, containing amounts – moderate to good – of quartz (Qtz), and a significant proportion – from good to abundant – of calcite (Cal) and chlorite (Chl), together with scarce amounts or traces of micas and of dolomite (Dol). In this group, SAI 92 stands out in part because of the abundant quartz content and scarce to moderate levels of calcite (Cal) and chlorite (Chl). The amounts of plagioclase (Pl), micas and dolomite (Dol) are also scarce in this sample.

Optical microscopy

Thin sections for petrographic analysis were prepared from 12 of the samples from the QPl group (SAI 78-88 and 93). The other four samples from the QCal group (SAI 89-92) were not analysed since the mineralogical analysis was already in itself enough to exclude any possibility of their being compatible with the ceramic petrofacies (D'Ercole *et al.* 2015).

The QPl group samples (SAI 78-88 and 92) all display unimodal grain size, and a framework composed predominantly of clasts of mono- and polycrystalline quartz, with straight and undulose extinction, grain size rarely greater than 0.5mm, sub-angular to sub-rounded (larger quartz clasts, as observed above for the ceramic samples, usually have well rounded edges). Plagioclase, K-feldspar, aggregates of iron oxide and muscovite mica make up the remaining mineral phases, along with micrite calcite, fragments of metamorphic and volcanic rocks, and traces of clinopyroxene, epidote and hornblende amphibole. Clay pellets and collophane (microcrystalline hydrated fluorapatite) occurred frequently. In none of the 12 samples examined was any trace of organic substance found.

Inside the macro-group QPl, samples SAI 78-80 and 83-88 all have grains of similar size (~ 0.125-0.25) **(Plate 4.5: a–b)**; while SAI 81 and 82 are distinguished by a finer texture (~ 0.062-0.125) as well as by a higher proportion of plagioclase (Pl), iron oxide (FeO), clay pellets and collophane **(Plate 4.5: c–d)**: these two samples can in fact be included in petrographic terms in the Fabric QPl-Col defined above. In sample SAI 93, the only sample in the QPl group to come from site 8-B-10C, no collophane or clay pellets were found; instead, aggregates of micrite calcite were more numerous (Fabric QPl-Cal) **(Plate 4.5: e–f)**.

X-ray fluorescence analysis

All 16 sediment samples were analysed using X-ray fluorescence (XRF). The resulting identification of just the major oxides was enough to allow a division into the two mineralogical groups defined above (QPl and QCal), but the chemical results also brought to light differences among the QPl samples which had not been seen in the petrographic or diffraction analyses **(Table 4.8)**.

The two mineralogical groups, QPl (quartz-plagioclase) and QCal (quartz-calcite) differ mainly in the levels of silicon dioxide (SiO_2) and of calcium oxide (CaO) that they contain, which derive respectively from the presence of silicate or carbonate minerals. The QPl group contains more SiO_2 (x=59.09; σ=6.54) than CaO (x=3.93; σ=0.78); while for the QCal group the values for CaO are x=26.87; σ=5.64, and for SiO_2, x=24.42; σ=8.30. Inside the QPl group, samples SAI 78-80 have the highest levels of SiO_2. The variability in the levels of aluminium oxide (Al_2O_3) and of magnesium oxide (MgO) is just as significant. In the QPl group, given the higher content of quartz and of silicates in general, the average value (x) for Al_2O_3 is

# Sample	XRPD Group	SiO_2	TiO_2	Al_2O_3	Fe_2O_3	MnO	MgO	CaO	Na_2O	K_2O	P_2O_5	LOI
SAI 78	QPl	67.36	1.18	11.46	5.90	0.16	2.70	4.32	1.26	1.56	0.30	3.80
SAI 79	QPl	71.32	1.03	10.00	4.91	0.10	2.29	4.13	1.23	1.37	0.23	3.39
SAI 80	QPl	69.29	1.08	10.86	5.41	0.12	2.46	4.00	1.23	1.40	0.22	3.94
SAI 81	QPl	52.35	2.00	16.46	10.82	0.18	3.49	4.13	1.01	1.23	0.17	8.16
SAI 82	QPl	51.46	2.14	16.70	11.64	0.21	3.59	4.08	0.98	1.18	0.16	7.87
SAI 83	QPl	55.95	1.71	14.28	9.04	0.16	3.27	3.27	3.96	1.04	0.13	7.20
SAI 84	QPl	55.53	1.86	14.91	9.61	0.16	3.50	3.40	2.76	1.10	0.13	7.04
SAI 85	QPl	58.61	1.90	14.32	9.10	0.15	3.29	3.28	2.18	1.08	0.13	5.97
SAI 86	QPl	56.92	1.90	14.54	9.47	0.15	3.40	4.00	1.88	1.09	0.13	6.51
SAI 87	QPl	58.05	2.14	14.80	9.76	0.19	3.15	3.14	1.94	1.12	0.13	5.59
SAI 88	QPl	55.26	2.07	15.14	10.43	0.24	3.56	3.35	1.74	1.10	0.12	7.00
SAI 93	QPl	57.00	1.46	12.61	7.13	0.15	3.50	6.02	1.36	1.04	0.12	9.61
x		59.09	1.71	13.84	8.60	0.16	3.18	3.93	1.79	1.19	0.16	6.34
σ		6.56	0.41	2.14	2.22	0.04	0.45	0.78	0.86	0.17	0.06	1.90
SAI 89	QCal	20.63	0.70	6.07	6.61	0.11	6.77	29.48	0.25	0.55	0.01	28.82
SAI 90	QCal	18.12	0.64	5.32	5.80	0.08	6.81	32.46	0.21	0.52	0.00	30.05
SAI 91	QCal	22.33	0.75	6.26	6.12	0.14	8.67	26.20	0.31	0.67	0.00	28.55
SAI 92	QCal	36.60	1.13	8.16	5.93	0.18	6.69	19.32	0.67	0.80	0.08	20.44
x		24.42	0.81	6.45	6.12	0.13	7.24	26.87	0.36	0.64	0.02	26.97
σ		8.30	0.22	1.21	0.36	0.04	0.96	5.64	0.21	0.13	0.04	4.40

Table 4.8 Major elements composition of the analysed local sediment samples in relation to the identified mineralogical/petrographic groups (x: average composition, σ: standard deviation). Major elements and LOI (Loss on ignition) in w%.

4. Archaeometric analysis

as high as 13.84, with a standard deviation (σ) of 2.14, while in the QCal group the level is lower (x=6.46; σ=1.21). The amount of Al_2O_3 is particularly high in the two samples SAI 81 and 82. This, together with the results of the loss on ignition test (LOI) may indicate a higher clay content in SAI 81 and 82. The quantity of MgO, on the other hand, is higher in the QCal group (x=7.24; σ=0.96) than in the QPl group (x=3.18; σ=0.45) and especially in samples SAI 78-80.

The two groups can also be distinguished by the quantities of oxides of titanium (TiO_2), sodium (Na_2O) and of potassium (K_2O) that they contain. The levels of the three oxides are always higher in the QPl samples **(Table 4.8)**. The Na_2O and K_2O are derived from the feldspar, while the TiO_2 is possibly connected to the presence of titanium bearing pyroxene, or more probably, derives from the composition of the clay size fraction itself, and from the presence of accessory phases such as ilmenite ($FeTiO_3$), identified in the SEM analysis also in the ceramic samples (see above).

The following scatter plots highlight the differences in composition between the samples from the two groups, QPl and QCal. In the first graph **(Figure 4.14)**, the relationship of CaO to Al_2O_3 allows a clear division of the calcium-rich sediments (QCal group, in grey) from the QPl group (in blue) with its greater quantities of Al_2O_3 but low levels of CaO.

The second graph, of MgO/K_2O vs. SiO_2/Al_2O_3, **(Figure 4.15)**, shows clearly the divergences between the QPl group (in blue) and the QCal group (in grey), above all in the pairing MgO/K_2O (x-axis), where the QCal sediment samples, with their significant quantities of MgO and lower levels of K_2O, lie out on the right of the graph. The three samples SAI 78-80 are detached from the other members of the QPl group because of the higher value of SiO_2/Al_2O_3 ($SiO_2>Al_2O_3$). It should be noted, in addition, how the sample SAI 92 is separated from the other members of the QCal group because of its higher SiO_2 and Al_2O_3 content.

In the scatter plot of Fe_2O_3/TiO_2 vs. SiO_2/Al_2O_3 **(Figure 4.16)**, the separation of the two groups is essentially the result of the relationship of Fe_2O_3/TiO_2: this gives a value above eight for the QCal samples (grey), poorer in TiO_2 while the balance is more even between the two oxides in the QPl group. The sample SAI

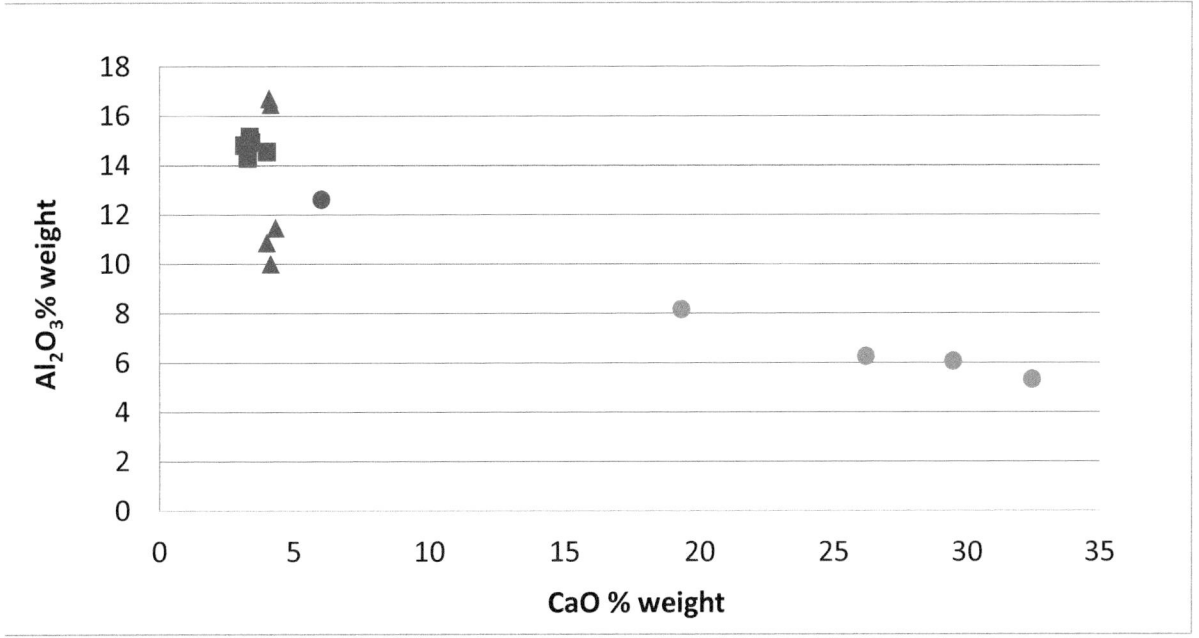

Figure 4.14 CaO vs. Al_2O_3 plot (wt %). The colour of symbols refers to the XRD groups: blue = QPl group; grey = QCal group. The shape of the symbols refers to the sites: triangles are samples from site 8-B-10A; squares are samples from site 8-B-76; circles are samples from site 8-B-10C.

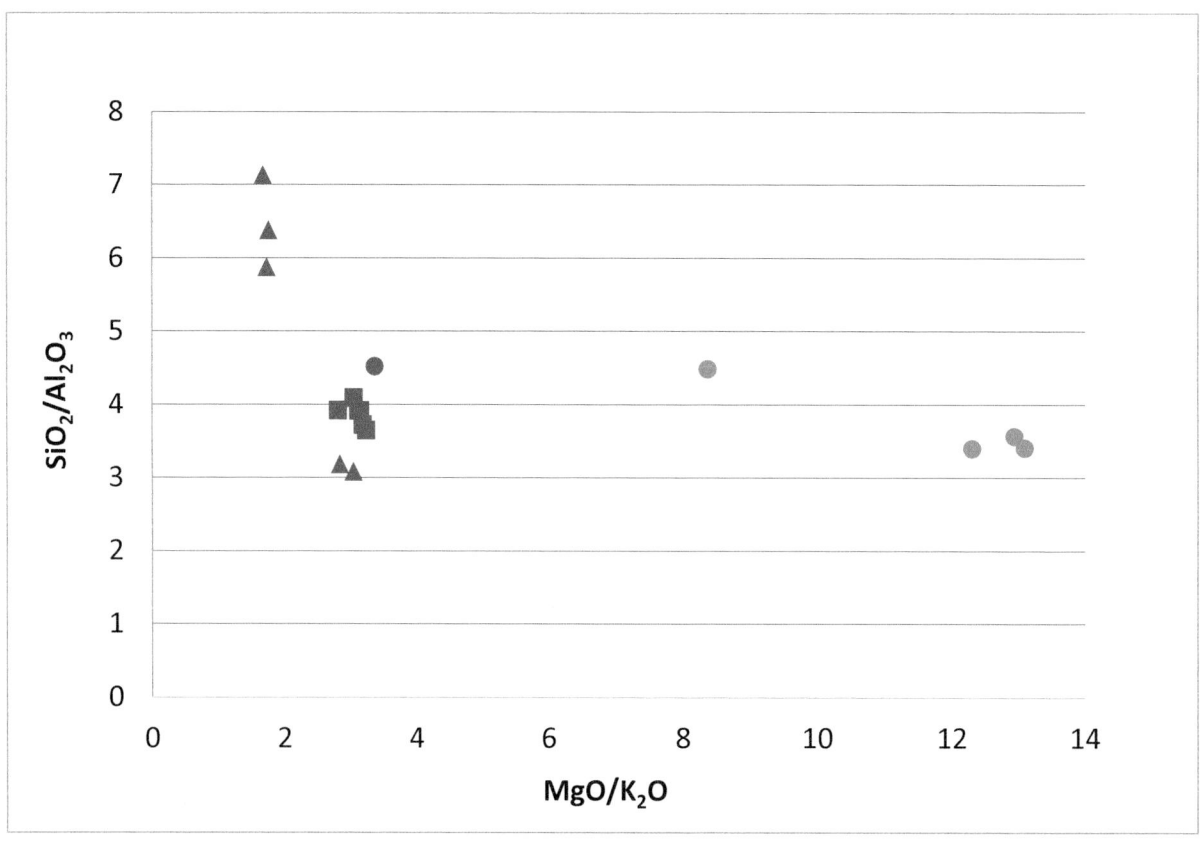

Figure 4.15 MgO/K_2O vs. SiO_2/Al_2O_3 plot. The colour of symbols refers to the XRD groups: blue = QPl group; grey = QCal group. The shape of the symbols refers to the sites: triangles are samples from site 8-B-10A; squares are samples from site 8-B-76; circles are samples from site 8-B-10C.

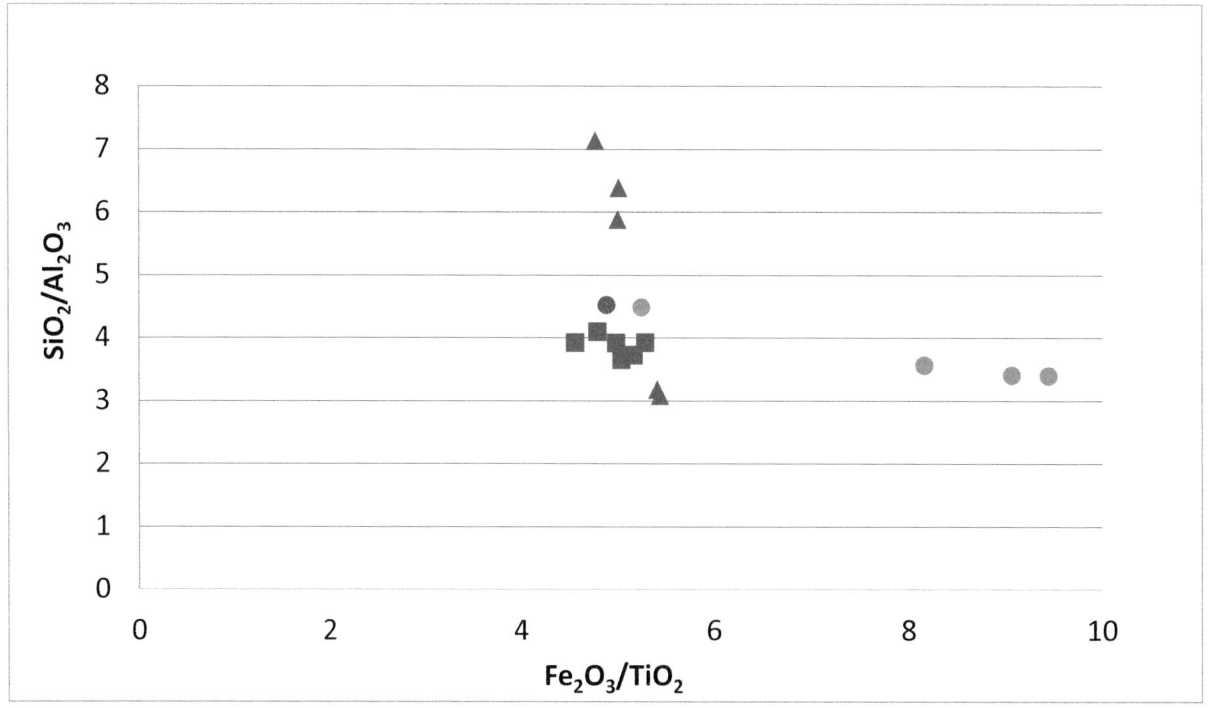

Figure 4.16 Fe_2O_3/TiO_2 vs. SiO_2/Al_2O_3 plot. The colour of symbols refers to the XRD groups: blue = QPl group; grey = QCal group. The shape of the symbols refers to the sites: triangles are samples from site 8-B-10A; squares are samples from site 8-B-76; circles are samples from site 8-B-10C.

# Sample	XRPD Group	Rb	Sr	Y	Zr	Nb
SAI 78	QPl	40	269	22	314	17
SAI 79	QPl	35	242	18	276	15
SAI 80	QPl	37	258	21	287	16
SAI 81	QPl	47	290	32	283	24
SAI 82	QPl	46	304	33	275	25
SAI 83	QPl	42	264	30	321	23
SAI 84	QPl	45	275	32	371	25
SAI 85	QPl	43	260	32	506	24
SAI 86	QPl	44	282	32	416	24
SAI 87	QPl	44	270	34	789	27
SAI 88	QPl	47	302	32	442	28
SAI 93	QPl	35	197	30	689	23
x		42	268	29	414	23
σ		4	29	5	170	4
SAI 89	QCal	15	240	10	86	4
SAI 90	QCal	14	225	9	68	3
SAI 91	QCal	18	308	11	128	6
SAI 92	QCal	24	312	19	353	18
x		18	271	12	159	8
σ		5	45	5	132	7

Table 4.9 Trace elements composition of the analysed local sediment samples in relation to the identified mineralogical/petrographic groups (x: average composition, σ: standard deviation). Trace elements in parts per million (ppm).

92 is an exception, the only member of the QCal group with a value for TiO_2 above zero and at the same time a higher percentage of SiO_2. The three samples, SAI 78-80 (QPl group), as noted for the previous graph, are separated from the rest of their group because of the higher value of SiO_2/Al_2O_3 ($SiO_2>Al_2O_3$).

Among the trace elements (Rb, Sr, Y, Zr and Nb), as was the case with the ceramics, it is mainly Rb, Y and Zr, and to a lesser extent, Nb, that show a more significant geochemical variability. All four occur in higher proportions in the QPl samples than in the QCal group **(Table 4.9)**.

This result can be seen very clearly in the scatter plots below **(Figures 4.17-4.19)**: on all three a good positive linear correlation is seen (particularly striking between Rb, Y and Nb), which allows a clean separation of the samples from the QPl group (blue), rich in Rb, Y and Nb from the QCal sediments (grey), with lower amounts of these elements. Samples SAI 78-80 (blue triangles) again occupy an intermediate position between the two groups. On the last graph **(Figure 4.19)**, the relationship between Rb and Zr again divides the two groups, QPl and QCal. Only samples SAI 87 and 93 (respectively the blue square and the blue circle in the upper right hand corner) contain a higher proportion of zirconium (Zr).

The ceramic and sediment samples: classification criteria and comparison of results

As described above, it was possible on the basis of mineralogical and petrographic data to make an initial division of the ceramic samples into three groups, defined in terms of the principal phase associations observed: QPl = quartz-plagioclase (SAI 01-25, 40-49, 51-56, 59-67 and 69; QKfs = quartz-K-feldspar (SAI 26, 28-30, 32-39, 50, 58, 68, 70-77); and Q = quartz (SAI 27 and 31).

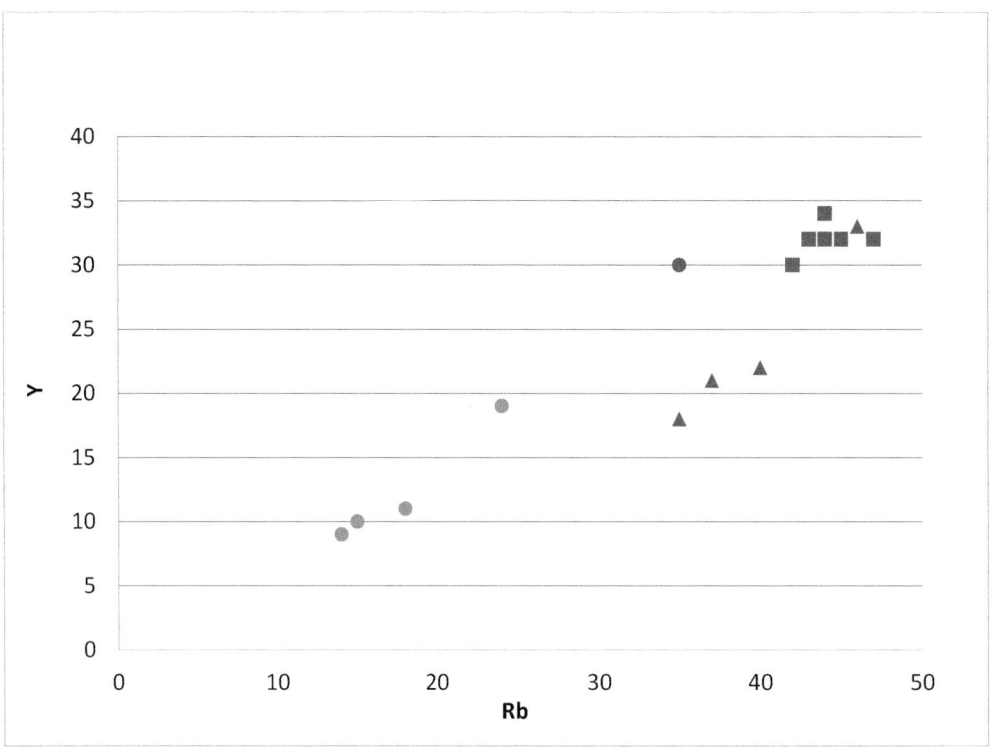

Figure 4.17 Rb vs. Y plot. The colour of symbols refers to the XRD groups: blue = QPl group; grey = QCal group. The shape of the symbols refers to the sites: triangles are samples from site 8-B-10A; squares are samples from site 8-B-76; circles are samples from site 8-B-10C.

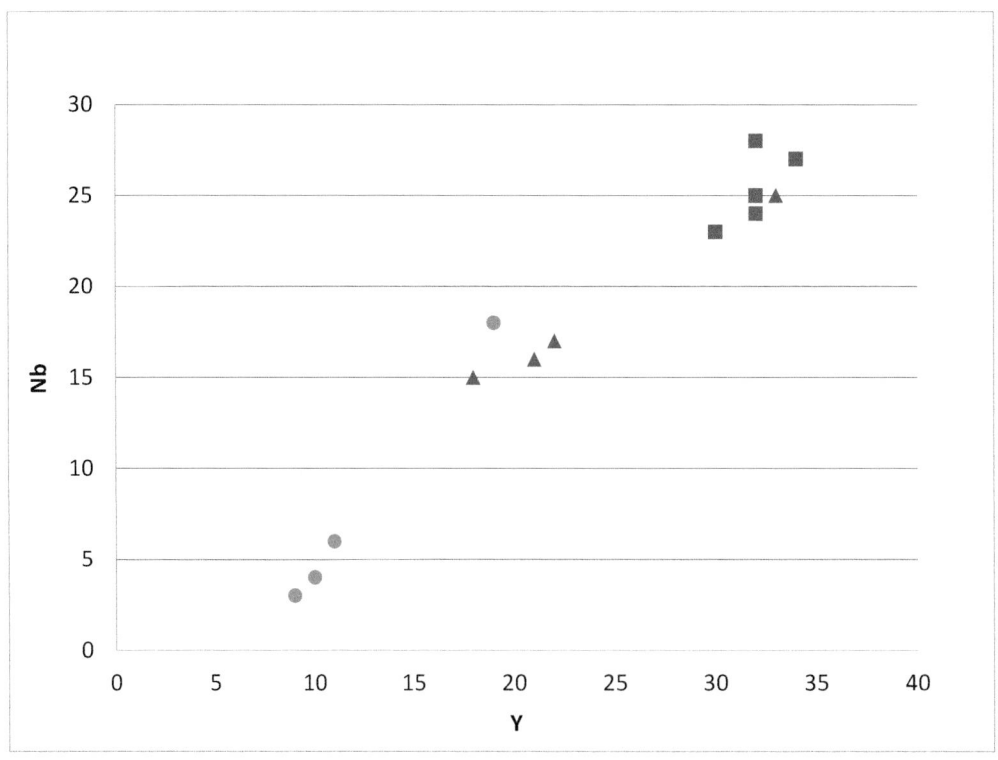

Figure 4.18 Y vs. Nb plot. The colour of symbols refers to the XRD groups: blue = QPl group; grey = QCal group. The shape of the symbols refers to the sites: triangles are samples from site 8-B-10A; squares are samples from site 8-B-76; circles are samples from site 8-B-10C.

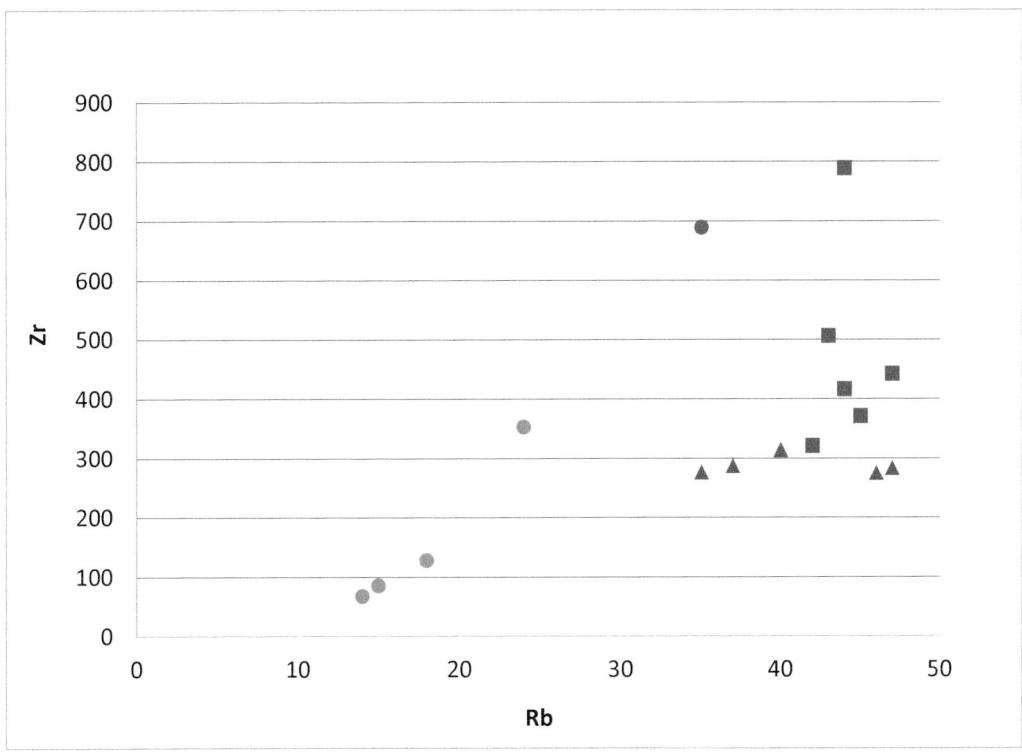

Figure 4.19 Rb vs. Zr plot. The colour of symbols refers to the XRD groups: blue = QPl group; grey = QCal group. The shape of the symbols refers to the sites: triangles are samples from site 8-B-10A; squares are samples from site 8-B-76; circles are samples from site 8-B-10C.

The first group, QPl, includes all the samples from the two most recent sites 8-B-10A (SAI 01-10, 40-42, and the clay lumps SAI 52-56) and 8-B-52A (SAI 11-21 and 43-48) and of most of the Abkan samples from site 8-B-76 (SAI 22-25, 49, 51, 59-67 and 69). This group was found to be substantially homogeneous in its composition, nevertheless, the petrographic analysis showed up certain differences – mostly derived from technological aspects of production – which led to a further subdivision of the samples into four different fabrics: QPl-Veg, QPl, QPl-Col and QPl-Cha **(Table 4.10)**. These differences are primarily the result of production methods varying with time, and reflect the samples provenances in successive Abkan and Pre-Kerma horizons. Finally, minor variations – mainly in grain-size – identified in individual fabrics (the sub-groups labelled on the table with small letters) are potentially related to the division into texture sub-groups proposed in the macroscopic classification (see for discussion Chapter 5).

The Khartoum Variant samples from site 8-B-10C (SAI 26, 28-30, 32-39 and 70-77) and the three Abkan samples from site 8-B-76 (SAI 50, 58 and 68) assigned to the QKfs group showed petrographic and chemical characteristics that were quite distinct from the ceramics of the QPl group. Specifically, these ceramics are characterized by significant amounts of K_2O and Rb, which are both related to the significant content of K-feldspars. This leads to the hypothesis that, besides the adoption of particular technological choices during manufacture, both a raw material and tempers with a different composition from those of the QPl samples were selected. In addition, analysis with the OM allowed a further subdivision of the QKfs group, identifying at least two different fabrics: QKfs-Bt and QKfs-Ms **(Table 4.10)**. Samples SAI 32 and 74-76, assigned (with the exception of SAI 76) to the fabric QKfs-Ms, also differ in part from the other members of their group through their composition: together with SAI 50, 58 and 68 (site 8-B-76), they occupy an intermediate position on the graphs between the samples from the QKfs and QPl groups (see above).

The Q group, composed of the two Khartoum Variant samples SAI 27 and 31, has a chemical make-up that is essentially similar to that of the QPl group, with, however, certain particular mineralogical and petrographic

XRPD Group	Mineral phases	Rock fragments	Fabrics	Subgroups	Grain size (mm)	Organics vegetal	Organics charcoal	Collophane	Clay pellets	Site
QPl (n=51)	Qtz, Pl, Kfs, Micas, FeOx, Cal, Px, Ep, Amp (tr)	M & V	QPl-Veg	1a	0.125-0.25	+		-	-	8-B-52A (no.=17)
				1b	0.062-0.125		?			
				1c	0.062-0.125 0.5-1					
			QPl	2	0.125-0.25	-		-	-	8-B-10A (no.=6)
			QPl-Col	3	0.062-0.125	-	?	+	+	8-B-10A (no.=12) 8-B-76 (no.=1)
			QPl-Cha	4a	0.125-0.25	-	+			8-B-76 (no.=15)
				4b	0.062-0.125 0.5-1	-	-			
QKfs (no.=23)	Qtz, Kfs, Micas, FeOx, Pl, Cal, Px, Ep, Amp (tr)	M (+/-)	QKfs-Bt	1a	0.5-1	?				8-B-10C (no.=20)
			QKfs-Ms	1b	0.125-0.25					8-B-76 (no.=3)
			QKfs-Bt	1c	0.062-0.125 0.5-1	?	-			
Q (no.=2)	Qtz, Pl, Ms, FeOx, Ep, Amp (tr)				0.25-0.5	?				8-B-10C (no.=2)

Table 4.10 Main groups, fabrics and sub-groups of the analysed ceramic samples and clay lumps. Main groups: QPl: quartz-plagioclase; QKfs: quartz-K-feldspar; Q: quartz. Fabrics: QPl-Veg: quartz-plagioclase-vegetal; QPl-Col: quartz-plagioclase-collophane; QPl-Cha: quartz-plagioclase-charcoal; QKfs-Bt: quartz-k-feldspar-biotite; QKfs-Ms: quartz-k-feldspar-muscovite. Mineral symbols mentioned are as in Kretz (1983). M: metamorphic rock fragments; V: volcanic rock fragments. Symbols (+), (-) e (--) express the abundance of the mineralogical phases; tr, traces (after D'Ercole et al. 2015).

XRPD Group	Mineral phases	Rock fragments	Fabrics	Grain size (mm)	Clay pellets	Collophane	Site
QPl (no.=12)	Qtz, Pl, Kfs, Micas, Cal, FeOx; Px, Ep, Amp (tr)	M & V	QPl	0,125-0,25	+	~	8-B-10A (no. = 3) 8-B-76 (no. = 6)
			QPl-Col	0,062-0,125	+	+	8-B-10A (no. = 2)
			QPl-Cha	0,25-0,5			8-B-10C (no. = 1)
QCal (no.=4)	Qtz, Cal, Chl, Micas, Dol, Pl						8-B-10C (no. = 4)

Table 4.11 Main groups and fabrics of the analysed local sediment samples. Main groups: QPl: quartz-plagioclase; QCal: quartz-calcite. Fabrics: QPl-Col: quartz-plagioclase-collophane; QPl-Cha: quartz-plagioclase-charcoal. Mineral symbols mentioned are as in Kretz (1983). M: metamorphic rock fragments; V: volcanic rock fragments. Symbols (+), (-) e (~) express the abundance of the mineralogical phases; tr: traces (after D'Ercole et al. 2015).

traits: SAI 27 and 31 have a framework made up mainly of well-rounded clasts of monocrystalline quartz, while K-feldspar and micas are completely or almost completely absent **(Table 4.10)**.

The mineralogical and compositional characteristics observed in the ceramics were subsequently compared to the results obtained from analysing 16 samples of sediment taken from three sites 8-B-10A (SAI 78-82), 8-B-76 (SAI 83-88) and 8-B-10C (SAI 89-93). As noted above, four of the sediment samples analysed from the oldest Khartoum Variant site 8-B-10C (SAI 89-92) had a completely different composition from that found in any of the island's ceramics: they were composed of calcareous sediments (group QCal) characterised by mineral phases normally absent from the diffraction spectrum of any of the ceramic samples, whether belonging to the groups QKfs, QPl or Q. On the other hand, the rest of the sediment samples, from site 8-B-10A (SAI 78-82) and 8-B-76 (SAI 83-88), and one sample, SAI 93, from site 8-B-10C, were very similar, mineralogically and petrographically, to the ceramics of group QPl, and so were assigned to this group **(Table 4.11)**.

This compatibility was subsequently confirmed for their chemical make-up by adding the results for the 12 sediment samples of the QPl group (SAI 78-88 and 93) to the scatter diagrams discussed above of the ceramic samples. Samples SAI 89-92 from the QCal group, were not entered on the graphs, since already known to be different in mineralogically and petrographically from the ceramics.

As can be seen on the two scatter plots below, showing the correlation of CaO with K_2O **(Figure 4.20)** and of MgO with K_2O **(Figure 4.21)**, the sediment samples, irrespective of where they came from, have a similar distribution on the graph to the samples from the QPl group (area circled in blue), with similar levels of CaO, MgO and K_2O to the Abkan and Pre-Kerma (QPl) ceramics, while the contrast with the Q group samples (in green) is not so clear-cut. The two graphs show a complete separation from most of the QKfs samples (in orange), which after all always contain higher levels of K_2O and lower levels of CaO and MgO.

The scatter plot of MgO/K_2O vs. SiO_2/Al_2O_3 **(Figure 4.22)** confirms this pattern, with the exception that the figures for the relationship SiO_2/Al_2O_3, detach three of the five samples from site 8-B-10A (SAI 78-80, triangles in the upper part of the graph) from the rest of the group of QPl sediments – making them differing from the ceramic samples.

Principal Component Analysis (PCA) of these results brings out even more clearly the divergences between the different groups suggested by mineralogical and petrographic analysis (D'Ercole *et al.* 2015). This analytical procedure, a part of multivariate statistics, aims to express significant variance within a given set of data by reducing the number of original n. variables (in this case the chemical values of the individual samples) through the generation of new principal components – linear combinations of the original variables – which could not be expressed otherwise in space (Baxter 1999, 2001.)

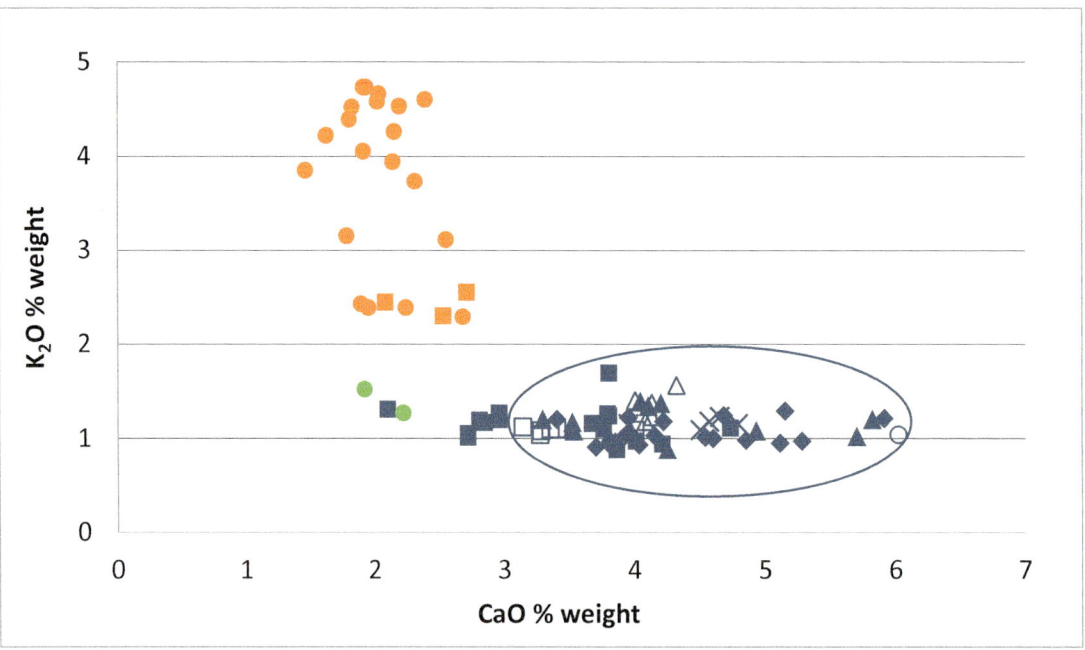

Figure 4.20 CaO vs. K$_2$O plot (wt %). The colour of symbols refers to the XRD groups: blue = QPl group; orange = QKfs group; green = Q group. The shape of the symbols refers to the sites: triangles are samples from site 8-B-10A; diamonds are samples from site 8-B-52-A; squares are samples from site 8-B-76; circles are samples from site 8-B-10C; crosses are the clay lumps from site 8-B-10A. Empty triangles are sediments from site 8-B-10A; empty squares are sediments form site 8-B-76; empty circles are sediments from site 8-B-10C.

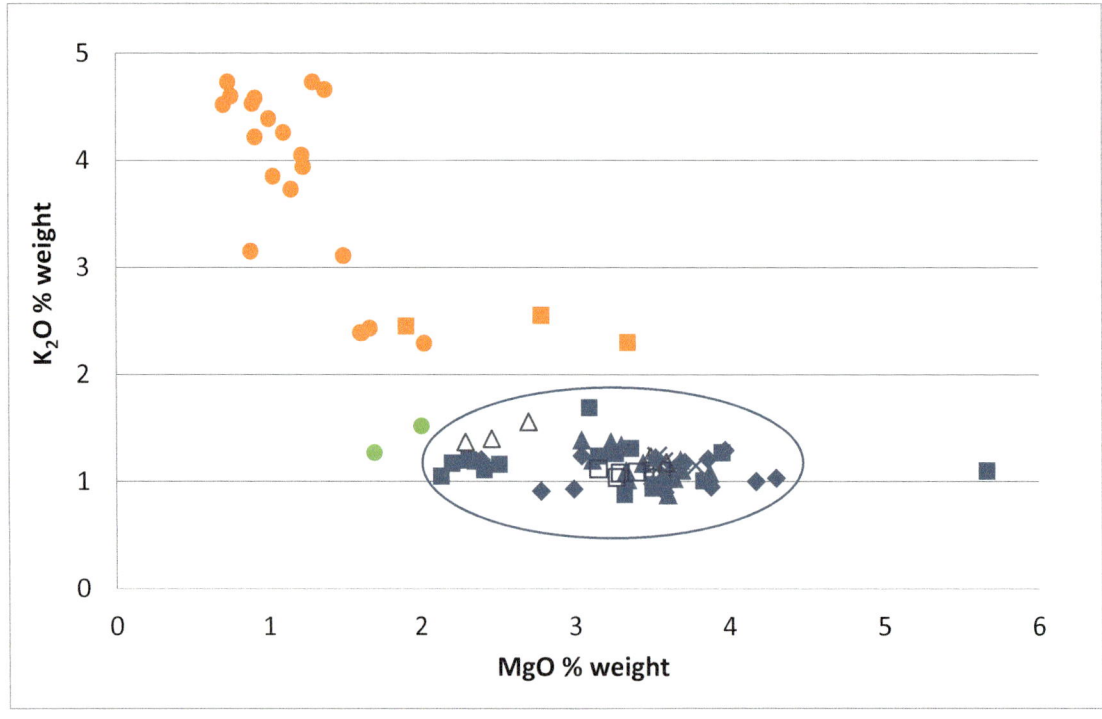

Figure 4.21 MgO vs. K$_2$O plot (wt %). The colour of symbols refers to the XRD groups: blue = QPl group; orange = QKfs group; green = Q group. The shape of the symbols refers to the sites: triangles are samples from site 8-B-10A; diamonds are samples from site 8-B-52-A; squares are samples from site 8-B-76; circles are samples from site 8-B-10C; crosses are the clay lumps from site 8-B-10A. Empty triangles are sediments from site 8-B-10A; empty squares are sediments form site 8-B-76; empty circles are sediments from site 8-B-10C.

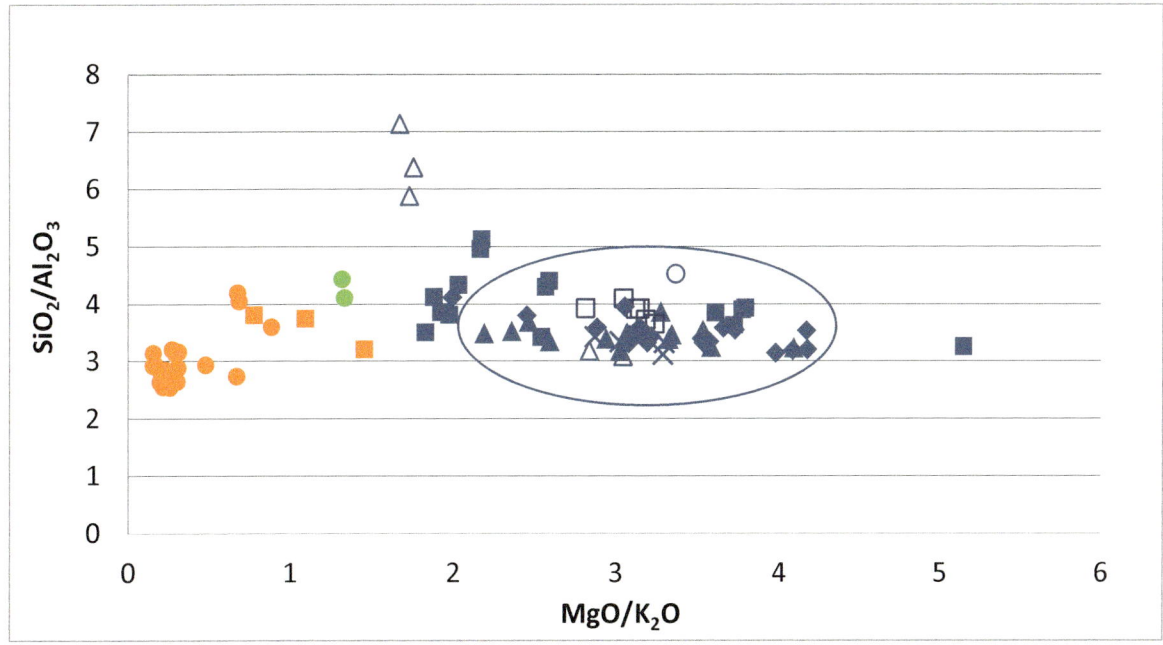

Figure 4.22 MgO/K$_2$O vs. SiO$_2$/Al$_2$O$_3$. The colour of symbols refers to the XRD groups: blue = QPl group; orange = QKfs group; green = Q group. The shape of the symbols refers to the sites: triangles are samples from site 8-B-10A; diamonds are samples from site 8-B-52-A; squares are samples from site 8-B-76; circles are samples from site 8-B-10C; crosses are the clay lumps from site 8-B-10A. Empty triangles are sediments from site 8-B-10A; empty squares are sediments form site 8-B-76; empty circles are sediments from site 8-B-10C.

For the data considered here, since the greatest variance is contained in the first two components calculated (PC1 and PC2), they were found sufficient in themselves to represent the distribution of the samples, and so are given on the graph below **(Figure 4.23)** as a new series of data for the different groups. It should be stated that before creating the new components, the original variables were standardized, and that it was decided to exclude certain chemical values such as silicon and phosphorus, the former (SiO$_2$) as too evenly spread over the different samples, and the latter (P$_2$O$_5$) as the possible result of post depositional contamination. On this graph, the vectors indicated with dashes vary in length in proportion to the variance of the different chemical variables, while their position relative to each other defines the level of correlation between them: if the lines are close together or superimposed (as with Fe and Ti) there is a positive linear correlation between the two components, if they are at 90° the correlation is weak, and if they are at 180° then the correlation is negative.

The graph shows the existence of two principal clusters of ceramic, the macro-groups QPl (blue symbols: 1a and 1b) and QKfs (orange symbols: 2a, 2b, and 2c), occupying clearly distinct positions in opposite quadrants of the graph. This result is produced essentially by the variance caused by the Fe$_2$O$_3$, TiO$_2$, MgO and CaO content – greater in the QPl samples – and the K$_2$O, MnO, Rb, and Zr content, characteristic of the chemical make-up of the QKfs group.

As shown above, in the QPl samples, the TiO$_2$ content is closely related to the Fe$_2$O$_3$ content, and is attributed not just to the presence of titanium bearing pyroxene or other heavy mineral accessory phases (such as rutile and/or ilmenite), but also to the composition of the matrix itself (i.e., the clay-size fraction of the samples). The higher levels of MgO come also from the pyroxene, while the CaO content is derived from the plagioclase. In the QKfs samples the dominance of K$_2$O and Rb – due to the potassium feldspars – indicates the metamorphic origin of the mineral components making up the ceramics. A similar origin should be considered for zircon too: this mineral is found in metamorphic rocks in fact

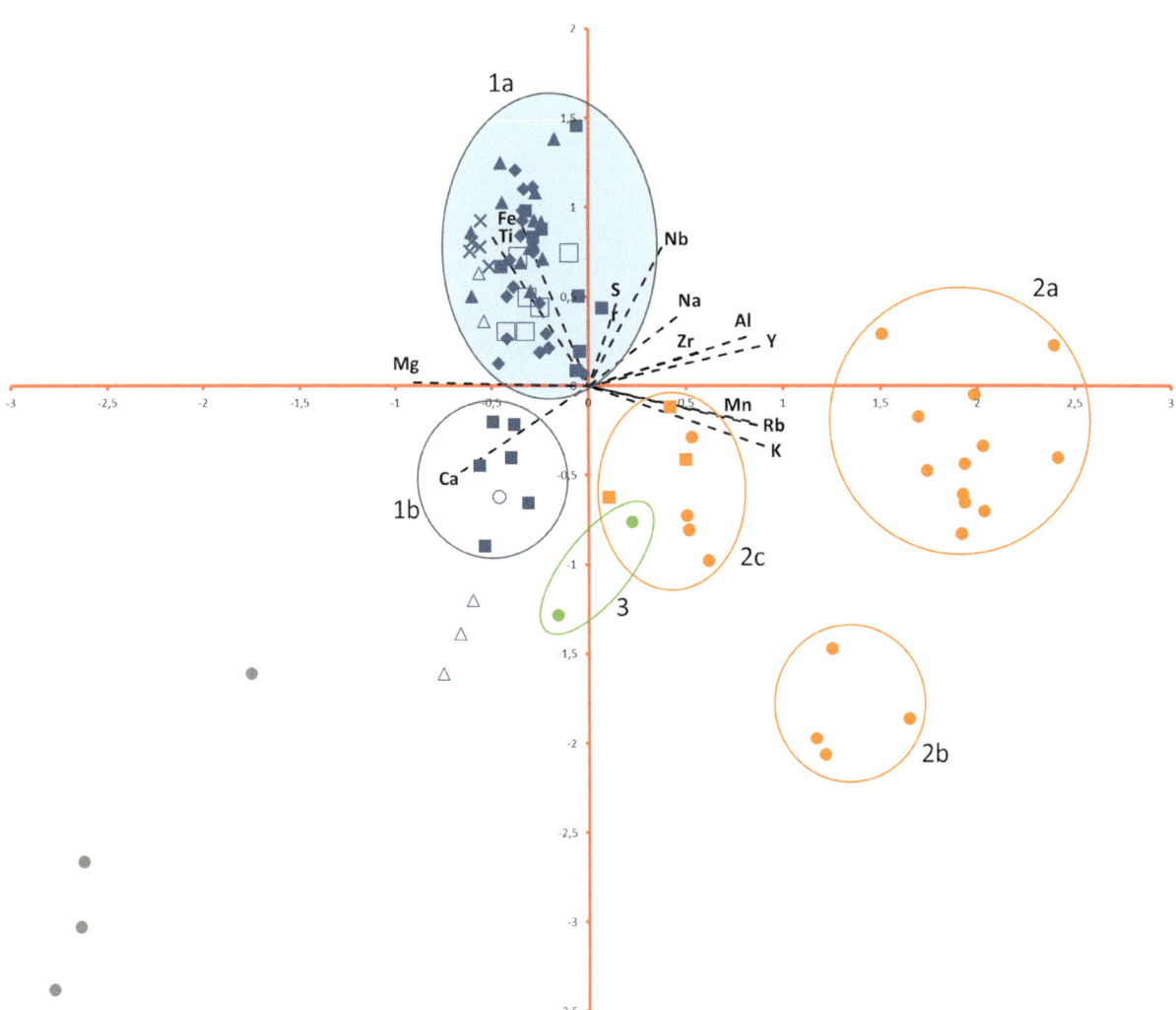

Figure 4.23 Biplot of the PCA analysis (PC1 versus PC2) for derived pottery and clayey sediments, showing the clusters identified. The colour of symbols refers to the XRD groups: blue = QPl group; = QKfs group; green = Q group. The shape of the symbols refers to the sites: triangles are samples from site 8-B-10A; diamonds are samples from site 8-B-52-A; squares are samples from site 8-B-76; circles are samples from site 8-B-10C; crosses are the clay lumps from site 8-B-10A. Empty triangles are sediments from site 8-B-10A; empty squares are sediments form site 8-B-76; empty circles are sediments from site 8-B-10C (modified after D'Ercole et al. 2015).

but is also found in intrusive quartz veins. It should be noted, in addition, that for both groups some of the samples are distributed very compactly on the graph (this is particularly the case with the most recent samples from sites 8-B52A and 8-B10A, blue diamonds and blue triangles), but that others are more spread out, forming minor sub-groups. From the QPl samples, a sub-group is formed by samples SAI 22, 49 and 60-62 (1b) which are richer in calcium. With the QKfs samples, the internal differentiation seems more obvious, with the formation, alongside the core group (2a), of two smaller clusters, one composed of samples SAI 26, 28, 73 and 77 (2b), the other of SAI 32, 50, 58, 68 and 74-76 (2c). The two Khartoum Variant samples making up the Q group (green circles) again stand out on their own.

Finally, the PCA confirms that the samples of sediment from the sites 8-B-76 and 8-B-10A (with the exception always of SAI 78-80) and most of the Abkan and Pre-Kerma ceramics of the QPl group (area shaded in blue) are chemically very similar to each other, and that the four calcareous sediment samples from site 8-B-10C (SAI 89-92; grey symbols) are different in their chemical composition from any of the ceramics.

4. Archaeometric analysis

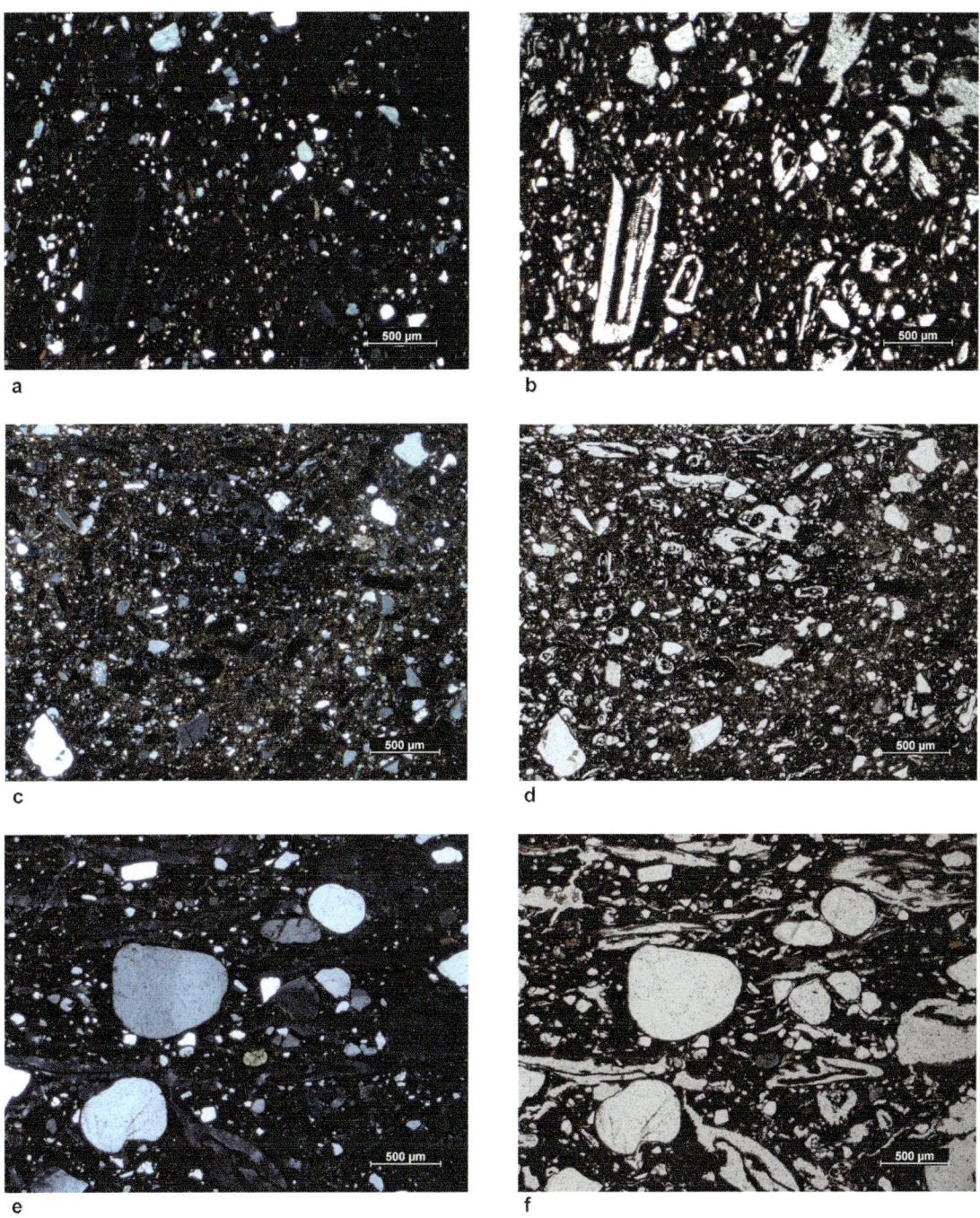

Plate 4.1 Microphotographs of examples from Pre-Kerma samples from fabric QPl-Veg. (a, c, e) Cross Polarized Light; (b, d, f) Plane Polarized Light. Photographs by G. Eramo.

Plate 4.2 Microphotographs of examples from Pre-Kerma samples (a-d) from fabric QPl (a-b) and QPl-Col (c-d) and from Abkan samples (e-f) from fabric QPl-Chr. (a, c, e) Cross Polarized Light; (b, d, f) Plane Polarized Light. Photographs by G. Eramo.

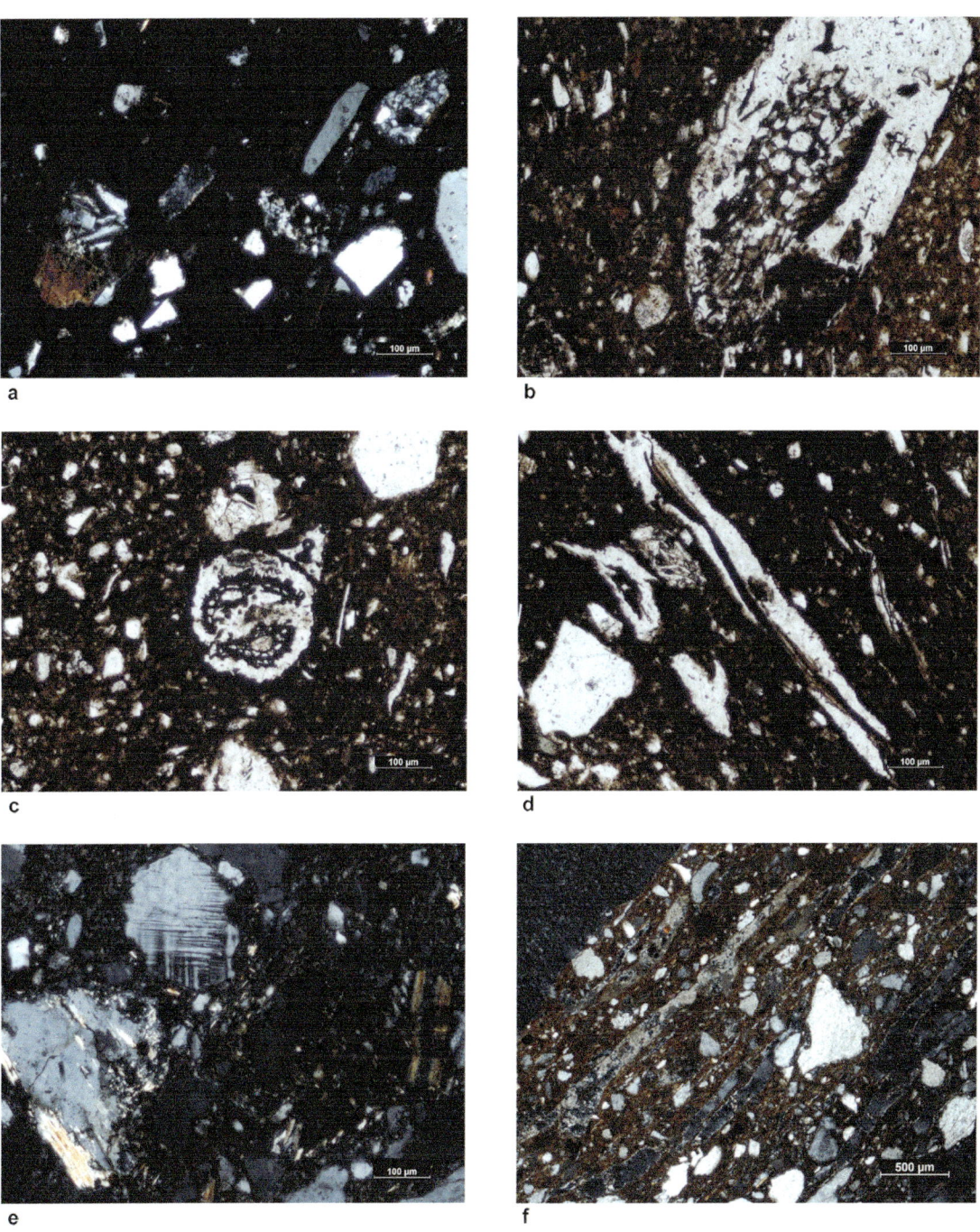

Plate 4.3 Microphotographs of examples from Pre-Kerma samples (a-d) and Khartoum Variant samples (e-f) showing mineral phases and organic structures recognized under the microscope. (a, e, f) Cross Polarized Light; (b, c, d) Plane Polarized Light. Photographs by G. Eramo.

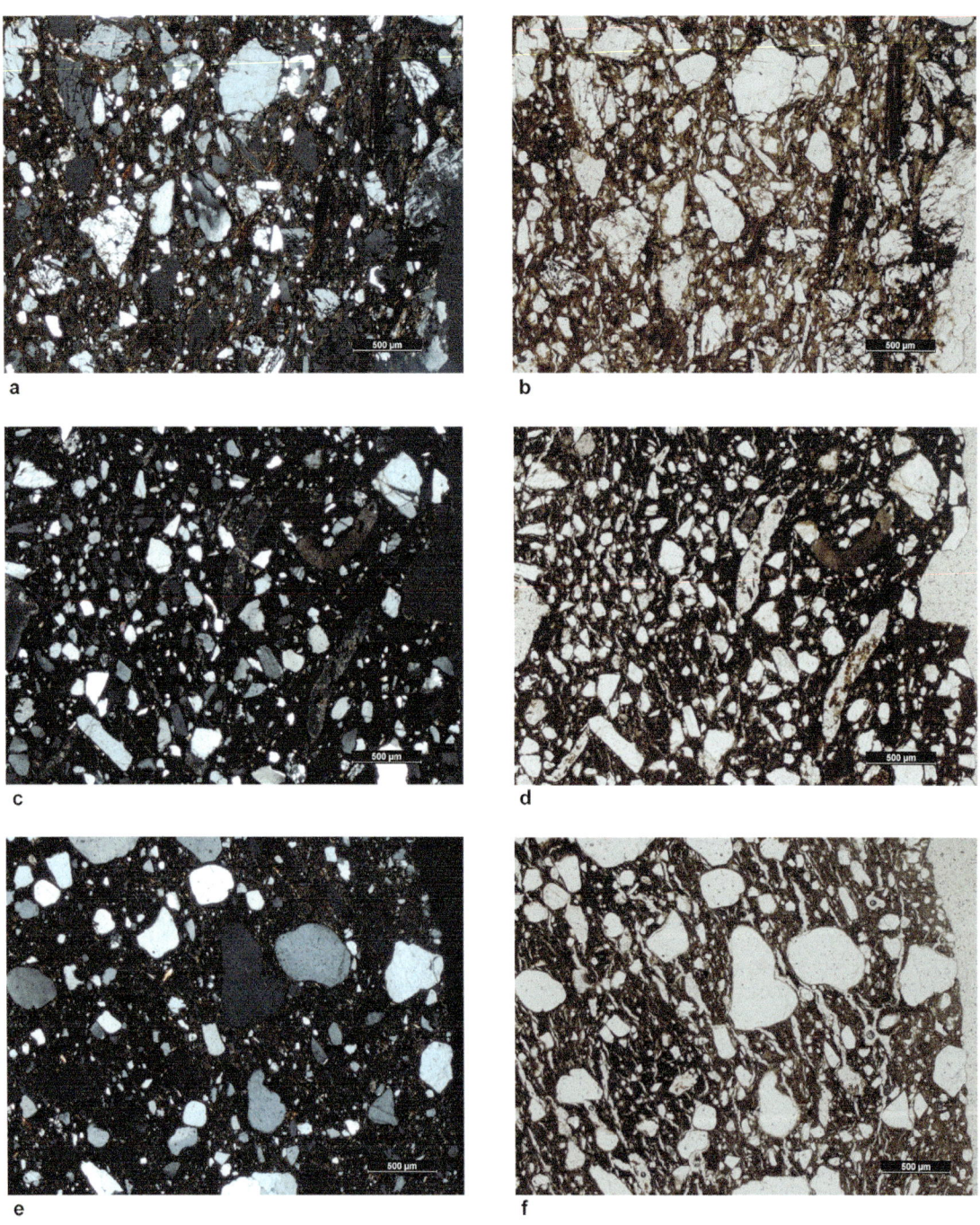

Plate 4.4 Microphotographs of examples from Khartoum Variant samples from fabric QKfs-Bt (a-b); QKfs-Ms (c-d) and Q (e-f). (a, c, e) Cross Polarized Light; (b, d, f) Plane Polarized Light. Photographs by G. Eramo.

Plate 4.5 Microphotographs of examples from local sediment samples from fabric QPl (a-d); and QPl-Cal (e-f). (a, c, e) Cross Polarized Light; (b, d, f) Plane Polarized Light. Photographs by G. Eramo.

5. Comparing chaînes operatoires: continuity and discontinuity in the ceramic assemblages of Sai Island

Introduction

This chapter discusses and compares the principal characteristics – technological and stylistic – of the ceramic productions on Sai Island in the four sites described above (8-B-10C, 8-B-76, 8-B-52A and 8-B-10A), that is over the course of the Khartoum Variant (*c.* 7600–4800 BC), Abkan (*c.* 5500–3700 BC) and Pre-Kerma (*c.* 3600–2500 BC) horizons. These characteristics are initially discussed with reference to the information obtained from the macroscopic analysis of the ceramics described in Chapter 3. Later, the ceramics are re-examined in the light of the archaeometric results (see Chapter 4), attempting to establish a possible correspondence between the categories determined macroscopically and the petrographic-mineralogical groupings that emerged from laboratory analysis.

A first specific aim is to assay the evolution of the ceramic manufacturing traditions on Sai Island from the earliest appearance of ceramic technology from the Early Holocene (Khartoum Variant horizon) to the Middle and Late Holocene (Abkan and Pre-Kerma horizons). Cultural watersheds or chronological 'breaks' have a crucial significance for the economical and cultural history of the local Nubian societies. The transition, at *c.* 5000 BC, from the Khartoum Variant to the Abkan period marks the passage from a fully foraging economy based on hunting, fishing and gathering to a 'pastro-foraging' economy (Riemer 2007), where herding was integrated in the former extractive subsistence system (Garcea 2016a); whereas the beginning of the Pre-Kerma period (*c.* 3600 BC) is accompanied by the progressive establishment of more complex societies, with a mixed agro-pastoral economy (see Chapters 1 and 2).

In this chapter, we set out to observe if any stylistic and/or technological change – or else continuity – occurred in the ceramic traditions of the island over the course of this time frame; when and why did the change eventually take place; and how 'discontinuity' and 'continuity' (Roux 2008) in pottery assemblages are related to and may reflect the economic and cultural development of human societies.

Finally, a more general point of interest is to evaluate (1) the compatibility of the two methodological approaches used for the macroscopic and the analytical/archaeometric analysis of the ceramic record, and (2) the increased effectiveness that could derive from their combined use when applied to archaeological issues.

Discussion and comparison of the macroscopic data

Preparation: clay processing and addition of non-plastic inclusions

The Khartoum Variant ceramics from site 8-B-10C are characterised by sandy pastes with abundant mica, composed of small (43.1%) or medium sized (33.1%) angular inclusions **(Figure 5.1)**. The ceramics are hard, only slightly porous, and rough to the touch. In fracture, the vessels are light brown or reddish-brown, due to the use of a raw material rich in iron oxides and poor in – or completely without – organic content. Ceramics with a dark grey or black core are rarer and present only in the few samples (13.9%) containing organic, vegetable substances **(Figure 5.2)**. These latter samples correspond mainly to the fine products and their paste contains fewer mineral inclusions, which are smaller, and often more rounded than in the samples of medium texture (see Chapter 3).

The Abkan ceramics from site 8-B-76 differ evidently from those of site 8-B-10C. The pottery is soft, brittle and relatively porous, with greyish-brown to light brown surfaces and dark cores because of

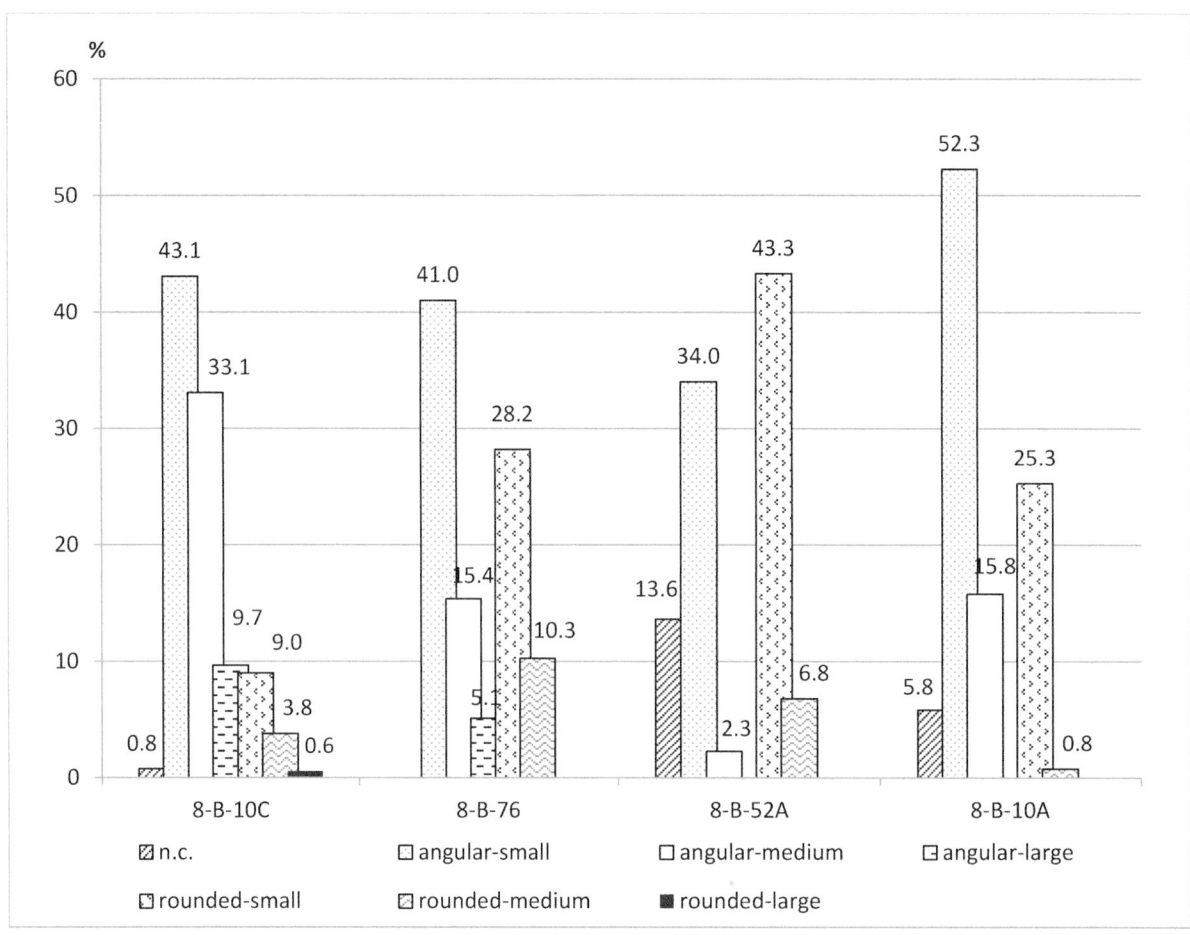

Figure 5.1 Percentages of types and dimensions of mineral inclusions from sites 8-B-10C, 8-B-76, 8-B-52A and 8-B-10A.

the presence of organic material in all the pastes. 79.5% of the ceramics contain thin, tubular organic inclusions, possibly derived from herbivore dung, which, considering their significant amount in the paste, have been likely deliberately added to the clay **(Figure 5.2)**. The most common mineral inclusions are angular in shape. However, medium sized inclusions become less common (15.4% against 33.1% recorded at site 8-B-10C), while the samples with mineral rounded inclusions, mainly smaller than 1mm, increase in number (28.2% against 9% from site 8-B-10C) **(Figure 5.1)**.

The use of organic matter as a temper marks the beginning of a new important technological tradition. Further, the use of dung as a tempering agent may also take on a cultural meaning. In this sense, the addition of dung inclusions to the ceramics may represent a new tradition connected to the development of a pastoral economy and to the growing economic and social importance of domestic animals for the inhabitants of the island. This tradition persisted through the following Pre-Kerma horizon, becoming to some extent a distinguishing characteristic of pottery production in this period. At both sites 8-B-52A and 8-B10A, most of the ceramic pastes contain tubular organic inclusions most likely derived from herbivore dung (88.4% and 76.8% respectively) and only a limited number of samples are tempered with flat organic inclusions, possibly indicating the addition of grass, chaff or other plant remains **(Figure 5.2)**.

The ceramics from site 8-B-52A and 8-B-10A differ however in the pastes' mineral content. In both cases, small inclusions (< 1mm) predominate, being 'common' (fine texture) or, more often, 'frequent' (medium texture); but at site 8-B-52A ceramics contain a higher proportion of rounded inclusions **(Figure 5.1)**. It

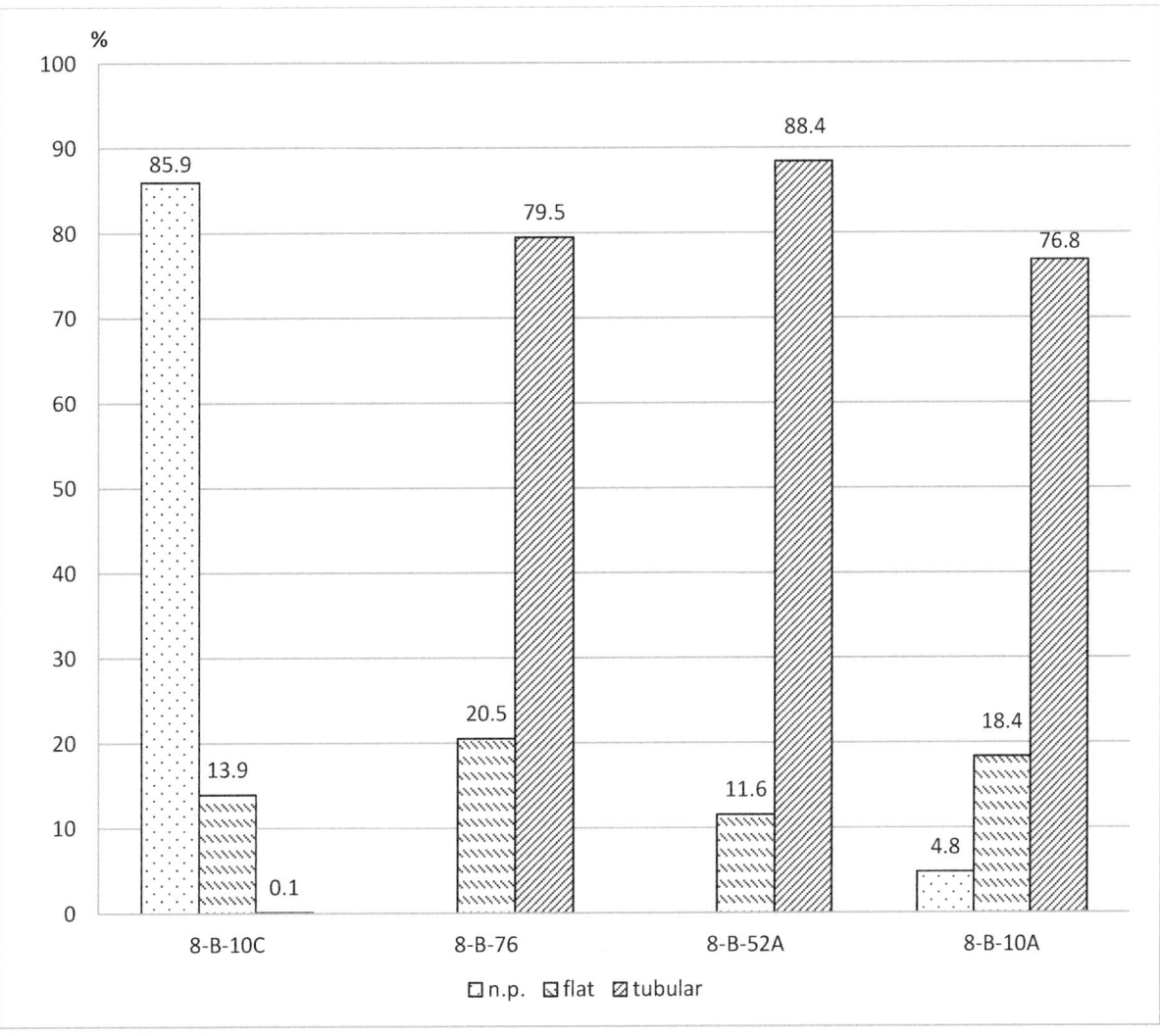

Figure 5.2 Percentages of organic inclusions from sites 8-B-10C, 8-B-76, 8-B-52A and 8-B-10A.

is reasonable to suppose that the 8-B-52A pottery, given that it was probably intended mainly to store foodstuffs in and not for domestic use, was the result of specific technological choices, through the careful sorting of the raw material and/or methods of preparation. The presence of rounded inclusions, as described in Chapter 3, may also indicate the intentional addition of aeolian sand to the paste.

Production: shaping and surface treatment

Ceramics from site 8-B-10C have walls between 6 and 10mm thick **(Figure 5.3)**. Only a small percentage of potsherds are burnished or polished, while most of the fragments have just plain or smoothed surfaces **(Figures 5.4 and 5.5)**. Some fragments were covered with a thin layer of clay, a self-slip, of the same composition as the rest of the vessel. It is possible that burnishing and polishing skills initially developed in response to functional rather than stylistic demands: at site 8-B-10C, when vessels are burnished or polished, it is more often on interior surfaces **(Figures 5.4 and 5.5)**.

During the Abkan horizon (site 8-B-76) walls become thinner **(Figure 5.3)** and surface treatments become more common. 10.3% of the samples are burnished on the inside, while in 23.1% of cases vessels are burnished on both surfaces. None of the ceramics, however, are polished **(Figure 5.5)**.

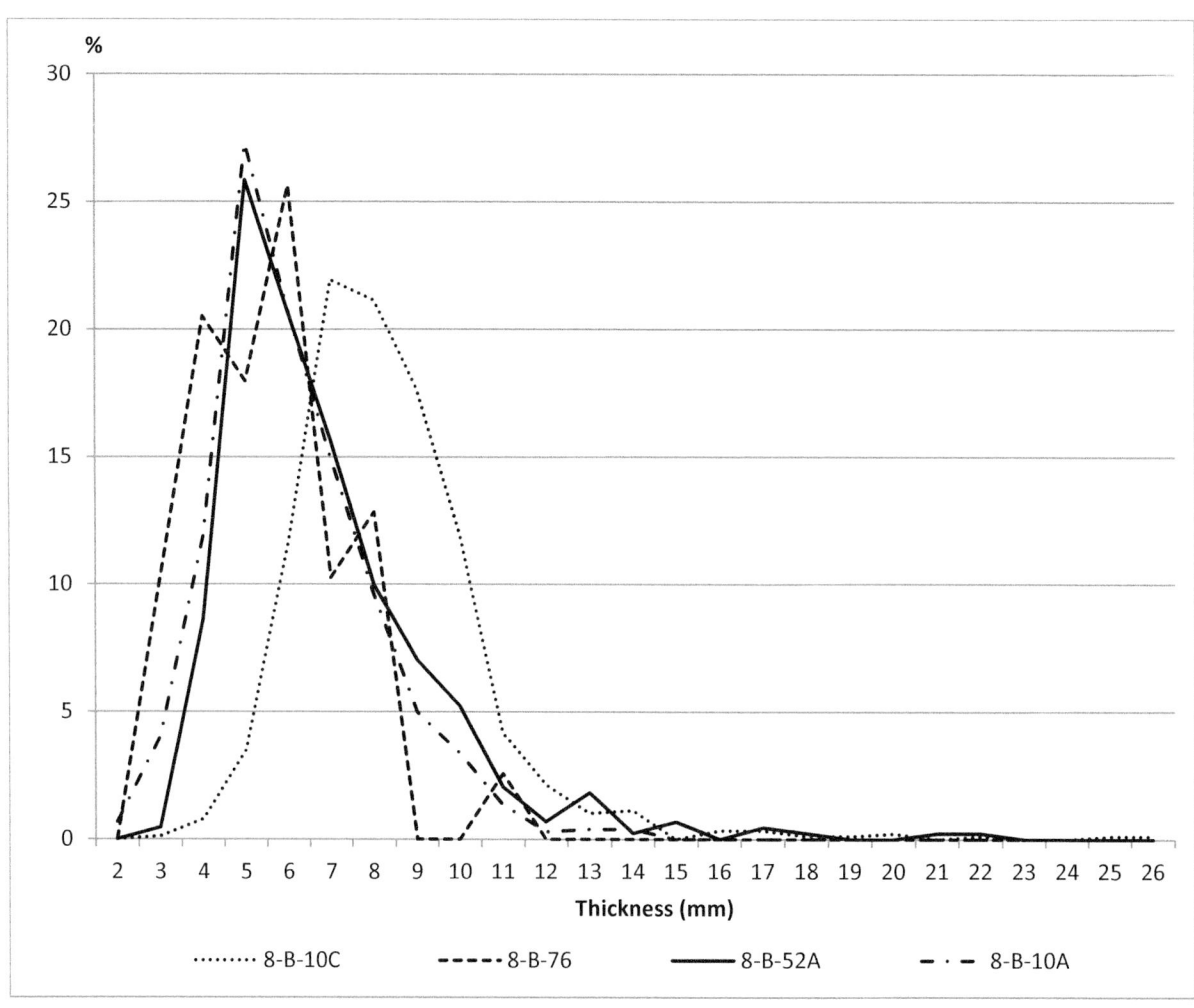

Figure 5.3 Sherd thicknesses from sites 8-B-10C, 8-B-76, 8-B-52A and 8-B-10A.

Burnishing is common to the products of both the most recent sites 8-B-52A and 8-B-10A **(Figure 5.4)**. In contrast, polishing is found only occasionally at site 8-B-10A (77.3% of the classified fragments show no sign of polishing) while at site 8-B-52A, over a third of the samples (38.8%) are polished on the inner surfaces, and 24% are polished both internally and externally **(Figure. 5.5)**. A possible explanation is that at site 8-B-52A polishing was used, in combination with other technological choices, to meet given functional demands resulting from the specific context, and the type of activity – storing foodstuffs – that was carried on there (see Chapter 3). Polishing, is recognized to be primarily an important 'visual performance characteristic', as may serve as decoration, however 'it also has an impact on 'utilitarian performance characteristics', reducing the permeability of the vessel (Skibo 2013: 102). This quality is commonly required for containers used for storing liquids, where high permeable walls serve to keep the content cool, by promoting cooling by evaporation (Schiffer 1988; cf., also Tite 2008), but permeability might represent a concern in vessels used for storing cereals and foodstuffs, as an excess of humidity could damage the food.

Stylistic traditions

Just as technological and functional choices vary over the course of the various phases of occupation, so too are stylistic traditions periodically adjusted through the adoption of different motifs and decorative techniques.

5. Comparing chaînes operatoires: continuity and discontinuity in the ceramic assemblages of Sai Island

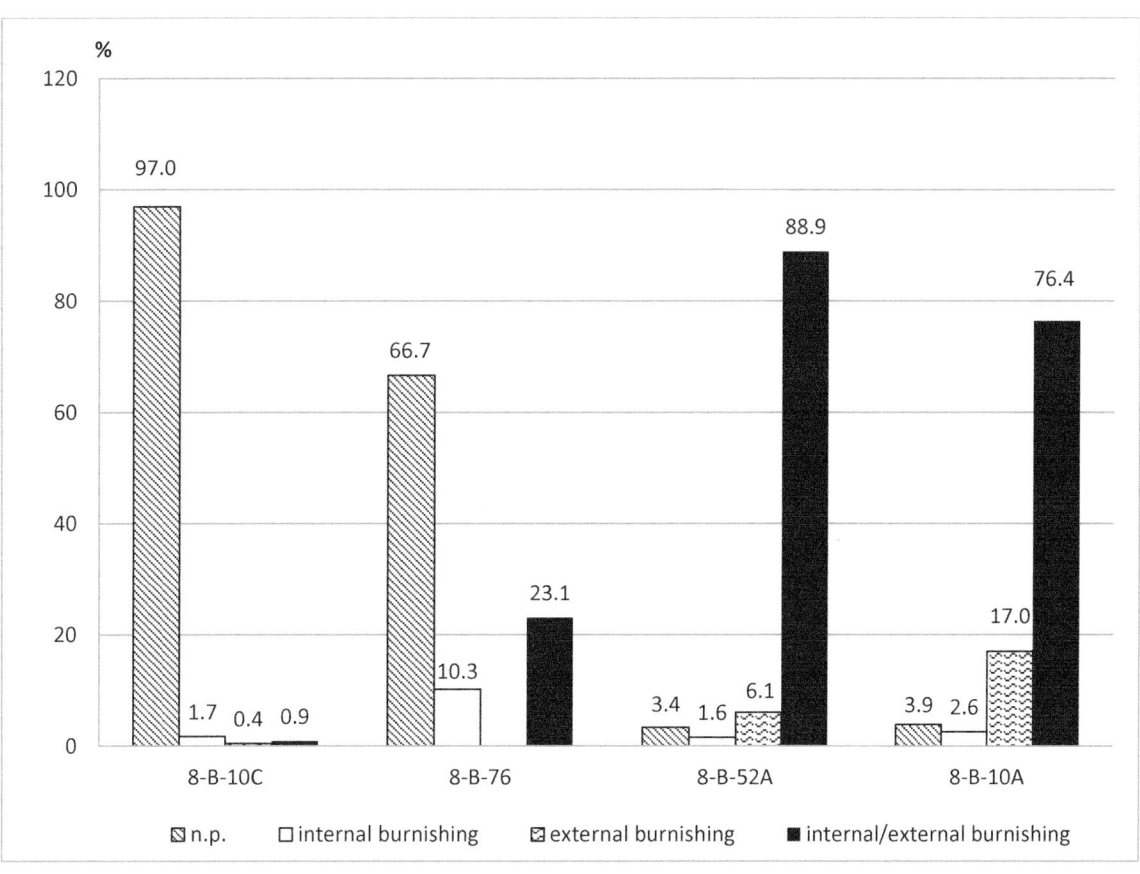

Figure 5.4 Percentages of burnishing from sites 8-B-10C, 8-B-76, 8-B-52A and 8-B-10A.

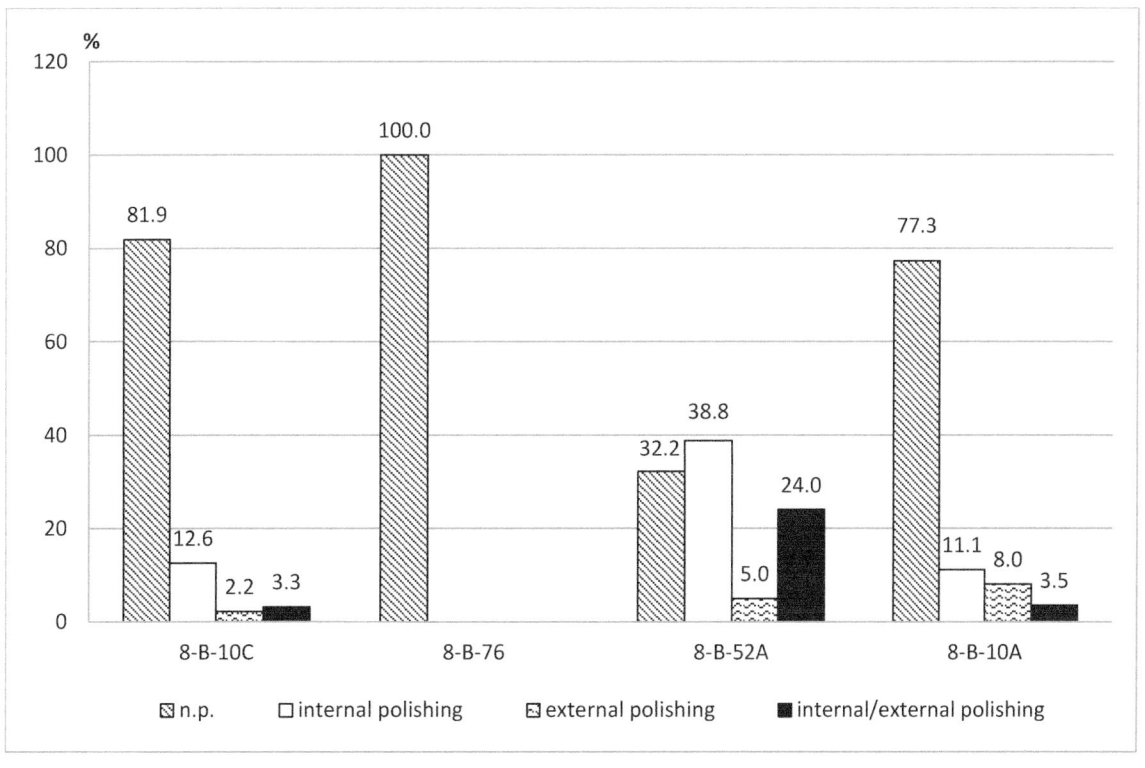

Figure 5.5 Percentages of polishing from sites 8-B-10C, 8-B-76, 8-B-52A and 8-B-10A.

SITE	8-B-10C		8-B-76		8-B-52A		8-B-10A	
	No.	%	No.	%	No.	%	No.	%
Not decorated	676	76.0	36	92.3	380	86.2	959	91.9
Alternately pivoting stamp	2	0.2	0	0.0	11	2.5	5	0.5
Incision	1	0.1	3	7.7	14	3.2	48	4.6
Rocker stamp impression	184	20.7	0	0.0	32	7.3	3	0.3
Rouletting	0	0.0	0	0.0	0	0.0	17	1.6
Simple impression	15	1.7	0	0.0	4	0.9	12	1.1
Unclassifiable	11	1.2	0	0.0	0	0.0	0	0.0
TOTAL	889	100.0	39	100.0	441	100.0	1044	100.0

Table 5.1 Percentages of decorative techniques from sites 8-B-10C, 8-B-76, 8-B-52A and 8-B-10A.

Pottery from the oldest site, 8-B-10C, is often not decorated though it is more frequently decorated than the pottery from the other three sites. Where they can be identified, the decorations consist mainly of various types of zigzag made with the rocker stamp technique (20.7% of the samples) **(Table 5.1)**.

A particular characteristic of the 8-B-10C ceramics is the dotted wavy line (DWL) motif, made with a three teeth comb, in a structure of alternating decorated and undecorated bands **(Figure 5.6)**. Equally common are packed zigzags of dots or dashes, made with a larger comb (between 7 and 25mm long) with four to 12 teeth. In some ceramics, the rocker stamp technique seems to have been applied using a roulette : in this case, the motif consists of partially overlapping, closely packed zigzags of horizontal dashes, which at times preserve the impression of the organic fibre the tools were made of (for an accurate description of the roulette tool, see Chapter 3) **(Figure 5.7)**. The effect in the vertical plane of the 'rocker' movement made to produce these motifs is less obvious: at first sight, the dominant impression is of lines of horizontal dashes. There are occasional incised decorations (single lines near the rim) or impressed dots, dashes or triangles, laid out in lines parallel to the rim, and, if the alternately pivoting stamp technique (APS) is used, in equidistant pairs **(Table 5.1)**. Rims are decorated with milled or notched impressions (see Chapter 3). The majority of decorated sherds come from layer 1 and just few single specimens are from the other layers of the stratigraphic sequence.

Figure 5.6 Sherd decorated with dotted wavy line motifs from site 8-B-10C (photo by R. Ceccacci).

Figure 5.7 Sherd decorated with a roulette by using the rocker stamp technique from site 8-B-10C (photo by R. Ceccacci).

During the following horizon, the variety of decorative techniques and motifs is more limited. More than 90% of the Abkan ceramics from site 8-B-76 are not decorated **(Table 5.1)**. The few decorations observed are usually found around rims **(Figure 5.8)**, sometimes on black-topped vessels or on ripple ware. There seems also to be a significant reduction in the quantity of pottery produced. It is possible that the introduction of a pastoral economy, and the effects that this change brought about over time on the organisation of human groups and the use they made of their ceramics led to pottery being produced in more limited quantities but with new technological characteristics – possibly designed to meet new functional demands (see below, final considerations).

Figure 5.8 Sherd decorated with milled impressions on the rim from site 8-B-76 (photo by R. Ceccacci).

Most of the decorated Pre-Kerma ceramics from site 8-B-52A have walls decorated with zigzags motifs produced using the rocker stamp technique (7.2%) **(Table 5.1)**. Unlike at site 8-B-10C however, the structure of the decoration is usually continuous rather than banded and the zigzags are almost always made with plain edged implements or combs, which can be as much as 31mm long – more suitable tools for decorating the wide surfaces of the storage jars discovered at this site. Incised motifs are less common, usually found around rims, as are simple impressions. Similarly, only a limited percentage of samples (2.5%) display decorations made with the alternately pivoting stamp technique **(Table 5.1 and Figure 5.9)**. These particular samples are almost exclusively associated with ceramics having a fine texture and surfaces polished both inside and outside. A similar result for a set of samples from the same site was reported previously by Garcea (Garcea and Hildebrand 2009). These fine ceramics also show the greatest stylistic and technical affinity with the Middle A-Group tradition (thin walled vessels, open forms, burnished or polished surfaces, and presence of black-topped) and seem to have a different function from typical local Pre-Kerma products, i.e., large food storage jars. Garcea and Hildebrand (2009) pointed out that the presence of A-Group pottery in the 8-B-52A's assemblage indicates 'well-established relationships' between Sai and A-Group peoples according a more robust north-south axis of contact.

A different decorative repertoire can be seen at site 8-B-10A. Only occasional potsherds have impressed decorations made using the alternately pivoting stamp or rocker stamp techniques. The most common technique instead is incision (4.6%) **(Table 5.1)**. The incised motifs form geometrical or herring-bone structures on the vessel surfaces, or panels which are sometimes produced by using a combination of techniques (e.g., incised triangles with single impressed dots or dashes inside them). The proportion of motifs laid out in a continuous structure is lower than at site 8-B-52A. Some of the potsherds from 8-B-10A bear a different type of impressed decoration not observed at the earlier sites or at site 8-B-52A. This is possibly made with an instrument – a roulette – similar to the tool mentioned for 8-B-10C, except that the decorations here are made by the rouletting technique and not the rocker stamp technique **(Figure 5.10)** (see Chapter 3).

Figure 5.9 Black-topped sherd resembling A-Group pottery decorated by using the alternatively pivoting stamp technique from site 8-B-52A (photo by R. Ceccacci).

Figure 5.10 Sherd decorated by using the rouletting technique from site 8-B-10A (photo by R. Ceccacci).

Vessels from site 8-B-52A and 8-B-10A differ also in shape. Although at both sites walls are of almost equal thickness (mainly varying between 4 and 8mm thick) **(Figure 5.3)**, at site 8-B-10A open forms predominate. Differently, at site 8-B-52A, there is a higher percentage of closed forms or vessels with straight walls – globular bowls and large neckless jars suitable for storing foodstuffs.

The stylistic and typological divergences between the assemblages of sites 8-B-52A and 8-B-10A, if considered along their technological differences, show a certain internal variability. Explaining this variability in chronological (as evolution of ceramic traditions) or functional terms (stylistic and technological diversification in response to different pottery uses at the two sites) is a very complex matter. The currently available dating from 8-B-52A indicate that the time-span of use of the site was actually much longer than suggested by the first dating of macro-botanical remains (Geus 1998), as it encloses a period of over 800 years from *c.* 3600 to 2500 BC (Hildebrand and Schilling 2016; see also Chapter 2). Silo 39, which is the feature in our sample returning more potsherds (268 classifiable sherds vs.119 from silo 44 and only 54 from silo7), gave a date between 2871 and 2626 BC (Geus 1998), however there are no dates from silos 7 and 44 for a direct comparison. Moreover, all the dates published from this site were obtained from charcoal and macro-botanical remains but no dates are presently available from the ceramics. Therefore, it is likely that the 8-B-52A ceramics, however homogeneous they seem, may not all date from the same period and be all associated to the same phase of use of the site. The settlement site 8-B-10A presently yielded three dates with overlapping ranges *c.* 1900-1650 BC (Hildebrand and Schilling 2016), which corresponds to a period of occupation about 1000 years younger than site 8-B-52A. However, as a direct dating of the ceramics is missing from this site too, it is tricky at the moment to draw any certain chronological interpretation concerning the stylistic and technological variability observed in the assemblages of the two sites.

A summary of the macroscopic technological and stylistic features of the different ceramic productions is provided in **Table 5.2**.

Discussion and comparison of the archaeometric data

Archaeometric characterisation analyses (see Chapter 4) have allowed defining in greater detail the technological and compositional nature of ceramics, making it possible to understand and explain more clearly the technological and stylistic variables, which have in part already been observed in the macroscopic classification described above. For the sake of simplicity, we can make a distinction between:

a) compositional variables, resulting from the selection of different raw materials;

b) technological variables connected with different phases of the manufacturing sequence, in particular the preparation of pastes, shaping, drying and firing.

It should be pointed out that both of these variables derive from decisions made by potters, and are therefore to be understood as socio-cultural variables. Nevertheless, they are also inevitably connected to the environmental and geological context, and to the nature of each site, and lastly to the possible different uses the ceramics were intended for.

Cultural horizon	Khartoum Variant	Abkan	Pre-Kerma	
Site	8-B-10C	8-B-76	8-B-52A	8-B-10A
Technological features				
Texture	medium; fine and coarse	fine; medium	medium; fine	medium; fine
Mineral inclusions	small and medium	small	small	small
	common; frequent	common	frequent; common	frequent; common
	angular	angular and rounded	rounded	angular
Organic inclusions	rare	common	frequent	frequent
	flat (vegetable fibres)	tubular (herbivore dung)	tubular (herbivore dung)	tubular (herbivore dung)
Surface treatment	not present; rare burnishing/internal polishing	not present; rare burnishing	burnishing; internal polishing	burnishing; rare polishing
Vessel type and morphology				
Body shape	large open bowls and jars	bowls with everted or straight walls	storage jars; rare open forms	open forms (bowls); rare closed forms
Thickness (mm)	6–10	6–8	4–8; up to 22 (heavy-duty jars)	4–8
Stylistic features				
Technique	rocker stamp	undecorated; incision; ripple ware	rocker stamp; alternately pivoting stamp	incision; rouletting; simple impression
Tool	combs with 3 teeth for DWL and combs with 3 to 12 teeth; roulette	stylus	plain edge tools; combs with 3 to 12 teeth; double pronged tools	stylus; roulette; combs with 6 to 8 teeth
Motif	DWL (small waves); dotted/dashes zig-zags	single lines	plain zigzags; dotted/dashes zig-zags; paired lines	single lines; paired lines
Structure	banded	-	continuous	panelled; geometric; herring-bone
Rim decoration	milled/notched impressions	milled impressions; black topped	black topped	black topped

Table 5.2 Synoptic description of the macroscopic technological and stylistic features of the different ceramic assemblages (after D'Ercole 2015).

Compositional variables

The ceramics from site 8-B-10C stand out from the later pottery not just macroscopically but also for their mineralogical, petrographic and chemical properties. The majority (20 out of 22) of these samples were assigned to the mineralogical-petrographic group QKfs (quartz-K-feldspar) while none of these samples belong to the QPl group (quartz-plagioclase), into which all the samples from the most recent sites 8-B-52A and 8-B-10A fall, along with most of the Abkan samples from site 8-B-76 (see Chapter 4).

The main difference between the QKfs and the QPl group consists in the varying proportions and types of feldspar present in the composition. That is, there is a significant proportion of potassium (K) feldspar in the QKfs group, and a usually more limited amount of sodium/calcium plagioclase (Pl) feldspar in the

QPl group. This difference is reflected in the chemistry of the pottery, allowing a clear distinction, just on the basis of certain major oxides, of the Khartoum Variant samples from the others (see Chapter 4). The presence of K-feldspar is associated with a frequently high mica content, principally biotite, and rock fragments of metamorphic origin: the latter vary in size but are always angular and irregular in shape. Also the quartz grains observed in thin section analysis usually have sharp rather than rounded edges. The inclusions are generally very poorly sorted and the fabric presents a seriate, inequigranular texture. Moreover, petrographic consistency between the fine and coarse fraction of the non-plastic inclusions suggests that they were naturally present in the clayey raw material (D'Ercole *et al.* 2015, in press).

All of these characteristics taken together for the ceramics from site 8-B-10C point to the use of a primary, residual clay as a raw material for pottery manufacture. The analyses of the local sediments indicate that there is no chemical or petrographic correspondence between this raw material and the natural samples taken from sites 8-B-10C, 8-B-76 and 8-B-10A (see Chapter 4). However, a more general petrographic compatibility with the geological substrate of the island makes it seem likely that the raw material for the Khartoum Variant pottery was nevertheless of local origin. It is a lean, non-calcareous clay, containing an high proportion of mineral inclusions of potassium feldspar and metamorphic rock fragments, but also rich in iron oxides and biotite, which lend it its characteristic reddish-brown coloration (D'Ercole *et al.* 2015). This suite of minerals and rock fragments is related to the Precambrian Basement (Hays and Hassan 1974). This old rock complex, consisting mainly of granites, gneiss and schists, in Sudanese Nubia is exposed on both banks of the Nile, in the stretch between Dongola and Wadi Halfa (Shang *et al.* 2010, see also for details Chapter 2). It emerges in part also on Sai even though for the most part overlaid with elements of the more recent Nubian Sandstone Formation, including the Jebel Adu outcrop, situated to the west of the settlement 8-B-10C (Geus 2000; Goossens *et al.* 1997). The clayey raw material used by the Khartoum Variant people to make their pottery may have been very likely collected from the eluvial deposits on the Nubian Precambrian Basement (D'Ercole *et al.* 2015, in press).

Certain 8-B-10C samples however (SAI 28, 32 and 73-77), though assigned to the same QKfs group, showed specific characteristics of composition (lower levels of mica and predominance of muscovite over biotite) and of texture (finer) which separate them partially from the site's other ceramics. These samples formed the sub-group Fabric QKfs-Ms (see Chapter 4). These characteristics, rather than being explained by some technological process (purification/washing of the natural clay with consequent sorting of the mineral element in the paste) seem more to suggest that the raw material – though still with a common petrofacies – was taken from a different clay bed. An unlike consideration applies to the samples SAI 27 and 31, which make up the mineralogical group Q (quartz). Their differences from the other samples are more distinct (their framework is composed almost exclusively of monocrystalline rounded quartz clasts, with traces of muscovite and heavy minerals) and chemically very clear (see Chapter 4).

The most interesting point is that most of these samples (SAI 27 and 31, but also SAI 28, 32 and 73-77) also stand out on the macroscopic level through their particular technological characteristics: burnished or polished interiors (SAI 28, 31, 73-75) and occasionally presence of organic inclusions in the paste (SAI 31-32, 74 and 75) (see Chapter 4, Table 4.2). This would support the interpretation proposed above based on the macroscopic analyses of the existence of a diversified production with a deliberate recourse to different sources/beds of clay or to different recipes. This diversification might be a response to a visual and/or functional need: the need to distinguish fine wares from the rest of the ceramics, whether by the selection of a particular raw material or through particular technological choices. If we consider the chronological sequence at site 8-B-10C within two distinct phases of occupation at 7600–7200 BC (level 2) and 5050–4800 BC (level 1) (see Chapter 2), this variability could also assume a chronological significance and indicate a potential change/evolution of the ceramic traditions over the time. A greater sampling from this site would help in confirming or not this preliminary reading.

The raw materials change during the following Abkan horizon. Most of the samples from site 8-B-76 (16 out of 19) belong to the mineralogical-petrographic group QPl (quartz-plagioclase). There are fewer mineral inclusions, and in particular, fewer metamorphic rock fragments in these ceramics. The proportion of potassium feldspar is reduced, replaced by plagioclase, though only by limited amounts. The quartz grains, as was observed in the macroscopic analysis, are more often small and rounded. These compositional features suggest the use of a raw material with new and different characteristics: a sediment of alluvial origin (i.e., a Nile alluvial clay) rather than a residual clay, more 'evolved' in composition, in that it is lacking those mineral components more liable to alteration (such as feldspar, rock fragments of metamorphic origin and biotite mica) and is better sorted in grain size. In this sediment there remain, however, certain stable mineral phases, such as zircon (zirconium silicate), which can be found in both metamorphic rocks and in intrusive veins of quartz (D'Ercole *et al.* 2015).

It should be emphasised that the raw materials used during the Abkan horizon are by no means uniform in composition. Three of the samples from site 8-B-76, for example, SAI 50, 58 and 68, have characteristics (higher feldspar content, metamorphic rock fragments, and biotite mica) which justify their attribution to the QKfs (quartz-K-feldspar) group, and so are associated with a raw material more similar in composition to that of the Khartoum Variant ceramics.

The totality of samples (30 out of 30) from the two most recent sites 8-B-52A and 8-B-10A were assigned to the QPl (quart-plagioclase) group. These ceramics include well-sorted, fine sand-sized quartz, plagioclase, volcanic rock fragments (tephra) and various type of heavy minerals, among them clinopyroxenes and amphiboles; whereas the content of K-feldspar, biotite mica and metamorphic rocks is much less important. These petrographic characteristics, as observed at site 8-B-76, point to the selection of sediments of alluvial origin. Further, the presence of collophanes (microcrystalline hydrated fluoroapatite), which seems to distinguish the petrographic content of a group of ceramics from site 8-B-10A (the QPl-Col group), suggests the use of a raw material taken not directly from the Nile banks but rather from inside or near an inhabited area (i.e., a paleo-alluvium). The formation of phosphates in archaeological deposits seems to indicate the presence of organic matter in soil, together with dissolved bone, which constitutes one of the main sources of phosphates (Weiner 2010). At site 8-B-10A, the higher levels of phosphates observed in certain sediment samples could derive from the accumulation of organic matter of animal origin (herbivore droppings) as a result of the repeated use of the area for pasture or penning, with the possible presence of animal enclosures/folds (D'Ercole *et al.* 2015).

In conclusion, a shift can be perceived from the use of residual sediments as the raw material for pottery production – during the Khartoum Variant horizon (site 8-B-10C) and to a very limited extent still in the Abkan period (site 8-B-76) – to an exclusive use, starting from the Abkan and then in the Pre-Kerma period, of alluvial, Nile clay, sediments. These alluvial sediments are geochemically very similar to the natural samples taken from archaeological deposits at both site 8-B-76 and 8-B-10A (see Chapter 4), and are therefore certainly of local origin. The strongest similarity has been observed with samples 81 and 82 from site 8-B-10A which were characterized by the highest clay content (see Chapter 4). All in all, this similarity also suggests, for both Abkan and Pre-Kerma periods, the selection of a raw material that lay to hand, if not necessarily inside the settlements, at least in the nearby deposits of sediment which, from the second half of the Holocene, formed around the original nucleus of the island, and on which the sites were established (D'Ercole *et al.* 2015).

Technological variables

The compositional variables discussed above are matched with evidence of a plurality of technological choices applied to individual stages in the manufacturing sequence. This variation, observed through petrographic analyses, led to a further subdivision of the three groups into fabrics and then into further sub-groups (see Chapter 4, Table 4.10). The groups and sub-groups are described below following again the different stages of the chaîne opératoire (preparation, production and firing).

Preparation: addition of organic inclusions

The oldest ceramics from site 8-B-10C are made of a sandy-micaceous clay and are tempered with mineral inclusions, probably occurring naturally in the sandy component, without addition of organics, or sorting of the mineral component through washing or purification of the clay. An exception is the limited number of samples (SAI 31-32 and 74-75) which are distinguished by an intentional addition of organic matter to the paste. These samples also contain less mineral inclusions, which have smaller size. Further, like other samples from the QKfs-Ms fabric sub-group (SAI 28, 73 and 76-77), they have a finer texture and more developed shrinkage porosity. This internal variability might be the result of changes taking place over time. Alternatively, it may support the idea of an internal division of production on the basis of function, between heavy-duty items mainly for storage use, made of medium to coarse pastes (QKfs-Bt fabric samples, rich in feldspar, metamorphic fragments and biotite), and ceramics with a fine texture and/or burnished or polished surfaces (essentially the QKfs-Ms fabric and the Q group samples) which were possibly intended for different uses within the settlement. The vessels made of fine pastes do not seem to be suitable for storing and were possibly used for serving or preparing the food.

Site 8-B-76 and sites 8-B-52A and 8-B-10 ceramics, in contrast, always contain organic matter, which, given the significant quantities, must have been intentionally added during preparation of the pastes. Petrographic analysis, allowed observing however, beyond this simple fact, a multiplicity of technological traditions.

Abkan pottery (site 8-B-76), independently of the compositional characteristics of the raw material, was tempered with organic inclusions possibly derived from herbivore dung, but also with charcoal fragments, which were not recognized macroscopically. In the QPl group, these samples are assigned to the QPl-Cha fabric (quartz-plagioclase-charcoal) precisely to indicate their technological peculiarity in comparison with the other samples of the same group (see Chapter 4). The presence of this organic component, consisting of particles of charcoal only a few millimetres large, seems to mark these ceramics out unambiguously, distinguishing them from the older Khartoum Variant artefacts and from the more recent assemblages of the sites 8-B-52A and 8-B-10A. Furthermore, given the amounts of these inclusions in the pastes, and their shape (the carbonized particles are usually angular), it can be assumed their presence is intentional, the consequence of some functional objective (D'Ercole *et al.* 2015, in press). At site 8-B-76, signs of internal variability in the amount of organic inclusions can be seen which seem to be connected to the differing textures identified in the macroscopic classification. There is abundant carbonized vegetable matter in most of the fine pastes (see for example SAI 24-25, 60-62, and 64). The ceramics with a medium texture however generally have lower levels of organic content and a more developed primary porosity, sign of a less carefully prepared paste (samples SAI 22, 49, 59, 63 and 67 from the QPl group and SAI 50, 58 and 68 from the QKfs group).

Laboratory analyses confirmed the divergences already observed between the assemblages of sites 8-B-52A and 8-B-10A. The discriminating factor in technological terms is the presence of organic tempers, which give the ceramics of the two sites a characteristic in common which distinguishes them from Khartoum Variant andAbkan pottery, but at the same time allows to differentiate the technological traditions of the two sites. The ceramics of site 8-B-52A contain distinctly higher levels of organic matter than the samples from 8-B-10A, so as to justify their being assigned to a special fabric, QPl-Veg (see Chapter 4). When organic matter is not completely burned up, it is possible to identify the structure of the vegetable fibres and estimate their relative size, which is between 0.5 and 3mm, though some inclusions are as much as 5 or more mm long. These dimensions, along with the morphology of the inclusions (thin and tubular, when the longitudinal section of the fibre is visible), indicate the addition of an organic component of animal origin (herbivore dung), made up of small pieces of grass, straw and other vegetable matter, resulting from herbivore digestion (D'Ercole *et al.* 2015, in press). Some samples from site 8-B-52A seem to contain seeds, fruit or berries, either intentionally added by the potter as temper, or happening to end up in the paste because they were present in the potter's immediate neighbourhood. Given the site's function as food storage area, it is likely that nearby there were abundant remains of various types of plants and/or cereals, however the space used as workshop for producing the ceramic containers could also be distant from the site (see Chapter 2).

The significant presence of organic inclusions is linked to certain technological characteristics such as greater plasticity and porosity in the paste, which qualities probably sought out intentionally by the potter both to facilitate manufacture and to improve vessels' performance (D'Ercole *et al.* 2015, in press). It is recorded in the literature (London 1981; Skibo *et al.* 1989; Velde and Druc 1999) that organic inclusions, given their hydrophilic nature, not only make the clay easier to work and improve the cohesiveness of coils during the shaping phase, but also contribute to the lightness of the ceramic body.

Petrographic analysis demonstrated a certain element of variation also within the produce of 8-B-52A. In the QPl-Veg fabric, three sub-groups were recognized on the basis of grain size (see Chapter 4, Table 4.10). The samples assigned to them correspond for the most part to the texture type groups established by the macroscopic analysis. The fine grained ceramics (samples SAI 11, 14, 17, 20-21, 43, 46 and 48, in sub-group 1b) contain not just smaller mineral inclusions, but also finer organic tempers. Further, these ceramics are usually polished, and often have stylistic as well as technological characteristics similar to those of A-group pottery (see above). This is reflected, in proportion, in the sample chosen for analysis: SAI 11, 43, 46 and 48 in particular show stylistic traits that are comparable to the A-group tradition. In addition, the fine paste ceramics at site 8-B-52A are frequently open forms, suitable for the serving and consumption of food, rather than for storage: this function seems therefore to be associated with the adoption of particular technological choices.

Production: shaping (manufacturing techniques)

Examination of samples under a petrographic microscope contributed to the identification of further information on the manufacturing techniques, and especially on the stage of shaping. Pastes containing organic fibres provide the best results: the way potters manipulate clay, and the use of different possible techniques for manufacture or surface treatment, act together to determine the position and orientation of organic inclusions within pastes. This stands out distinctly in thin section analysis, in the varying orientation of the voids left by combustion of organic matter (Velde and Druc 1999).

The examination of a number of typical ceramic samples showed that hand building by coiling seems to be the most common techniques in all the horizons. Where the longitudinal section of vegetable fibres is visible, the effects of how the potters stretched, squeezed and smoothed the clay coils are clearly visible **(Figure 5.11: a)**. There are also areas of uneven oxidization inside the matrix at the point where the clay coils touch. There are rarer samples, in which the disposition of the organics – parallel to the vessel's surface – suggests the use of a different building technique, by drawing out a lump of clay (D'Ercole *et al.* 2015) **(Figure 5.11: b)**.

Firing

In general, all the samples, irrespective of their chronology or of their mineralogical group (QPl, QKfs or Q), seem to have experienced similar firing conditions. Temperatures were probably no higher than 800°C (hematite out, clay minerals in, in the XRPD spectra), which excludes the use of some sort of combustion chamber such as a kiln, and suggests the use of bonfires or pitfires (D'Ercole *et al.* 2015, in press).

Petrographic examination gave further evidence for this, since most of the samples had a medium or high index of birefringence of the matrix as a consequence of incomplete sintering. This characteristic can be especially observed above all in the Abkan ceramics. Site 8-B-76 samples seem to be marked by highly developed

Figure 5.11 Photos of two thin sections in which it is recognizable the manufacturing technique used by the potter can be seen. Image a: coiling technique; image b: drawing out a lump of clay (figure by G. D'Ercole).

shrinkage porosity, which suggests particularly short firing periods at low temperatures of 600–700°C. On the macroscopic level, this is reflected in the softness and crumbly appearance of the pastes. The Khartoum Variant ceramics from site 8-B-10C also have a medium or high index of birefringence, which indicates similar firing technology and temperatures, even though the higher percentage of sand makes these pastes harder. In the case of the Pre-Kerma ceramics, though the degree of birefringence of the minerals is masked by the organic matter spread throughout the matrix, incompletely closed shrinkage porosity and only partially burned up vegetable matter suggest that again this pottery experienced short firing and/or firing at low temperatures (D'Ercole *et al.* 2015, in press).

Most of the Khartoum Variant ceramics are a uniform light brown or reddish-brown in fracture. This coloration is obtained by firing in an oxidizing atmosphere pastes rich in iron oxides and lacking organic content (D'Ercole *et al.* 2015, in press) **(Figure 5.12: a)**. The samples with vegetable inclusions (SAI 31-32 and 74-75) show a characteristic pattern in cross-section of their dark grey or black core, resulting from the partial combustion of vegetable matter in the paste, with oxidized surfaces. Certain 8-B-10C samples have irregular patches of differing coloration spread over the surface of the edges of the fragments. These patches were produced after firing and could be linked to post-depositional factors, or could be a sign of the vessels' possible re-use. However, as they occur on the edges of the sherds, they must be a feature of the fragment itself and not of the original complete vessel (D'Ercole *et al.* 2015).

There are some completely oxidized pieces among the Abkan ceramics from site 8-B-76, but 'sandwich' structure's fragments with a dark core and oxidized surfaces are more common. In these cases, firing was not intense enough for all the organic matter to be burned off and be completely dispersed. In consequence, a firing process can be hypothesised which involved an oxidizing atmosphere only in the last stage and at a low temperature (D'Ercole *et al.* in press; cf. also Quinn 2013; Velde and Druc 1999) **(Figure 5.12: b)**. Similar evidence comes from the analysis of the ceramics of sites 8-B-10A and 8-B-52A. Given the high organic content, they are usually dark grey or black in fracture, with a particularly thin (1mm or less) oxidized outer layer, often visible on the convex surface. Vessels were fired upside down, with the opening in direct contact with the fuel, but at temperatures not high enough or for periods not long enough to ensure complete oxidization of the entire thickness of the walls. Certain samples from the settlement 8-B-10A are distinguished by a thicker layer of oxidized clay (more than 1mm) and occasionally by oxidized cores. In these cases, the oxidizing process occurred while the heat was

the most intense (D'Ercole *et al.* 2015, in press) **(Figure 5.12: c)**.

A summary of the compositional and technological features of the different ceramic productions is provided in **Table 5.3**.

Final considerations

The comparison of analytical and macroscopic data allows us to perceive numerous variables affecting the production of pottery on Sai Island, to be ascribed both to the varying composition of the raw materials selected to make pots with, and to the successive choices made by potters at the moment of manufacture. These factors, along with changes in style, may be interpreted first on a diachronic level, where technological choices join stylistic variables to reflect changes in the products of the different chronological horizons (Khartoum Variant, Abkan and Pre-Kerma), and, second, on a synchronic level, which is in a certain sense unconnected to the diachronic level.

Figure 5.12 Examples of the different effects of oxidation-reduction resulting from firing. Image a: completely oxidized fracture for a sample lacking organic content; b: fracture with a dark core and oxidized surfaces; c: completely oxidized fracture for a sample with organic content (figure by G. D'Ercole).

Over the course of the whole cultural sequence, the first moment which stands out and is associated with the greatest number of changes in the production of ceramics is at *c.* 5000 BC, from the Khartoum

Cultural horizon	Khartoum Variant	Abkan	Pre-Kerma	
Site	8-B-10C	8-B-76	8-B-52A	8-B-10A
Compositional features				
Raw material	residual clay	silty alluvial clay	silty alluvial clay	silty alluvial clay
Non-plastic inclusions	quartz; K-feldspar; metamorphic rock fragments; micas	quartz; feldspar; micas	quartz; plagioclase; micas; volcanic rock fragments	quartz; plagioclase; micas; volcanic rock fragments
Technological features				
Organic inclusions	rare vegetable fibres for the fine ware	charcoal; herbivore dung	frequent vegetable fibres and herbivore dung (charcoal for the fine ware)	herbivore dung and vegetable fibres
Manufacturing technique	coiling	coiling	coiling	coiling; drawing out a lump of clay
Porosity	low; medium	high	medium	low; medium
Firing temperature	600-800 °C	600-700 °C	600-800 °C	600-800 °C
Firing atmosphere	oxidizing	oxidizing in the final stage	oxidizing in the final stage	oxidizing; oxidizing in the final stage; oxidizing

Table 5.3 Synoptic description of the compositional and technological features of the different ceramic assemblages (after D'Ercole 2015).

Variant to the Abkan horizon. At this time, several significant innovations have been recorded involving the following phases of pottery manufacture:

A. <u>raw material procurement</u>: change from residual clays with coarse K-feldspar, quartz, metamorphic rocks and biotite mica to Nile alluvial clays with fine quartz, plagioclase, volcanic rock fragments and heavy minerals;

B. <u>preparation</u>: change from mineral-tempered to organic-tempered pastes;

C. <u>production</u>: reduction in the thickness of vessel walls, disappearance of certain techniques and decorative motifs (i.e., dotted wavy line decoration), use of treatments for surfaces (burnishing, black-topped and ripple ware), different firing atmosphere.

The second important moment of economic and social transition, from the Abkan to the Pre-Kerma horizon, at c. 3600 BC, is characterised by a different pattern. Here, a substantial continuity can be seen in the stages of:

A. <u>raw material procurement</u>: continued use of Nile alluvial clays;

B. <u>preparation</u>: continued use of organic-tempered pastes;

C. <u>production</u>: continued use of black-topped and ripple wares, burnishing and polishing as treatments of vessel surfaces, similar firing temperatures and atmosphere.

Change between the Khartoum Variant and the Abkan horizon

There are many possible reasons for the changes observed in ceramic manufacture between the Khartoum Variant and the Abkan period. There might be a socio-cultural or functional explanation, or it could be a response to ecological factors deriving from the climatic and environmental changes that took place on the island. This involves the concepts of 'context of production' and of mutual relationship between it and technological choices. As pointed by Sillar and Tite (2000: 5), 'natural environment, technological knowledge and economic system influence technological choice, first, via the availability of raw materials, tools, energy sources and techniques, and, second, via the properties and performance characteristics that the options chosen possess in procuring, processing, forming, surface treatment, and firing'. Further, beside ecological, economic and functional factors, also cultural and ideological variables can justify the adoption of given technological choices and contribute to the change (or alternatively to the continuity) of certain ceramic traditions (see e.g., Arnold 1985, 2011; Gosselain 1992, 1998, 2000; Livingstone-Smith 2000; Stark 1998; Stark *et al.* 2000).

Khartoum Variant people were hunter-fisher-gatherers and both their subsistence economy and settlement pattern were dependent on the Nile regime and on the resources seasonally offered by the river. At that time, the Nile's level was higher than during the Middle Holocene and the river would represent to some extent a factor of risk and unpredictability for the people. Site 8-B-10C, located on a gravel bar on the eastern side of the island, because of the seasonal floods of the Nile, was characterized by a swampy environment and the gravel bars periodically became separate islands (see Chapter 2). Under these conditions, we might suppose that the Khartoum Variant people would not make themselves even more dependent on the Nile by deriving not only their food supply but also raw materials from it. Moreover, we could expect that these periodic landscape changes would impose a seasonal pattern on their supply activities, with a possible seasonal (and mental?) separation between the places intended for food supply (fluvial terraces/Nile alluvia) and those (pediment, Pleistocene sediments) dedicated to the procurement of the clayey raw material of their ceramics, and possibly other activities. Ethnographic studies recognized that pottery production in traditional societies is often concentrated in specific periods of the year, in correspondence with festivals and traditional celebrations (Van der Leeuw 1984), or according to a seasonal organization of communal activities (Arnold 1985, 2011; cf. also Santacreu 2014). Finally, the availability and workability of the clay itself might also vary according to the season, and is strongly influenced by changes in ecological and natural cycles (DeBoer 1984; Rice 1984). On Sai, as in the northern-most parts of Sudan, during the Early Holocene precipitation occurred during the winter months instead of during the summer, as in central Sudan (Garcea and Hildebrand 2009: 318). Following on from this, we might reasonably suppose that the Khartoum Variant potters collected the clay for making their vessels immediately after the rainy season and before the dry season.

It could be that at a given point, during the Abkan horizon (*c.* 5500-3700 BC), supplies of the residual sediments preferred in the earlier Khartoum Variant horizon ran out, and provoked a changeover – to some extent 'imposed' – to the use of alluvial sediment. It needs to be borne in mind here that the configuration of the island gradually changed. If during the Early Holocene, Sai was limited to just the pediment around Jebel Adu, after the fifth millennium BC, Sai took on its present morphology, and wide floodplains, rich in alluvial sediments, formed all around its inner pediment (see Chapter 2).

The change in the raw material source might also be due to choices dictated by technological and functional needs. The first pots produced by the Khartoum Variant people were mainly intended to be used as storage facilities within the household; therefore, they first needed to be resistant to mechanical stress (D'Ercole *et al.* 2017). The selection of a raw material with a high content of alkali feldspar and quartz inclusions would best meet this performance requirement (cf., Dal Sasso *et al.* D'Ercole *et al.* 2017; Schiffer and Skibo 1987; Schiffer *et al.* 1994).

Later, during the Abkan horizon, the introduction of the new herding activities in the subsistence system inevitably generated a transformation both in the social organization of the group and in the function of pottery, which progressively took a more central role in the domestic sphere. Pots became necessary not just for storage but also for other functions, and potters possibly started to experiment with different raw materials and technical options which are better suited to this new and diversified production (D'Ercole *et al.* 2017).

The innovations introduced with the Abkan period may be explained in functional terms, as co-dependent technological choices, required by the change in the clayey raw material, but also as new options made possible thanks to better technical knowledge on the part of potters. It is worth noting that the shift from the Khartoum Variant to the Abkan period corresponds paradoxically to a decrease in pottery produced. This could indicate the existence, at least at an initial phase, of a 'trial stage', during which the new raw material and technological choices were tested and first implemented in order to check their efficiency and compatibility with the function of the pots.

An important issue is the means of transmission of these innovations. How did the change happen? Was it the result of a process of 'technical acculturation' within the same group or did the change involved the arrival of an external population who brought along its own technical knowledge and technological and stylistic traditions?

V. Roux and M.A. Courty (2013), referring to evolutionary theory and to the mechanisms of cultural transmission and its effects on variability in cultural traits, distinguish between 'continuous' and 'discontinuous' change. The first indicates 'a continuous social learning between generations and among peers', while 'the change is discontinuous when there is a complete cessation of transmission', which means that the population is replaced (Roux and Courty 2013: 189). The authors argue that: 'Population replacement can be demonstrated when there is a replacement of material culture (break in a pattern of phylogenetic continuity)' (Roux and Courty 2013: 191).

The changes observed in the ceramic assemblages between the Khartoum Variant and the Abkan period, which involve all the steps of the manufacturing sequence and entail both technological and stylistic variability, seem to point to a 'cessation of transmission' and to a possible replacement of population.

This scenario is made particularly intriguing by the latest chronological and stratigraphic evidence provided from the site 8-B-76. This site, including Khartoum Variant and Abkan materials, was occupied, at first by Khartoum Variant groups of hunter-fisher-gatherers (*c.* 6400–6250 BC), and at a later time (*c.* 5500–3700 BC) by Abkan pastoral groups, who occupied the same site as the Khartoum Variant people, although their occupation was closer to the river (Garcea 2016a). During the same period (*c.* 5050–4800 BC), Khartoum Variant groups were still living at site 8-B-10C on the other side of the island. Therefore, for a certain time, the Khartoum Variant foragers and the new Abkan occupants lived side by side, surrounded by the same environment and exploiting similar resources. The newcomers, however, produced from the very beginning a completely different pottery. In some way, it seems that the two populations, although coexisting for some years, did not really interact or communicate with each others.

Continuity between the Abkan and the Pre-Kerma

In a different way, a consolidation and development of most of the Abkan ceramic traditions can be recognized through the Pre-Kerma. During this latter horizon, some changes and innovations occurred in the stylistic dimension, with the appearance of a wider variety of vessel forms, decorative techniques, motifs and structures; however, the 'technological style' (Lechtmann 1977) remains unvaried. In particular, the stages of raw material procurement and preparation were not changed. What we observe is instead a 'technical development' (Roux and Courty 2013) of certain specific technological choices (i.e., addition of organic tempers to the paste, burnishing and polishing of the surfaces, production of black-topped vessels). This picture points to a strong 'continuity' in ceramic traditions between the Abkan and the Pre-Kerma horizon and to a probable cultural transmission between the occupants of the two periods. In other words, it seems that the Pre-Kerma potters have learned well from their predecessors (the Abkan) the well established 'art' of making pots, so that they did not make significant changes in the chaîne operatoire but substantially repeated and improved the same 'recipe' (D'Ercole *et al.* 2017).

6. The Sai Island sequence and the Nubian and Sudanese traditions

Introduction

In this chapter, the pottery of Sai Island is compared with the Nubian, Egyptian and Sudanese productions introduced in Chapter 1. This discussion focuses on both a comparison of the stylistic features of the ceramics (first section) and their technological and compositional aspects (second section).

The following areas are included in the discussion: the district of Abka – Wadi Halfa, Lower Nubia (Adams and Nordström 1963; Carlson 1966; Mills 1965, 1968; Mills and Nordström 1966; Myers 1958, 1960; Nordström 1972; Shiner 1968a, 1968b; Williams 1989), the regions of Kerma (Honegger 1995, 1997, 1999, 2003, 2004a, 2004b, 2004c, 2005, 2006, 2010, 2014; Honegger and Williams 2015) and Mahas – Third Cataract, Upper Nubia (Edwards and Osman 2012; Sadig 2010), and the region of Nabta Playa and Bir Kiseiba, in the Egyptian Western Desert (Nelson and Associates 2002).

Additionally, to facilitate discussion of the stylistic and technological traits of ceramics production, areas which are on the fringes, geographically and culturally, of these principal areas will be considered, such as the regions of Laqiya and the Wadi Howar, west of Sai Island, in the desert (Jesse 2004, 2006, 2008; Lange 2004, 2006; Lange and Nordström 2006), and the province of Khartoum, in Central Sudan (Arkell 1949, 1953; Caneva 1983; Caneva and Marks 1990; Salvatori 2012; Salvatori and Usai 2006–2007; Salvatori *et al.* 2014) (see Chapter 1 for a description of these regions).

It should be pointed out that the data available for these different contexts are not always homogeneous. Although many analytical and technological studies of ceramics from Egyptian and Sudanese sites have been carried out over the years, only a few of them have included archaeometric characterisations with petrographic, mineralogical and chemical analyses associated with a macroscopic classification of the ceramic evidence. Consequently, while detailed information is available for some of these sites which can be compared with the results obtained for the Sai sample, for other sites much less is available, consisting more of technological than of compositional data, since obtained by means of macroscopic observation of ceramics or, at most, through examination under a stereoscopic microscope.

Section 1: Comparison of decorative styles (Figure 6.1)

Khartoum Variant horizon

Dotted Wavy Line

The earliest pottery at Sai Island was mainly decorated by impression with the rocker technique; producing dotted wavy line motifs and various types of packed zigzags. The presence of the characteristic dotted wavy line decoration among the products of site 8-B-10C is one of the principal features confirming its attribution to the Khartoum Variant tradition (Garcea 2006-2007; Garcea and Hildebrand 2009). The dotted wavy line motif is common to the Khartoum Variant sites of the Abka –Wadi Halfa region, the materials from which were first described by Shiner (1968a) and Nordström (1972) and later re-examined by Gatto (2006a). These sites share with settlement 8-B-10C on Sai two particular versions of the dotted wavy line: the 'Dotted Wavy Line-Arch-shaped' motifs, which 'only resemble the real technique of dotted wavy line impressions' (Garcea and Hildebrand 2009: 314), and the proper 'Dotted Wavy Line' motif (Gatto 2006a). At Sai, this latter was produced using a small comb, with a maximum of three teeth, and forming on the vessel a typical banded structure resulting in an alternation of decorated and undecorated zones (see Chapters 3 and 5). Gatto (2006a: 60), in her re-exam of the assemblages of the Khartoum Variant sites in the district of Abka –Wadi Halfa, defines this motif as 'Zonal Dotted Wavy Line' and observes that it was exclusively related to Ware Group A and Fabric III: 'medium coarse mineral tempered ware' (= coarser version of Fabric IA in Nordström 1972: 48) (see technological section below). Ware A also includes many undecorated sherds but, as suggested by Gatto (2006a: 62), this could depend 'on the common use of zonal decorative patterns in this ware'.

At site 8-B-10C, a single sherd (0.5%) decorated with a dotted wavy line motif was found in level 2 (*c.* 7600-7200 BC) and more sherds (9.4%) with this type of decoration come from level 1 (*c.* 5050-4800 BC) (see Chapter 3).

In Lower Nubia, ceramics decorated with dotted wavy line motifs were found in stratigraphic context at site Abka IX, in proximity of the bottom part of level 6, which is dated at 8260±400 uncal years bp (8291-6396 BC) (Gatto 2006a; Myers 1960).

This version of the dotted wavy line motif, also known in the literature as dotted wavy line of the type 'short waves' because of the short dimension of the comb and, consequently, of the waves obtained with the rocker technique (Jesse 2002: 80), is also characteristic of the ceramics of Nabta-Kiseiba (Gatto 2002a). Here, it first appears during the El Nabta phase (*c.* 7100-6700 BC) in the assemblage of site E-91-1 (Gatto 2002a, 2006a).

Finally, this motif is distinctive of the ceramic assemblage of the site of Wadi el-Arab in the region of Kerma. In this site, the dotted wavy line decoration is found in association with the alternately pivoting stamp herring-bone pattern and is dated to the phase Wadi el-Arab III-IV/Mesolithic III (*c.* 7200-6300 BC) (Honegger 2014: 23, Plate 1).

The dotted wavy line 'short waves' is not found in the regions of Laqiya and in the Wadi Howar or further south, in the area of Khartoum , where it is replaced by similar motifs which are however produced with a larger comb, arranged in a continuous structure on the vessel (dotted wavy line 'long waves') or by incised motifs ('incised wavy line') (Jesse 2002).

Packed zig-zags of dots and dashes

At site 8-B-10C, the rocker technique is also frequently associated with zig-zag motifs consisting of lines of dots or dashes.

The former (packed zig-zags of dots) is more frequent than the latter and was found on the surface and in level 1 (*c.* 5050-4800 BC) (see Chapter 3).

Packed zig-zags of dashes were found only in level 1 and were occasionally produced with a roulette instead of a comb, though still using the rocker technique (see Chapters 3 and 5).

The rocker dotted packed zig-zags is very common at the Khartoum Variant sites around Abka –Wadi Halfa and at the sites in the region of Nabta-Kiseiba, especially during the Al Jerar phase (*c.* 6700-6200 BC) (Gatto 2002a, 2006a). This decoration is also found in the region of Kerma during the phase Wadi el-Arab V/Mesolithic IV (*c.* 6300-6000 BC) (Honegger 2014: 23, Plate 1).

At Nabta-Kiseiba, among the rocker decorations, motifs – including the so-called Rocker 4 (R4) or 'Halina' type – are similar to the packed zig-zags of dashes observed on Sai, which Gatto hypothesizes were produced with a 'cord-wrapped implement of perishable material' (Gatto 2002a: 68). The R4 or 'Halina' motif appears earlier than the packed zig-zags of dots, that is in relation to the El Nabta phase (*c.* 7100-6700 BC) (see Gatto 2002a: 70, Table 5.5).

A further comparison comes from the region of Mahas in the Third Cataract, where ceramics have recently come to light at Mesolithic/Neolithic settlements with impressed decorations akin to the rocker motifs on Sai and in Lower Nubia (Sadig 2010; Edwards and Osman 2012).

In the region of the Wadi Howar, Jesse (2002: 83) reports the presence of impressed zig-zag motifs made with an instrument she calls a 'cord-wrapped tool', corresponding to a similar tool referred to at site 8-B-10C as roulette. In the Wadi Howar, this decoration was found during the earliest phase of occupation of the region, which starts however much later than along the Nile Valley (*c.* 5000 BC).

Unlike the 'dotted wavy line short waves' motif, which is a stylistic prerogative of the Khartoum Variant sites in Nubia (from Abka to Kerma) and of the sites in the region of Nabta-Kiseiba, rocker-stamp zig-zags are a tradition shared with ceramics from the sphere of Khartoum, where similar decorations have been found both at sites dating from a Mesolithic horizon (Early Khartoum) and at Neolithic sites. Here, again these motifs were impressed with the rocker technique, for the variant with horizontal dashes, by means of a 'cord-wrapped tool', not a comb (Caneva and Marks 1990: 17, Pl. III).

Alternately Pivoting Stamp

At site 8-B-10C, only two sherds respectively from level 2 (*c.* 7600-7200 BC) and from the pit 1E, were decorated with the alternately pivoting stamp technique (see Chapter 3). This decoration constitutes a further element of contact with the pottery traditions of the El Nabta phase at Nabta-Kiseiba (*c.* 7100-6700 BC) (Gatto 2002a), and of the Khartoum Variant sites around the Second Cataract, where different variants of alternately pivoting stamp were found in association with different fabrics (Gatto 2006a). The alternately pivoting stamp technique is also found in the Mesolithic sites of the region of Kerma, starting from the phase El-Barga I/Wadi el-Arab II/Mesolithic II (*c.* 7800-7200 BC) so almost contemporary to its first appearance at Sai Island and earlier than the dotted wavy line and rocker zig-zags motifs (Honegger 2014: 23, Plate 1).

The alternately pivoting stamp technique is not found in the Early Khartoum ceramics of the Wadi Howar (Jesse 2002) or in the region of Laqiya (Lange 2006) or at sites of the Mesolithic Khartoum horizon (Caneva 1983).

Simple impression

Decorations obtained with the simple impression technique are also attested at site 8-B-10C, in level 1 (*c.* 5050-4800 BC), although they were not very common (see Chapter 3). The same technique is found at site Abka IX (Lower Nubia), in the two variants of horizontal and vertical impressions and in association with more fabric types (see Gatto 2006a). In the region of Nabta-Kiseiba, oblique simple impressions are attested exclusively during the earliest El Adam phase (*c.* 8500-7850 BC) (Gatto 2002a).

Decorations on the rim

At site 8-B-10C, notched and milled impressions are found as characteristic decorations of rims in level 1 (*c.* 5050-4800 BC) (see Chapters 3 and 5). These decorations are well documented in the assemblages of the Khartoum Variant sites around the Second Cataract. Here, Gatto (2006a) noticed that they were mainly related to Ware Group B and Fabric I: 'high-grade fine sand tempered ware' (= Fabric IA in Nordström 1972: 48). In the region of Nabta-Kiseiba, milled impressions on the rims are mostly found during the Al Jerar phase (*c.* 6700-6200 BC) (Gatto 2002a).

Abkan horizon

The ceramic sample analysed from site 8-B-76, on Sai Island, presents stylistic traits typical of the Abkan tradition (Lange and Nordström 2006; Nordström 1972).

Vessels are typically burnished or polished, and are either not decorated or are decorated only near the rim with milled impressions (see Chapters 3 and 5). Further, black-topped and ripple wares are both recorded and will continue to be produced over the course of the following Pre-Kerma horizon (see Chapters 3 and 5).

The ceramics of site 8-B-76 therefore share stylistic characteristics with those of other Nubian sites, such as the settlements SJE 365, 369, and 371, and CPE 604 and 629 in the region of Abka – Wadi Halfa, and the site 5-S-25 on the island of Shagir, at the Second Cataract (Adams and Nordström 1963: 17–18). Similar decorative features are found further south, along the valley, at site 11-I-16 in the district of Murshid (Carlson 1966: 56, Pl III) and in the locality of Adindan, Cave East (Williams 1989: 5–6 and 8, Fig. 2) (see Chapter 1 for a description of these contexts).

In addition, as was the case with the Khartoum Variant horizon, pottery traditions of the Abkan period seem to evolve in a similar way to those of geographically marginal sites, outside the core area of the Abkan culture in Lower Nubia. These include the sites of Nabta Playa and Bir Kiseiba (Nelson and Associates 2002), and the Early Nubian sites of the area of Laqiya and of the Wadi Shaw, situated at about the same latitude as Sai, west of the Nile Valley (Lange and Nordström 2006).

At Nabta-Kiseiba, the pottery of the first phase of the Middle Neolithic (*c.* 6100–? BC) is still technologically similar to the one of the Early Neolithic of El Nabta/Al Jerar (*c.* 7100–6200 BC). Starting from the Second Phase of the Middle Neolithic (?–5500 BC), 'there are changes in clay, vessel wall thickness and surface treatments' (Nelson 2002b: 9) (see technological section below). Further, during the Second Phase of the Middle Neolithic of Nabta-Kiseiba, Nelson (2002c: 32) documented 'the disappearance of rocker-stamped surfaces and any sign of rim decoration, so typical of the Early Neolithic'. The ceramic assemblage from site E-75-8 is representative of this transition (Nelson 2002a).

The ceramics of site 8-B-76, on Sai Island, are also comparable to the Neolithic assemblages from the area of Kerma and Kadruka, in Upper Nubia.

At the Eastern Cemetery of Kerma, Honegger (2014: 23, Plate 1) reports the presence, during the Neolithic II phase (*c.* 5000–4000 BC), of ceramics made of fine pastes, with burnished or polished surfaces, undecorated or decorated with single impressions near the rim, along with black-topped and ripple wares, which resemble certain vessel types seen on Sai and at other Abkan sites in Lower Nubia.

On the other hand, no significant stylistic affinity has been observed with ceramics from the area of Khartoum attributed to the 'esh-Shaheinab Neolithic' (*c.* 4800–4500 BC) (Arkell 1953; Caneva 1983).

Pre-Kerma horizon

The ceramics produced at sites 8-B-52A and 8-B-10A retain many of the techniques and stylistic motifs employed during the Abkan horizon, (e.g., the black-topped and ripple wares common to both horizons), but also introduce other features (e.g., the geometric incised patterns) which indicate a new cultural tradition, the Pre-Kerma.

Many of the decorative motifs observed at the two Sai sites find matches in the typology proposed by Honegger (2004b, 2014) for the region of Kerma, in particular in the various combinations of incised

6. The Sai Island sequence and the Nubian and Sudanese traditions

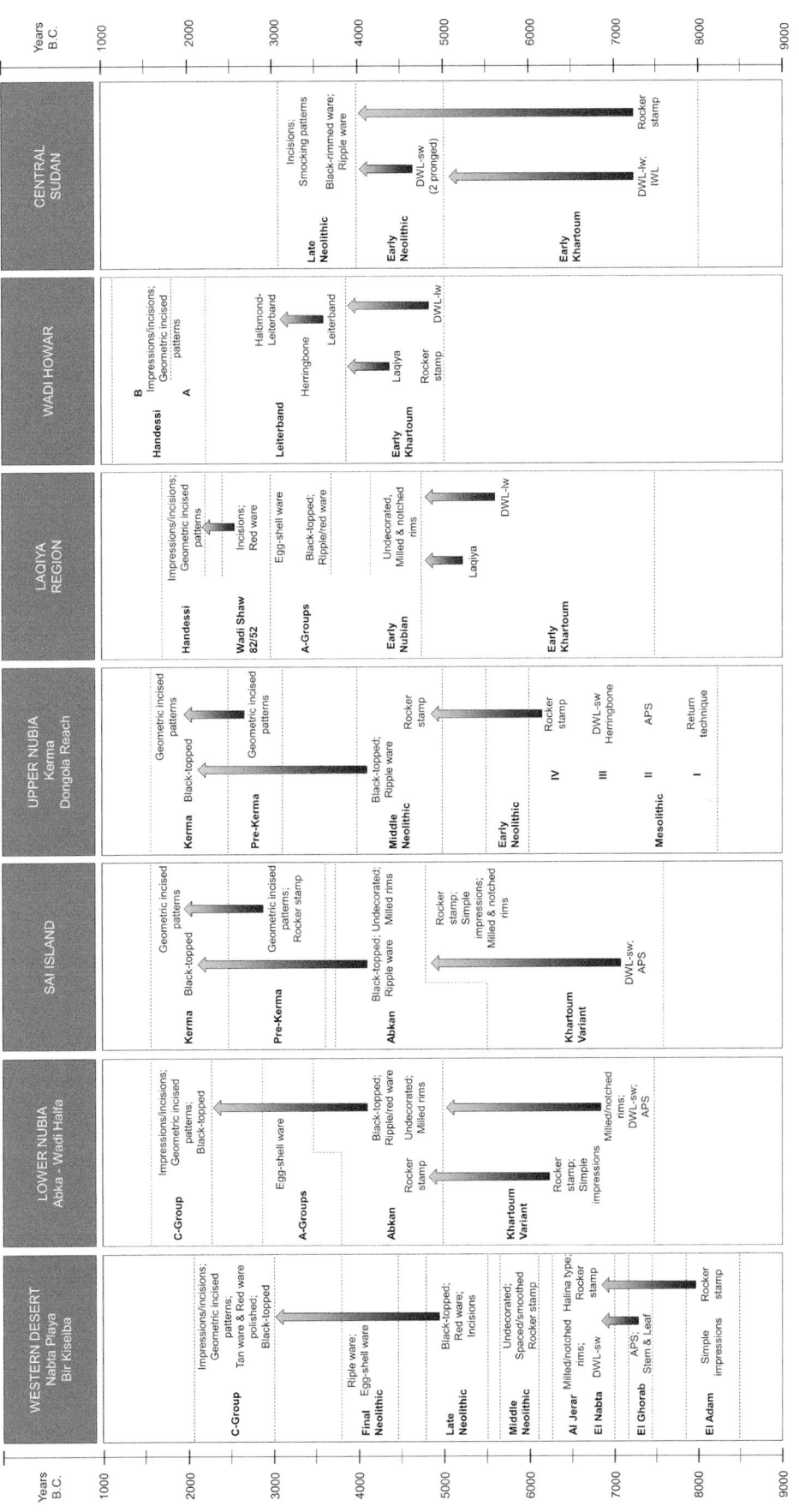

Figure 6.1 Evolution of the main techniques and decorative motifs of ceramics in the regions of Egypt and Sudan mentioned in the text. (Acronyms: DW-sw; DW-lw; APS; IWL stand for: Dotted Wavy Line-Short Waves; Dotted Wavy Line-Long Waves; Alternately Pivoting Stamp; Incised Wavy Line). Symbols > stand for an increase of mineral/organic inclusions; >> a significant increase of mineral/organic inclusions; < a decrease of mineral/organic inclusions. (figure by G. D'Ercole).

decorations organized in geometrical, criss-cross and herring-bone structures, typical principally of site 8-B-10A (see Chapter 3 and 5). These motifs, characteristic of the Recent Pre-Kerma phase (*c.* 2700–2600 BC) are found at Kerma at the site of Boucharia II, along with other variants of geometrical designs which continue to be used into the Kerma period (Honegger 2014: 23, Plate 1). They are also found in the region of Mahas, Third Cataract, at sites dated between the Pre-Kerma horizon and Early Kerma, and further north, at sites originally attributed to the so-called 'B-Group' (Gratien 1995: 52, Fig. 1).

In the region of Laqiya, a stylistic repertoire with many affinities comes from Wadi Shaw site 82/85, dated to *c.* 2500 BC, a period of transition between the A-Group and the 'Handessi' horizon (Lange 2004: 318, Pl. 1.2-6, 2006: 112). In the Wadi Howar, similar designs, in particular, incised herring-bone patterns, are observed in the dune settlements near Abu Tabari, in the lower wadi (Jesse 2008: 63, Fig. 11.2: 5).

Pre-Kerma traditions, unlike those of the Khartoum Variant and Abkan horizons, seem to extend even at southern sites along the valley, including the area of the Fourth Cataract (Smith and Herbst 2008: 209, Fig. 5) and the region of Kassala, in Eastern Sudan, where a correspondence has recently been proposed between Butana Group ceramics and Pre-Kerma products (Manzo *et al.* 2010: p. 5 of the on-line report).

The assemblage from site 8-B-52A includes a percentage of ceramics (the fine production, usually open forms, including red and/or black polished bowls, black-topped vessels and ripple ware) which show several points of contact with the A-Group tradition of Lower Nubia. This suggests some type of interaction/exchange between the two cultural spheres during one or more phases of occupation (see Chapter 5).

One of the characteristics of the settlement 8-B-10A is the presence of a group of samples decorated with motifs possibly produced by the technique of rouletting (see Chapters 3 and 5). This technique is amply documented in the Sahara right from the first appearance of pottery there (Haour and Manning 2010) but is at present unknown at Pre-Kerma sites in the Nile Valley.

Section 2: Comparison of the technological traditions (Figure 6.2)

Khartoum Variant horizon

Macroscopic examination (Chapter 3) and archaeometric analyses (Chapter 4) found that the ceramics of the earliest site, 8-B-10C, are of a sandy, gritty consistency, made of a Pleistocene residual clay characterised by a ferruginous oxidized matrix, and are tempered with mineral inclusions, containing sub-angular to angular grains of quartz, K-feldspar, mica and rock fragments of metamorphic origin. Within the main Group QKfs (quartz-K-feldspar), two different fabrics have been distinguished: Fabric QKfs-Bt (quartz-K-feldspars-biotite) and QKfs-Ms (quartz-K-feldspar-muscovite) which differ in grain-size (coarser in fabric QKfs-Bt than in QKfs-Ms) and in particular compositional characteristics (i.e., biotite *vs* muscovite mica, different amount of K-feldspar).

The two main fabrics identified in the ceramics of site 8-B-10C on Sai Island find a good correspondence with Fabrics IA and IB described by Nordström (1972: 48-49) at the Khartoum Variant sites at the Second Cataract (especially sites 18A and 428 from the area of Wadi Halfa). Although Nordström's classification was based essentially on macroscopic analysis of the samples, it is still possible to highlight certain common characteristics of the two sets of pottery.

Nordström defines both fabrics as 'Sand-Tempered Ferruginous Fabrics' marked by a dense homogeneous matrix and by the presence of numerous mineral inclusions, from angular to sub-rounded in form, including quartz, feldspars and other rock fragments (Nordström 1972: 48). In particular, Fabric IA ('fine sandy Khartoum Variant type') contains 'abundant quartz, consisting of angular, sub-angular or sub-rounded grains in sizes up to 500μ. 'Feldspar is present generally in the same proportion as quartz.

Mica is presents in small amount only. It is made up of scales and scaly aggregates, mainly of biotite in size up to 1mm.' (Nordström 1972: 48). Fabric IB ('micaceous Khartoum Variant type') is distinguished from 1A by differing mica content, in which muscovite dominates, as it does in fabric QKfs-Ms on Sai Island (Nordström 1972: 49). From a technological point of view, these ceramics are characterised by a hard consistency and low porosity which is the result of significant mineral inclusions and an absence of organic matter. Nordström (1972: 48-49) suggests that they were fired at between 700° and 800° C, in an oxidising atmosphere during the final stages. Both fabrics IA and IB are typical of Ware Family K: Khartoum Variant wares.

A re-examination of certain Khartoum Variant ceramic assemblages from the district of Abka –Wadi Halfa substantially confirmed the paste groups proposed by Nordström. Gatto (2006a) distinguished between Fabric I (= Nordström fabric IA) and Fabric II (= Nordström fabric IB). Additionally, she had a third fabric (Fabric III = coarser version of Nordström fabric IA). For Fabric II ('micaceous tempered'), Gatto demonstrated the presence of vegetable remains and of small, rounded grains of sand. Both of these characteristics recall what was observed in some of the 8-B-10C samples assigned to the Fabric QKfs-Ms (quartz-K-feldspar-muscovite), which stood out already at the macroscopic level for their finer grain-size and the presence of scattered organic inclusions (see Chapters 4 and 5).

A further comparison can be made with the Early Neolithic horizon sites of the region of Nabta-Kiseiba, within the El Nabta/Al Jerar phase (*c.* 7100-6200 BC). The ceramics produced during this period were made of playa clays, tempered only with mineral inclusions, and distinguished after firing by their characteristic red colour. These ceramics were shaped by coiling; their surfaces were smoothed inside and outside, and were then fired in open fires at low temperatures in oxidising conditions (Nelson 2002b, 2002c). The composition of the mineral inclusions would indicate the use of local raw materials, referring to outcrops of basement granite (Zedeño 2002). According to Zedeño (2002: 54), who realized the petrographic and chemical analysis on a sample of these ceramics, 'the poor sorting observed in most of samples suggests that sand inclusions may have come from reworked deposits within the playa basin and, therefore, they could have been naturally mixed with the clay and silt deposits rather than artificially added to obtain a workable paste.'.

In the analysis of Gatto (2006a), the different ware groups and fabrics identified in the cultural sequence of the Early Neolithic of Nabta-Kiseiba and at the Second Cataract sites correlate with different stylistic categories, suggesting a technological evolution parallel to the evolution of decorative techniques and motifs which mark successive phases over the course of the Khartoum Variant horizon (see above). In particular, fabric III ('medium coarse mineral tempered') would be typical of the earlier El Nabta phase (*c.* 7100-6700 BC) with decorative motifs as the dotted wavy line, both 'Arch-shaped' and 'Wide Zonal', rocker packed dotted zig-zags and the alternately pivoting stamp (APS) herring-bone patterns. Fabric I ('high-grade fine sand tempered') would include mostly rocker decorations applied all over the body of the vessel and rim band decorations, typical of the Al Jerar phase (*c.* 6700-6200 BC) (Gatto 2006a).

At Sai Island, most of the ceramics analysed from site 8-B-10C (from both levels 1 and 2) were assigned to fabric QKfs-Bt (coarse wares with high content of K-feldspar and biotite, similar to Gatto fabric III) and a limited number of samples belonged to fabric QKfs-Ms (medium, fine ware with lower content of K-feldspar and muscovite, similar to Gatto fabric II). Finally two samples were assigned to group Q (medium, fine ware containing only quartz). Some internal variations and a correlation between fabrics and macroscopic features were observed, but they seem to reflect more a functional rather than a chronological tendency (see Chapters 3 and 5).

The technological properties of the earliest products of the Khartoum Variant tradition seem not to be too different from what has been observed so far even in areas stylistically foreign to the sphere of influence of the Khartoum Variant culture, such as the regions of Laqiya and the Wadi Howar. The pottery of the first hunter-gatherer groups to settle in the area in the 6th millennium BC is in both cases made of pastes containing only mineral inclusions (mainly large, angular quartz inclusions and granite tempers), used to produce vessels by the coiling method, with walls that are burnished or simply smoothed, decorated with dotted wavy line motifs ('long waves' variant) or the characteristic 'Laqiya' style (Keding 2006; Klein *et al.* 2004; Lange 2006-2007).

Further south, in the region of Kadada and in the province of Khartoum, the first ceramic assemblages attributed to the so-called 'Wavy Line' horizon (Arkell 1949) seem to have many technological similarities to Nubian ceramics. Earlier archaeometrical studies conducted in the area (De Paepe 1986; Francaviglia and Palmieri 1983; Hays and Hassan 1974) indicate that many Early Khartoum assemblages, despite stylistic differences, have technological affinities with the Khartoum Variant of Lower Nubia. The Mesolithic pottery of the sites of Saggai, Umm Marrahi, Sorurab and el-Qoz, as well as the earliest ceramics of el-Kadada (before the Neolithic occupation of the site) are all produced in the same way, with pastes rich in quartz, micas and feldspar, and granitic tempers, all deriving from the petrofacies of the Precambrian Basement, which outcrops in the region of the 6th Cataract over an area of about 3000km^2 (De Paepe 1986).

The available chemical data, though obtained with different instruments and analytical methods from those adopted in this study, can also provide material for comparison. The Mesolithic assemblages of Saggai, Umm Marrahi, Sorurab and el-Qoz (Francaviglia and Palmieri 1983) and the earliest ceramic nucleus of el-Kadada – the I group in De Paepe's classification (1986: 118) – show average measurements for certain chemical elements comparable to those for site 8-B-10C (quartz-K-feldspar group). The most significant similarities, besides those for aluminium and iron, are for percentages of potassium and rubidium, both contained in K-feldspar, which are among the principal mineral phases to characterise the pastes of both Khartoum Variant ceramics from Nubia and Early Khartoum pottery from the area of the 6th Cataract **(Table 6.1)**.

Finally, alkali concentrations, related to the presence of K-feldspar and granite inclusions, characterized the earliest Mesolithic assemblage of the site of Al-Khiday, 17km south of Khartoum (petrographic group I: alkali-feldspar-and quartz-rich potsherds) (Dal Sasso *et al.* 2014). Here, the authors explain the presence of alkali-feldspar as the result of an intentional addition of granite tempers possibly in order to improve the resistance of this pottery to mechanical stress (Dal Sasso *et al.* 2014).

This overall picture demonstrates a tendency common to all the first pottery producing cultures of the Early Holocene settled along the course of the Nile (Lower Nubia, Upper Nubia and Central Khartoum) and in its current desert hinterland (regions of Nabta-Kiseiba, Laqiya and Wadi Howar): the choice as a raw material for their ceramics of clays rich in alkali feldspar and fragments of granitic and/or metamorphic rocks, taken from Precambrian outcrops, which despite variation in details of composition connected to local geology, display very similar macroscopic characteristics and technological features.

Abkan horizon

On Sai Island during the Abkan horizon (*c.* 5500-3700 BC), a significant change can be seen in methods and techniques of ceramic production. Most of the samples from site 8-B-76 are made of a raw material different in composition from the clays of site 8-B-10C. This is a silty alluvial sediment, rich mainly in fine quartz and plagioclase, and to a lesser extent in muscovite mica and heavy minerals, assigned consequently to the group QPl (quartz-plagioclase) (see Chapters 4 and 5). In the transition

Contexts	SiO2	TiO2	Al2O3	Fe2O3	MnO	MgO	CaO	Na2O	K2O
Sai (KV)	56.57	0.74	19.02	6.47	1.16	1.18	2.04	2.06	3.83
el-Kadada (EK)	61.60	0.80	17.70	5.36	0.06	0.90	1.25	1.51	3.99
Saggai (EK)	60.39	0.69	18.88	4.39	0.14	1.00	1.88	1.28	4.78
Umm Marrahi (EK)	62.91	0.65	16.83	3.34	0.04	0.89	2.42	1.33	3.42
Sorurab (EK)	58.96	0.56	19.36	3.16	0.08	1.38	1.83	1.04	4.87
el-Qoz (EK)	61.29	0.65	18.16	3.55	0.09	1.10	1.85	1.40	4.55

Table 6.1 Major elements composition of the Khartoum Variant (KV) ceramics analysed from Sai Island (group QKfs: 20 samples) in comparison with the Early Khartoum (EK) productions from el-Kadada (20 samples), Saggai (17 samples), Umm Marrahi (4 samples), Sorurab (5 samples) and el-Qoz (4 samples). Major elements in weight percent (w%).

from the Khartoum Variant to the Abkan horizons, the introduction of new technological choices can be distinguished, such as a new procedure for preparing pastes consisting in the addition of organic tempers (small charcoal fragments and herbivore droppings), or frequent burnishing of vessels, inside and outside (see Chapter 5).

During the same period, a similar evolution in pottery traditions is recorded in Lower Nubia in the region of the Second Cataract in the so-called 'Ware Family M' from the Abka settlements 365, 369, 370, 371 and 414 and from site 303 at Faras and 440 at Gamai, which Nordström (1972: 49-50) suggested should be assigned to a new category of paste – Fabric IC: 'sandy Abkan type' - defined by a dense homogeneous matrix and a higher proportion of silt than was observed in the fabrics of the preceding horizon. These ceramics, as at site 8-B-76, are usually grey-brown or dark grey, and contain both mineral tempers and angular, irregular particles of charcoal with a fibrous surface texture. This pottery is distinguished from ceramics of the Khartoum Variant period (Ware Family K) by a soft or only moderately hard consistency and by higher porosity, the result of firing at lower temperatures, between 600° and 700° C (Nordström 1972: 49).

What was observed at Nubian sites in the valley finds a parallel in changes which took place over the course of the chronological sequence at Nabta-Kiseiba. Here, during the transition from the Early Neolithic horizon to the second phase of the Middle Neolithic (c. 5500 BC) some important innovations in pottery production occur, indicative both of a possible change in raw material, and a progressive development of technological skills. The pastes are now no longer red after firing but grey to dark grey, and contain larger amounts of organic matter, for the first time intentionally added as a tempering agent. From this period on, procedures aimed at making surfaces smoother become more common, leading to the introduction at Late Neolithic sites (c. 5500-4800 BC), of the first burnished vessels and black-topped wares (Nelson 2002a: 7). Over the course of the Late and Final Neolithic phases, the variety of ceramic types increases, with the appearance of at least three different classes of product, each characterised by particular technological and compositional traits: 'Qussier Clastic Yellow Ware', made of geological clay/marl from the complex of clastic rock at El Qussier, 'Red Ware', made with iron-rich alluvial clays available locally on lake or river banks, and the co-called 'Olive Ware', probably not a local product, exclusive to Final Neolithic sites (c. 4400-3800 BC) (Nelson 2002b).

This type of technological development is a feature also of the pottery traditions of the region of Laqiya, which around 4500 BC seem to change completely, precisely as they do in neighbouring regions. Lange (2006-2007: 244) associates the ceramics of this new horizon, called 'Early Nubian' precisely because of affinities with the cultures of the valley, with a type of paste similar to Nordström's fabric IC ('sandy Abkan type') in which for the first time not only mineral inclusions can be seen but also vegetable tempers – hence the definition 'sandy fabric with a limited amount of plant material as tempering agents' (Fabric SP1). In general, pastes appear less coarse during this phase and walls grow thinner; and, as observed of the Late Neolithic phase at Nabta-Kiseiba, burnished surfaces and ripple ware make their first appearance (Lange 2006-2007).

It is interesting to note however how the pottery traditions of the adjacent Wadi Howar region follow their own line of development, technologically and stylistically, independently of the Nubian cultural sphere. 'Leiterband-style' ceramics, typical of the pastoral communities settled in the middle wadi at the beginning of the 4th millennium BC, like earlier dotted wavy line and Laqiya ceramics (c. 5200-4000 BC), are distinguished by pastes which contain only mineral inclusions, and by technological properties similar to those observed in the previous horizon. Keding (2006) reports the use of the same building technique – coiling – and the use of analogous techniques of surface smoothing and burnishing. An archaeometric characterisation of a sample of pottery revealed however a certain discontinuity in the selection of raw material over the course of this period. This correlates in part with other changes in cultural traditions and decorative styles; more generally it reflects the geomorphological complexity of the region, and the availability for local production of a wide range of raw materials with varying compositional characteristics (Klein *et al.* 2004).

Further south, in the province of Khartoum also, with the development of the first pastoral cultures of the 'esh-Shaheinab Neolithic' (5th millennium BC), certain significant technological innovations are introduced into pottery production. In this period, a substantial change in the raw material can be deduced from a comparison of Arkell's macroscopic descriptions (1953) with the analytical data available for this area (Francaviglia and Palmieri 1983; Hays and Hassan 1974). And the Neolithic complex of el-Kadada, dating from the first half of the 4th millennium BC (Late Neolithic) differs completely as far as composition is concerned from the Mesolithic ceramics found in the same area as a result of using a different raw material, characterised by a lower content of feldspar and granite, and a higher proportion of fine sand (De Paepe 1986: 119).

The Neolithic ceramics from el-Qoz and el-Kadada were prepared therefore, like Abkan pottery in Lower Nubia, with alluvial sediments collected near settlements, in a region whose geological substrate is dominated by Nubian sandstone. Ceramics containing large amounts of feldspar are typical of areas with Precambrian soils, while those tempered mainly with grains of quartz come from regions where Nubian sandstones are exposed (De Paepe 1986: 132).

It is interesting to note that the chemical composition of alluvial deposits on the Nile does not vary significantly even over large distances. The sediments analysed by De Paepe (1986: 126) from the part of the valley between el Ghaba and Doshein (Central Sudan, 6th Cataract) have a chemical make up for the most part compatible with the composition of the archaeological deposits of the Abkan and Pre-Kerma periods analysed in the Nubian context of Sai, and so with the ceramics produced there in the same periods **(Tables 6.2-6.4)**. Petrographically too, these raw materials have many characteristics in common: a matrix rich in iron oxides and a sandy fraction containing, besides grains of quartz, accessory minerals from the group of amphibole, epidote and zircon, mica, augite and lithic fragments of volcanic origin.

A similar picture comes from the site of Al-Khiday, 17km south of Khartoum. Here, all the Neolithic pottery (c. 4600-4000 BC) is characterized by fine sand-sized quartz inclusions (petrographic group 3: 'fine sand-sized quartz-rich potsherds) (Dal Sasso *et al.* 2014).These ceramics are more standardised in comparison with the Mesolithic pottery and are mainly composed of small sub-angular and sub-rounded quartz grains. Further, the proportion of alkali feldspar is definitively lower in respect to the Mesolithic pastes.

Contexts	SiO2	TiO2	Al2O3	Fe2O3	MnO	MgO	CaO	Na2O	K2O
Sai (ABK)	59.81	1.57	15.12	9.77	0.18	3.16	3.42	1.54	1.16
el-Kadada (Neolithic)	64.24	1.26	11.14	6.59	0.11	2.07	2.74	1.67	1.89
el-Qoz (Neolithic)	68.40	0.98	14.53	5.11	0.22	1.19	1.71	1.18	1.28

Table 6.2 Major elements composition of the Abkan (ABK) ceramics analysed from Sai Island (group QPl: 16 samples) in comparison with the Neolithic productions from el-Kadada (90 samples) and el-Qoz (6 samples). Major elements in weight percent (w%).

Pre-Kerma horizon

On Sai, the production of ceramics in the Pre-Kerma period (c. 3600–2500 BC) seems to have developed directly from Abkan traditions, having inherited not just certain stylistic and technological aspects (e.g., , black-topped and ripple wares and burnishing and polishing processes) but also in general a similar manufacturing technology based above all on the use of the same raw material: silty alluvial sediments, available locally since deposited by the Nile, possibly tempered with sand and organic matter (ceramics attributed to the common mineralogical-petrographic group QPl = quartz-plagioclases). What distinguishes Pre-Kerma pastes from ceramics of the Abkan period however is the greater organic content added by potters during manufacture, consisting not just of ash and charcoal fragments but of vegetable fibres, either present as vegetable fibres or else after digestion by herbivores, i.e., in dung (see Chapter 5).

Nordström (1972) also mentions the presence of organic tempers in pastes as a technological characteristic of ceramics from the A-Group complexes of Lower Nubia. He identifies two main fabrics: fabric IIA, 'ash-tempered', containing particles of charcoal smaller than 0.25mm; and fabric IIB, 'dung-tempered', containing dung which itself contains very small fragments of straw and grass, varying in length from 0.25mm to 2 or 3mm (Nordström 1972: 51–52). He assigns both fabrics to the 'Ware Family H', i.e. to the macro-family of the so-called Nubian tradition. Nevertheless, fabric IIA is used exclusively for ripple ware, typical of the Classical A-Group horizon, while fabric IIB has a longer life, lasting through the period of the A-Group into the following period (Terminal A-Group) up until the occupation of the area by C-Groups and even into the New Kingdom (Nordström 1972: 52).

A-Group pottery found in the region of Laqiya, western of Sai Island in the desert, from the 4th millennium BC on also contains organic tempers. Lange (2006–2007: 149, Fig. 4) associates two fabrics with this horizon, fabrics SP2 and SP3 ('sandy fabrics with a limited amount of plant material as tempering agents'), corresponding respectively to Nordström's paste types ID and IIA. The pottery is dark grey or black, of moderately hard consistency, containing varying proportions of organic matter, which can be of vegetable (ash with carbonized particles) or animal (herbivore dung). The following phase (Facies 'Wadi Shaw 82/52') represents in certain ways a continuation of the technological tradition inherited from the A-Groups (red vessels with polished walls or black mouthed vessels). However, in this period ceramics tempered with a higher content of organic matter also start to appear (fabrics PS1 and PS2 with a high amount of plant material and sand as tempering agents) decorated with incised motifs which recall the Pre-Kerma style.

In the area of Laqiya and nearby in the Wadi Howar this tendency becomes more pronounced in the course of the Handessi horizon (c. 1800–1500 BC), hand in hand with the formation of a fully nomadic, pastoral socio-economic model and a renewal of close contacts between the two regions. During the last phase of occupation (both areas were later largely or completely abandoned because of increasing aridity) the pottery is uniformly distinguished by the addition of organic tempers to the pastes (vegetable fibres and/or dung) and by the use of similar decoration (different variants of incised motifs and mat impressions) (Keding 2006; Lange 2006–2007).

At the moment there are few technological data available for the Pre-Kerma complexes at Kerma, in Upper Nubia. Nevertheless, on the basis of an archaeometrical characterisation carried out on a collection of ceramics from the eastern cemetery and from the later settlement at Kerma (De Paepe and Brysse 1986) we can deduce that similar technological traditions became established along the valley. The Kerma period ceramics analysed from the site at Kerma are made of a mixture of alluvial Nile sediments and of organic matter added intentionally to the paste during preparation. The chemical composition of these ceramics is very similar to that of the Pre-Kerma ceramics from Sai (samples assigned to the QPl group – quartz-plagioclases – from sites 8-B-52A and 8-B-10A) **(Table 6.3)**. Furthermore, a similar compatibility is found between the samples of sediment analysed from the Abkan and Pre-Kerma sites on the island of Sai and alluvial Nile deposits from the region of Kerma **(Table 6.4)**.

The two sets of samples also display important petrographic similarities: prevalence among the mineral tempers of quartz, but also the presence of feldspar, occasional lithic fragments of metamorphic and volcanic origin, mica and accessory phases such as hornblende, epidote, opaque minerals and titanian augite (De Paepe and Brysse 1986).

A number of conclusions can be drawn from these comparisons. First of all, the confirmation of a certain compositional homogeneity in the Nile's alluvial deposits, since in at least three locations along the valley – as far apart as Sai, el-Kadada and Kerma – they display throughout the period a similar petrographic and chemical composition. Secondly, there is evidence of a common technological background to ceramic production which is shared both by agro-pastoral Nubian cultures in the valley (first the A-Group and Pre-Kerma, then the Kerma complexes) and by pastoral groups in the desert (cultures of the Wadi Howar and the region of Laqiya and C-Groups in Lower Nubia and in the area of Nabta-Kiseiba).

Final considerations

In the course of the Holocene, Nubian and Sudanese pottery traditions, in the valley and in the desert, progressively changed both decorative techniques and motifs **(Figure 6.1)** and the technology related to selection of raw materials and other phases of manufacture **(Figure 6.2)**. At individual sites however these changes sometimes took place at different rates and in different ways, and stylistic developments did not always coincide with technological innovations.

Contexts	SiO_2	TiO_2	Al_2O_3	Fe_2O_3	MnO	MgO	CaO	Na_2O	K_2O
Sai (PK)	54.92	1.87	15.84	10.67	0.20	3.49	4.33	1.86	1.10
Kerma Town	55.55	1.49	14.22	9.63	0.16	3.20	4.08	2.26	1.28
Kerma Cemetery	55.90	1.46	14.51	9.31	0.14	2.40	3.15	1.75	1.57

Table 6.3 Major elements composition of the Pre-Kerma (PK) ceramics analysed from Sai Island (group QPl: 30 samples) in comparison with the Kerma productions from Kerma: cemetery (51 samples) and town (16 samples). Major elements in weight percent (w%).

Contexts	SiO_2	TiO_2	Al_2O_3	Fe_2O_3	MnO	MgO	CaO	Na_2O	K_2O
Sai	59.28	1.73	13.95	8.74	0.17	3.15	3.74	1.83	1.21
VI Cataract	49.42	2.23	14.13	11.25	0.18	2.61	4.23	1.26	1.23
Kerma	51.03	1.6	13.91	9.99	0.16	3.01	4.57	1.89	1.19

Table 6.4 Major elements composition of the local sediment samples analysed from the Abkan and Pre-Kerma sites on Sai Island (11 samples) in comparison with the Nile alluvia from the regions of the VI Cataract (27 samples) and Kerma (7 samples). Major elements in weight percent (w%).

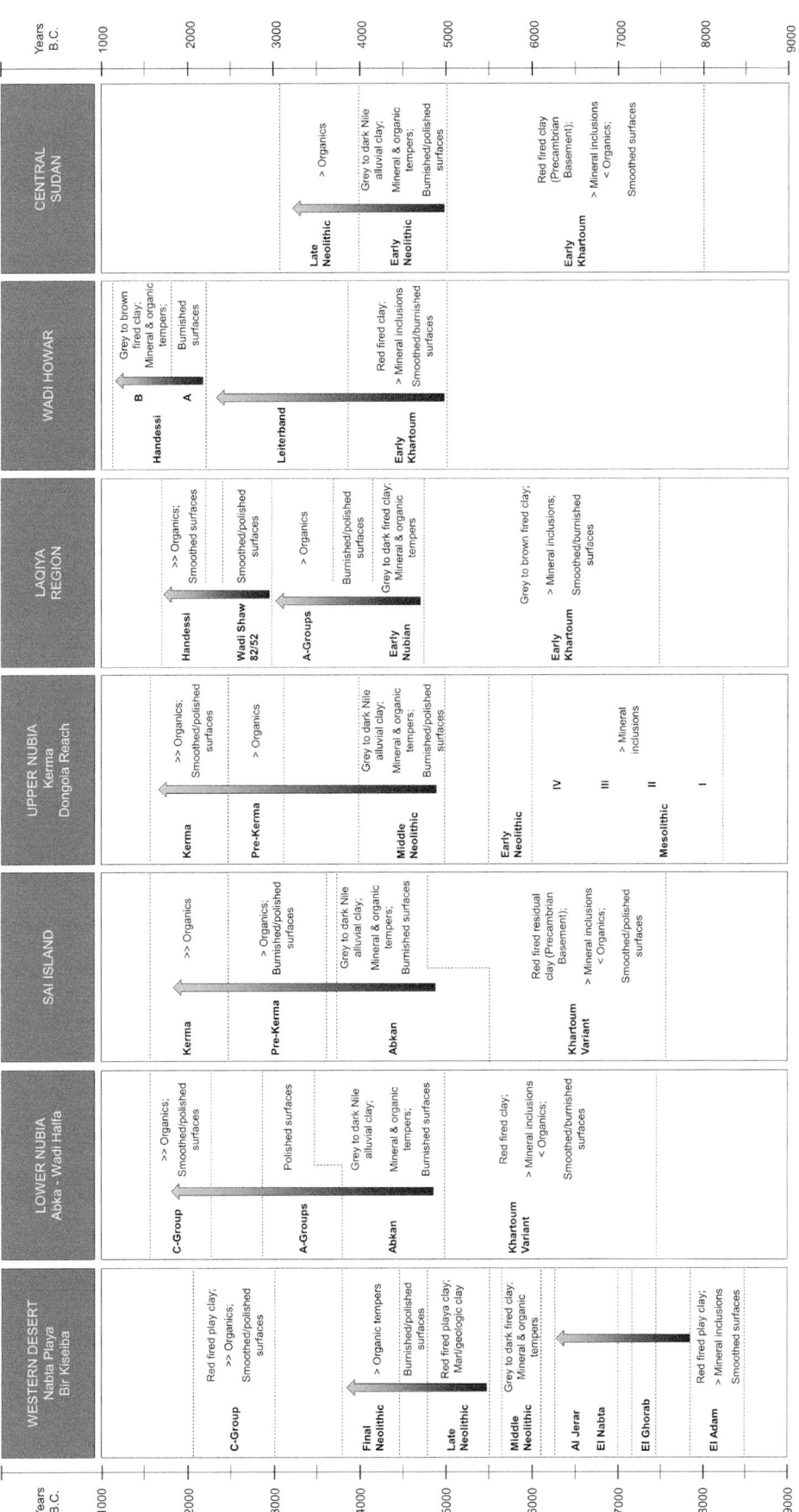

Figure 6.2 Evolution of the main technological features of ceramic in the regions of Egypt and Sudan mentioned in the text. Symbols > stand for an increase of mineral/organic inclusions; >> a significant increase of mineral/organic inclusions; < a decrease of mineral/organic inclusions. (figure by G. D'Ercole).

Analysis of stylistic characteristics of pottery has made it possible to identify in chronological sequence the formation of different cultural spheres or traditions, each one distinguished by a shared repertoire of decorative motifs and techniques. From a cultural point of view, this is the result of more or less complex dynamics of interaction and exchange between human groups, involving entire communities, or sub-groups within the community or even simply individuals. Forms of indirect communication, such as mechanisms of imitation or, conversely, rivalry/competition between groups that were not necessarily in physical contact with each other, could have acted over time as a stimulus promoting or limiting the transmission of stylistic traits.

During the earliest horizon (Early Holocene, c. 8000–5000 BC), the distribution of decorative motifs and techniques suggests the coexistence in this area of two main cultural spheres: first, the Khartoum Variant, which includes the Nubian complexes spread along the Nile from to the district of Abka –Wadi Halfa, at the Second Cataract, to Sai Island, in northern Upper Nubia; and second, the Early Khartoum tradition, defined by the stylistic characteristics of pottery from the site of Khartoum Hospital, and covering also the regions of Dongola Reach, Butana and the area of Atbai. In this period, the first pottery producing cultures of the Wadi Howar and the region of Laqiya, though chronologically more recent than the Nile Valley sites, seem to fall into the Early Khartoum sphere of influence. The Mesolithic sites (i.e., Wadi el Arab and El Barga) around Kerma, in Upper Nubia, and the Early Neolithic sites of Nabta-Kiseiba, in the Egyptian Western Desert show, instead, more important affinities (i.e., zonal dotted wavy line of the type 'short waves', milled and notched impressions on the rims) with the Khartoum Variant tradition.

This picture remains much the same during the following horizon, when in most of the area under consideration, gradually and at different rates in different regions, an economy based on the exploitation of animal resources supplemented and then completely replaced the economy of hunting-gathering-fishing typical of the earlier horizons.

The stylistic innovations observed through the course of the Middle Holocene (c. 5000–3500 BC) suggest again the simultaneous presence in the valley of two traditions that were culturally and geographically separate: the Abkan tradition, whose distribution coincides closely with the orbit of the Khartoum Variant culture, including the whole of Lower Nubia – from the latitude of Abka to Sai Island; and the culture of the so-called 'esh-Shaheinab Neolithic' horizon, spread out to the south, in the province of Khartoum. The ceramics from the region of Nabta-Kiseiba, in the desert, and from the area of Kerma, in Upper Nubia, display again major stylistic similarities with the Abkan sphere. Further, it is interesting to note that Laqiya, which in the previous period had been part of the sphere of influence of Khartoum, from the second half of the Holocene (c. 4500 BC) has ceramics more like those produced in Nubia and in the complexes of the Late Neolithic phase of Nabta-Kiseiba.

The people of Late Holocene cultures (c. 3500–2500 BC) seem not to have ranged so widely over their territories, and so opportunities for contact between groups are reduced and with that the possibilities for diffusion of pottery traditions, which now seem to follow shorter trajectories (Garcea and Hildebrand 2009). Nevertheless, a 'Nubian identity', formed over the centuries following the period of the Khartoum Variant, remains intact through the A-Group and Pre-Kerma horizons, finding expression in a decorative repertoire which spread not only along the valley but also into the pastoral communities in the desert on either side. Analysis of the stylistic traits of the pottery from Sai Island confirms the existence of periodic contacts and perhaps of prescribed cycles of exchange along the Nile between the A-Group cultures of Lower Nubia and the populations in the Pre-Kerma sphere settled further south.

Unlike the stylistic choices, which developed very different cultural traditions in different parts of the territory, technological innovations seem to take the same form in all the regions under consideration (see above).

The two key moments of technological innovation – a change in the type of raw material (alluvial silty quartz rich clays *vs.* earlier alkali feldspar rich clays), a new procedure for preparing pastes (addition of organic inclusions) besides the consolidation of particular techniques for treating surfaces – are common to all the sites examined. Irrespective of the cultural identity of each region, they coincide with the principal phases of economic and social transformation: first (at *c.* 5500 BC), the passage from an economy based on hunting, gathering and fishing to one based on animal husbandry; and then, two millennia later (*c.* 3500 BC), the emergence of more complex but also geographically more limited social nuclei, that along the Nile became nearly sedentary, with an agro-pastoral economy, and in the deserts were fully and nomadic pastoralist.

Over this period, as economic and social structures changed, so also did people's needs, and consequently it is hardly surprising to observe significant changes in the technological sequence of pottery making.

In the earliest period – the Khartoum Variant and Early Khartoum complexes – pottery technology was at an initial stage and the first ceramic containers were likely produced in a rather opportunistic way and were limited to few specific functions (i.e., storing and possibly cooking). During this period, the selection of the raw material (a lean, ferruginous clay, red in colour, found in granitic-metamorphic outcrops of Precambrian origin, which might either contain feldspar inclusions or be intentionally tempered with feldspar sand) would probably require a substantial effort. Much less care was taken at later stages of the production sequence. Clay was probably used as it was found, without undergoing a process of purification; and vessel walls were smoothed but rarely underwent any further surface treatment.

From the Abkan period (*c.* 5500–3700 BC) on however – and in the area of Upper Nubia and Central Sudan, from the first appearance of Neolithic pastoral cultures – a new tendency can be observed. Now the choice of raw materials seems to depend on the convenience of being able to exploit sources near to settlements (a silty alluvial clay is used, taken from the banks of the Nile or of other local water courses) and greater weight is given to the phase of preparation of the paste (addition of tempers) and finishing off (burnishing and polishing, the appearance of the first black-topped and ripple wares).

This tendency can be interpreted in terms of the different functions ceramics must have had for the first Khartoum Variant and Early Khartoum communities in comparison with communities in the second half of the Holocene (Abkan and Neolithic pastoral cultures) and later A-Group and Pre-Kerma societies.

Silty alluvial clays, less rich in feldspar and rock inclusions than the clays used previously, allowed vessels to be made with thinner walls that were therefore lighter, easier to transport and suitable for a larger range of uses. Alluvial clay was also available locally, and when needed could be found comparatively effortlessly close to settlements.

Equally, cultural considerations of the intrinsic significance of ceramic objects, beyond their purely practical, material functions, should not be excluded from this analysis. And in this light, even certain technological choices, just like stylistic decisions, would have had a possible cultural significance. This could be the case for the addition of organic tempers to pastes. This practice, still documented in modern Nubian and Sudanese ethnographic manufactures, was uniformly characteristic of most ceramic production in societies of the Middle and Late Holocene. In these cultures, pastoralism played a central role in the groups' economy, and it is not surprising that this fact – forced cohabitation and mutual dependence of humans and domesticated animals – should be reflected in ceramic traditions, through adding a certain amount of organic matter of animal origin (herbivore droppings) to the raw material.

On this subject, it is interesting to note how in the region of Nabta-Kiseiba, although there is good evidence of the presence of domestic cattle from the time of the Early Neolithic (*c.* 6500 BC) and of domestic caprines from *c.* 6100 BC (Gautier 2001, 2002), a substantial change in the technology of pottery can be seen only in the succeeding period (Second phase of the Middle Neolithic': *c.* 5500 BC) and during the Late Neolithic, at the time of a final transition to a pastoral economy in the region. In the area of Laqiya, west of Sai Island in the desert, many of the principal technological innovations in pottery production, such as the introduction of pastes with organic tempers, appeared at *c.* 4500 BC, again with the adoption of a fully pastoral economy. In the nearby Wadi Howar, the main technological changes seem to occur much later, during the latest Handessi phase (*c.* 1800 BC), after a definitive deterioration in climatic/environmental conditions which may have coincided with immigration into the area of a new population with different pottery traditions.

At Sai Island and in the other contexts along the Nile Valley, important changes and technological discontinuities in the way of producing pottery are first registered starting with the second half of the Holocene (*c.* 5300 BC), with the shift to semi-arid and drier climatic conditions, and with the adoption of cattle and small livestock.

During this period, which coincides with the 'Mid-Holocene regionalization phase' in the Eastern Sahara (Kuper and Kröpelin 2006), a hiatus in the human occupation is recorded in the region of Kerma, which from *c.* 5300 BC on is re-occupied with new settlements located on the alluvial plain and inhabited by Neolithic pastoral societies (Honegger and Williams 2015; see also Chapter 1).

In Lower Nubia and at Sai Island, there are no gaps in occupation between the Khartoum Variant and the Abkan period, however the location of the new Abkan settlements is slightly different from that of the earlier Khartoum Variant with sites closer to the Nile, on floodplains and river terraces (see Chapter 1).

All in all, the changes observed in pottery technology and style (i.e., use of alluvial silty clay, addition of organic tempers, new surface treatments, new wares and decorative techniques) together with the data coming from settlement evidences, seem to indicate a strong incoherence both in material culture and in economic and occupation strategies starting from the mid-Holocene. At Sai Island, this scenario was tentatively interpreted with a break in a pattern of phylogenetic continuity (*sensu* Roux and Courty 2013) corresponding to a gradual replacement in population at the passage from the Khartoum Variant to the Abkan period (see conclusions in Chapter 5). This interpretation might also work in explaining the analogous changes occurring, almost contemporarily, in the adjacent regions of Nubia and Sudan.

A deeper knowledge of the technological and compositional characteristics of the ceramic assemblages of northern Sudan and southern Egypt through the course of the Holocene, together with a better understanding of the human occupation strategies and of the 'timing' and 'routes' (Garcea *et al.* 2016b) of the animal domestication in these areas would shed new light on this question and could further support this interpretation.

Bibliography

Adams, W. Y. and Nordström, H. - Å. 1963. The Archaeological Survey on the West Bank of the Nile: Third Season, 1961-1962. *Kush* XI: 10–46.

Applegate, A. and Zedeňo, N. 2001. Site E-92-8: A Late Prehistoric C-Group Component at Nabta Playa (second author with A. Applegate). In F. Wendorf, R. Schild and Associates (eds.) *Holocene Settlement of the Egyptian Sahara, Volume 1: The Archaeology of Nabta Playa*: 529-533. New York, Kluwer Academic/Plenum Publishers.

Arkell, A. J. 1949. *Early Khartoum*. Oxford, Oxford University Press.

Arkell, A. J. 1953. *Esh Shaheinab*. Oxford, Oxford University Press.

Arnold, D. E. 1985. *Ceramic Theory and Cultural Process*. Cambridge, Cambridge University Press.

Arnold, D. 2011. Ceramic Theory and Cultural Process after 25 years. *Ethnoarchaeology* 3 (1): 63–69.

Azim, M. 1975. Quatre campagnes de fouilles sur la Forteresse de Saï, 1970-1973. 1ère partie: l'installation pharaonique', *Cahiers de Recherches de l'Institut de Papyrologie et d'Égyptologie de Lille* 3: 91-125.

Azim, M. and Carlotti J.-F. 2011-2012. Le temple à de l'île de Saï et ses abords. *Cahiers de Recherches de l'Institut de Papyrologie et d'Égyptologie de Lille* 29: 11-63.

Bailloud, G. 1969. L'évolution des styles céramiques au Ennedi (République du Tchad). In J. P. Lebeuf (ed.), *Actes du 1er Colloque international d'Archéologie Africaine (Fort Lamy 1966)*. Fort Lamy: 31-45.

Barclay, K. 2001. *Scientific Analysis of Archaeological Ceramics. A handbook of resources*. Oxford, Oxford Books.

Baxter, M. J. 1999. Detecting Multivariate Outliers in Artefact Compositional Data. *Archaeometry* 41 (2): 321–338.

Baxter, M. J. 2001. Multivariate Analysis in Archaeology. In D. R Brothwell and A. M. Pollard (eds.), *Handbook of Archaeological Sciences*: 685-694. Chichester, John Wiley & Sons, Ltd.

Bonnet, C. 1988. Les fouilles archéologiques de Kerma (Soudan): rapport préliminaire sur les campagnes de 1986–87 et 1987–88. *Genava* XXXVI: 5–20.

Bronk Ramsey, C., 2013. OxCal 4.2 Manual. Oxford, Radiocarbon Accelerator Lab. http://c14.arch.ox.ac.uk/oxcal.

Budka, J. 2014. The New Kingdom in Nubia: New results from current excavations on Sai Island. *Egitto e Vicino Oriente* 37: 55–87.

Budka, J. 2015. The Pharaonic town on Sai Island and its role in the urban landscape of New Kingdom Kush. *Sudan & Nubia* 19: 40–53.

Budka, J. and Doyen, F. 2013. Living in New Kingdom towns in Upper Nubia—New evidence from recent excavations on Sai Island. *Ägypten & Levante* 22/23, 2012–2013: 167–208.

Camps, G. 1969. *Amekni: Néolithique ancien du Hoggar*. Mémoires du C.R.A.P.E. 10. Paris, Arts et Métiers graphiques.

Camps-Fabrer, H. 1966. *Matière et art mobilier dans la préhistoire nord-africaine et saharienne*. Mémoires du C.R.A.P.E. 5. Paris, Arts et Métiers graphiques.

Caneva, I. 1983. 'Wavy Line' decoration from Saggai I: an essay of classification. In I. Caneva (ed.), *Pottery-using gatherers and hunters at Saggai (Sudan). Preconditions for Food Production, Origini* 12: 155–190.

Caneva, I. 1987. Pottery decoration in prehistoric Sahara and Upper Nile: a new perspective. In B. E. Barich (ed.), *Archaeology and Environment in the Libyan Sahara: The Excavations in the Tadrart Acacus, 1978-1983*: 231–254. Oxford, British Archaeological Reports 368.

Caneva, I. (ed.) 1988. *El Geili: The History of a Middle Nile Environment 7000 B.C.-A.D. 1500*. Oxford, British Archaeological Reports 424.

Caneva, I. 1995. New methods of data collection and analysis in Sudanese prehistoric archaeology. *Cahiers de Recherches de l'Institut de Papyrologie de Lille* 17 (1): 69-96.

Caneva, I., Garcea, E. A. A. and Gautier, A. 1993. Pre-Pastoral cultures along the central Sudanese Nile. *Quaternaria Nova* 3: 177-252.

Caneva, I. and Marks, A. E. 1990. More on the Shaqadud pottery: evidence for Saharo-Nilotic connections during the 6th–4th millennium B.C. *Archéologie du Nil Moyen* 4: 11-35.

Carlson, R. 1966. A Neolithic site in the Murshid district, Nubia. *Kush* XIV: 53-62.

Chaix, L. 2011. Contribution à l'étude de l'économie de la période Pré-Kerma. Premiers résultats sur la faune du site 8-B-10A sur l'Ile de Saï (Nord Soudan). In V. Rondot, F. Alpi and F. Villeneuve (eds.), *La pioche et la plume: Autour du Soudan, du Liban et de la Jordanie. Hommages archéologiques à Patrice Lenoble*: 207-223. Paris, Presses de l'université Paris-Sorbonne.

Chaix, L. and Honegger M. 2015. New Data on the Animal Exploitation from the Mesolithic to the Neolithic periods in Northern Sudan. Climate and Ancient Societies. In S. Kerner, R. Dann, P. Bangsgaard (eds.), *Climate and Ancient Societies*: 197-214. Museum Tusculanum Press.

Childe, V. G. 1936. *Man makes himself*. London, Methuen.

Childe, V. G. 1956. *Society and knowledge*. London, Allen and Unwin.

Cresswell, R. 1972. Les trois sources d'une technologie nouvelle. In J. Thomas and L. Bernot (eds.), *Langues et techniques, nature et société II: Approche ethnologique, approche naturaliste*: 21-27. Paris, Klincksieck.

Cuomo di Caprio, N. 2007. *La ceramica in archeologia 2. Antiche tecniche di lavorazione e moderni metodi d'indagine*. Roma, L'Erma di Bretschneider.

Dal Sasso, G., Maritan, L. Salvatori, S., Mazzoli, C. and Artioli, G. 2014. Discriminating pottery production by image analysis: a case study of Mesolithic and Neolithic pottery from Al Khiday (Khartoum, Sudan). *Journal of Archaeological Science* 46: 125-143.

David, N. and Kramer, C. 2001. *Ethnoarchaeology in Action*. Cambridge, Cambridge University Press: 168-224.

DeBoer, W. R. 1984. The last pottery show: System and sense in ceramic studies. In S. E. van der Leeuw and A. C. Pritchard (eds.), *The Many Dimensions of Pottery*: 527-571. University of Amsterdam, Amsterdam.

deHeinzelin, J. 1968. Geological history of the Nile valley in Nubia. In F. Wendorf (ed.) *The Prehistory of Nubia* Vol. 1: 19-55. Dallas, Fort Burgwin Research Center and Southern Methodist University Press.

deMenocal, P. B, Ortiz, J., Guilderson, T., Adkins, J., Sarnthein, M., Baker, L., and Yarusinski, M. 2000. Abrupt onset and termination of the African Humid Period: Rapid climate response to gradual insulation forcing. *Quaternary Science Reviews* 19: 347-361.

De Paepe P. 1986. Etude minéralogique et chimique de la céramique néolithique d'El Kadada et ses implications archéologiques. *Archéologie du Nil Moyen* 1: 113-140.

De Paepe, P. and Brysse, Y. 1986. Analyse microscopique et chimique de la céramique de Kerma (Soudan). *Genava* XXXIV: 41-45.

D'Ercole, G. 2015. The ceramic assemblage from Sai Island (northern Sudan): connecting technological choices to cultural traditions between the 6th and the 3rd millennium BC. In J. Kabaciński, M. Chłodncki and M. Kobusiewicz (eds.), *Hunter-Gatherers and Early Food Producing Societies in Northeastern Africa, Studies in African Archaeology* 14: 177-194. Poznan, Poznań Archaeological Museum.

D'Ercole, G., Budka, J., Sterba, J. H., Garcea, E. A. A. and Mader D. 2017. The successful 'recipe' for a long-lasting tradition: Nubian ceramic assemblages from Sai Island (northern Sudan) from prehistory to the New Kingdom. *Antiquity* 91: 24-42.

D'Ercole, G., Garcea, E. A. A., Eramo, G. and Muntoni, I. M. Variability and continuity of ceramic manufacturing of prehistoric pottery from Upper Nubia, Sudan: An ethnographic comparison. *Journal of Archaeological Science: Reports*. Available online 29 May 2017. In Press, Corrected Proof. https://doi.org/10.1016/j.jasrep.201

D'Ercole, G., Eramo, G., Garcea, E. A. A., Muntoni I. M. and Smith, J. R. 2015. Raw material and technological changes in ceramic productions at Sai Island, northern Sudan from the 7th to the 3rd millennium BC. *Archaeometry* 57 (4): 597–616.

Doyen, F. 2014. Sai Island New Kingdom Town (Northern Sudan): 3rd and 4th Seasons (2009-2010). In J. R. Anderson and D. A. Welsby (eds.), *The Fourth Cataract and Beyond. Proceedings of the 12th International Conference for Nubian Studies*: 367-375. Leuven, British Museum Publications on Egypt and Sudan 1.

Drake, N. A., Blench, R. M., Armitage, S. J., Bristow, C. S. and White, K. H. 2011. Ancient watercourses and biogeography of the Sahara explain the peopling of the desert. *Proceedings of the National Academy of Sciences* (PNAS) USA 108: 458–462.

Dunne, J., Evershed, R. P., Salque, M., Cramp, L., Bruni, S., Ryan, K., Biagetti, S., and di Lernia, S. 2012. First dairying in the green Saharan Africa in the fifth millennium BC. *Nature* 486: 390–394.

Edwards, D. N. and Osman, A. 1992. *Mahas Survey Reports 1. (The Mahas Survey 1991, Interim Report and Site Inventory)*. University of Cambridge.

Edwards, D. N. and Osman, A. 1993. *Mahas Survey Reports 2. (The Mahas Survey 1990, Interim Report and Site Inventory)*. University of Cambridge.

Edwards, D. N. and Osman, A. 2000. The Archaeology of Arduan Island - the Mahas Survey 2000. *Sudan & Nubia* 4: 58–70.

Edwards, D.N. and Osman, A. (eds.) 2012. *The Archaeology of a Nubian Frontier. Survey on the Nile Third Cataract, Sudan*. Leicester, Mauhaus Publishing.

Edwards, D. N., Osman, A., Tahir, Y. F, Sadig, A. M. and el-Zein I. S. 2012. On a Nubian frontier — landscapes of settlement on the Third Cataract of the Nile, Sudan. *Azania: Archaeological Research in Africa* 47 (4): 450–487.

Florenzano, A., Mercuri, A. M., Altunoz, M. and Garcea, E. A. A. 2016. Palynological evidence of cultural and environmental connections in Sudanese Nubia during the Early and Middle Holocene. *Quaternary International* 412, Part B: 65–80.

Francaviglia, V. and Palmieri, A. 1983. Petrochemical analyses of the 'Early Khartoum' pottery. A preliminary report. In I. Caneva (ed.), *Pottery-using gatherers and hunters at Saggai (Sudan). Preconditions for Food Production. Origini* 12: 191–205.

Franzini, M., Leoni, L. and Saitta, M. 1972. A simple method to evaluate the matrix effects in X-ray fluorescence analysis. *X-ray Spectrometry* 1: 151–154.

Franzini, M., Leoni, L. and Saitta, M. 1975. Revisione di una metodologia analitica per fluorescenza X, basata sulla correzione completa degli effetti di matrice. *Rendiconti della Società Italiana di Mineralogia e Petrologia* 31: 356–378.

Gasse, F. 2000. Hydrological changes in the African tropics since the last glacial maximum. *Quaternary Science Reviews* 19: 189–211.

Gabriel, B. 1976. Neolithische Steinplätze und Palaökologie in den Eben der österlichen Zentralsahara. In E. M. van Zinderen Baker (ed.), *Paleoecolgy of Africa*: 25–40. Cape Town, A. A. Balkema.

Gabriel, B. 1986. *Die Östliche Libysche Wüste im Junquartar*. Berlin, Berliner geographische Studien Bd. 19.

Garcea, E. A. A. 2005. Comparing *chaînes opératoires*: technological, cultural and chronological features of pre-pastoral and pastoral ceramic and lithic productions. In A. Livingstone Smith, D. Bosquet and R. Martineau (eds.), *Pottery Manufacturing Processes: Reconstitution and Interpretation*: 215–228. Acts of the XIVth UISPP Congress, University of Liège, 2001, Symposium 2.1. Oxford, British Archaeological Reports 1349.

Garcea, E. A. A. 2006a. Semi-permanent foragers in semi-arid environments of North Africa. *Word Archaeology* 38 (2): 197–219.

Garcea, E. A. A. 2006b. The endless glory of a site: esh-Shaheinab in the Sudanese prehistory. In I. Caneva and A. Roccati (eds.), *Acta Nubica: Proceedings of the X International Conference of Nubian Studies, Rome 9-14 September 2002*: 95–102. Roma, Istituto Poligrafico e Zecca dello Stato.

Garcea, E. A. A. 2006–2007. The Holocene prehistory at Sai Island, Sudan. In B. Gratien (ed.), *Mélanges offerts à Francis Geus. Cahiers de Recherches de l'Institut de Papyrologie et d'Egyptologie de Lille* 26: 107–113.

Garcea, E. A. A. 2008. The ceramics from Adrar Bous and surroundings. In J. D. Clark and D. Gifford-Gonzalez (eds.), *Adrar Bous: Archaeology of a Central Saharan Granitic Ring Complex in Niger*: 245–289. Tervuren, Royal Museum for Central Africa.

Garcea, E. A. A. 2011–2012. Revisiting the Khartoum Variant in its environment. *Cahiers de Recherches de l'Institut de Papyrologie et d'Egyptologie de Lille* 29: 139-150.

Garcea, E. A. A. 2012. Pottery manufactures at Sai Island, Sudan. *Prehistory of Northeastern Africa. New Ideas and Discoveries. Studies in African Archaeology* 11: 153-166. Poznań, Poznań Archaeological Museum.

Garcea, E. A. A. 2013. Manufacturing Technology of the ceramic Assemblage. In E. A. A. Garcea (ed.), *Gobero: the No-Return Frontier. Archaeology and Landscape at the Saharo-Sahelian Borderland. Journal of African Archaeology Monograph Series* Vol. 9: 209-240. Frankfurt am Main, Africa Magna Verlag.

Garcea, E. A. A. 2016a. Multi-stage dispersal of Southwest Asian domestic livestock and the path of pastoralism in the Middle Nile Valley. *Quaternary International* 412, Part B: 54–64.

Garcea, E. A. A. 2016b. Semi-permanent foragers in North and West Africa – An archaeological perspective. In T. Tvedt and T. Oestigaard (eds.), *Water and Food: From hunter-gatherers to global production in Africa, A History of Water*: 29–54. Series III, Vol. 3, I.B., London, Tauris.

Garcea, E. A. A. and Caputo, A. 2004. Outils statistiques pour l'étude de la production et de l'utilisation de la céramique au Sahara et au Soudan. *Préhistoire Anthropologie Méditerranéennes* 13: 87–96.

Garcea, E. A. A. and Hildebrand, E. A. 2009. Shifting social networks along the Nile: Middle Holocene ceramic assemblages from Sai Island, Sudan. *Journal of Anthropological Archaeology* 28: 304–322.

Garcea, E. A. A., Wang, H. and Chaix, L. 2016a. High-Precision Radiocarbon Dating Application to Multi-Proxy Organic Materials from Late Foraging to Early Pastoral Sites in Upper Nubia, Sudan. *Journal of African Archaeology* 14 (1): 83-98.

Garcea, E. A. A., Karul, N. and D'Ercole G. 2016b. Southwest Asian domestic animals and plants in Africa: Routes, timing and cultural implications. *Quaternary International* 412, Part B: 1–10.

Gardiner, A. 2010. Nubian sandstones of NW Sudan. *Nubian sandstone. Working Group abstracts volume* 13. Manchester, North Africa Research Group, University of Manchester.

Gatto, M. C. 2002a. Early Neolithic Pottery of the Nabta-Kiseiba Area: Stylistic Attributes and Regional Relationships. In K. Nelson and Associates (eds.), *Holocene Settlement of the Egyptian Sahara. Volume 2: The Pottery of Nabta Playa*: 65–78. New York, Kluwer Academic/Plenum Publishers.

Gatto, M. C. 2002b. Ceramic Traditions and Cultural Territories: the Nubian Group in Prehistory. *Sudan & Nubia* 6: 8–19.

Gatto, M. C. 2006a. The Khartoum Variant Pottery in Context: Rethinking the Early and Middle Holocene Nubian Sequence. *Archéologie du Nil Moyen* 10: 57–72.

Gatto, M. C. 2006b. The Nubian A-Group: a reassessment. *ArchéoNil* 16: 61–76.

Gautier, A. 2001. The Early to Late Neolithic Archaeofaunas from Nabta and Bir Kiseiba. In F. Wendorf, R. Schild and Associates (eds.), *Holocene Settlement of the Egyptian Sahara. Volume 1: The Archaeology of Nabta Playa*: 609–635. New York, Kluwer Academic/Plenum Publishers.

Gautier, A. 2002. The Evidence for the Earliest Livestock in North Africa: Or Adventures with Large Bovids, Ovicaprids, Dogs and Pigs. In F. A. Hassan (ed.), *Droughts, Food and Culture. Ecological Change and Food Security in Africa's Later Prehistory*: 195–208. New York, Kluwer Academic/Plenum Publishers.

Geus, F. 1994. L'île de Saï à travers l'histoire du Soudan. *La Nubie, L'archéologie au Soudan. Les Dossiers d'Archéologie* 196: 22-27.

Geus, F. 1995a. Archaeology and History of Sai Island. *The Sudan Archaeological Research Society Newsletter* 8: 27-34.

Geus, F. 1995b. Saï 1993-1995. *Archéologie du Nil Moyen* 7: 79-98.

Geus, F. 1996. Saï: fouilles anciennes et récentes. *Egypte, Afrique and Orient* 4: 2-7.

Geus, F. 1998. Saï 1996-1997. *Archéologie du Nil Moyen* 8: 85–126.

Geus, F. 2000. Geomorphology and prehistory of Sai Island (Nubia): report on a current research project. In L. Krzyzaniak, K. Kroeper and Kobusiewicz (eds.), *Recent Research into the Stone Age of Northeastern Africa*: 119-128. Poznan, Poznan Archaeological Museum.

Geus, F. 2002. Saï 1998-1999. *Archéologie du Nil Moyen* 9: 95–134.

Geus, F. 2004a. Sai. In D. A. Welsby and J. R. Anderson (eds.), *Sudan. Ancient Treasures. An Exhibition of recent discoveries from the Sudan National Museum*, London, British Museum Press: 114–116.

Geus, F. 2004b. Pre-Kerma Storage Pits on Sai Island. In T. Kendall (ed.), *Nubian Studies: Proceedings of the Ninth Conference of the International Society for Nubian Studies, Boston 21-26 August 1998*: 46–51. Boston, Museum of Fine Arts and Northeastern University.

Gifford-Gonzalez, D. and Hanotte, O. 2011. Domesticating animals in Africa: implications of genetic and archaeological findings. *Journal of World Prehistory* 24: 1–23.

Goossens, R., De Dapper, M. and De Paepe, P. 1997. SAIGAIS: A Geo-Archaeological Information System for the Island of Sai (Nubia, Sudan). Actes de la VIIIe Conférence Internationale des Etudes Nubiennes, II-Découvertes archéologique. *Cahiers de Recherches de l'Institut de Papyrologie et d'Egyptologie de Lille* 17 (2): 13–20.

Gosselain, O. P. 1992. Technology and style: potters and pottery among the Bafia of Cameroon. *Man* 27 (3): 559–586.

Gosselain, O. P. 1998. Social and technical identity in a clay crystal ball. In M. Stark (ed.), *The Archaeology of Social Boundaries*: 78-106. Washington, DC, Smithsonian Institution Press.

Gosselain, O. P. 2000. Materializing Identities: An African Perspective. *Journal of Archaeological Method and Theory* 7 (3): 187–217.

Gratien, B. 1976. Les nécropoles de l'île de Sai, IV. *Cahiers de Recherches de l'Institut de Papyrologie et d'Egyptologie de Lille* 4: 105-130.

Gratien, B. 1986. *La nécropole Kerma*. Paris, Editions du Editions du Centre national de la recherche scientifique.

Gratien, B. 1995. La Basse Nubie à L'Ancien Empire: Egyptiens et Autochtones. *The Journal of Egyptian Archaeology* 81: 43–56.

Guiraud, R., Bosworth, W., Thierry, J. and Delplanque, A. 2005. Phanerozoic geological evolution of Northern and Central Africa: an overview. *Journal of African Earth Sciences* 43: 83–143.

Haaland, R. 1987. *Socio-economic Differentiation in the Neolithic Sudan*. Oxford, British Archaeological Reports 350.

Hafsaas-Tsakos, H. and Tsakos, A. 2012. A Second Look into the Medieval Period on Sai Island. *Beiträge zur Sudanforschung* 11: 75-91.

Hakem, A. M. A. and Khabir A. M. 1989. Sarourab 2: a new contribution to the Early Khartoum tradition from Bauda site. In L. Krzyzaniak and M. Kobusiewicz (eds.), *Late Prehistory of the Nile Basin and the Sahara*: 381-385. Poznan, Poznan Archaeological Museum.

Haour, A. and Manning, K. 2010. *West African pottery decorated using roulettes*. (http://ads.ahds.ac.uk/catalogue/archive/wafrican_lt_2010/index.cfm?CFID=4215359&CFTOKEN=93794867, accessed December 2016).

Haynes, JR., C. V. 1985. Quaternary Studies, Western Desert, Egypt and Sudan - 1979-1983 field studies. *National Geographic Society Research Reports* 19: 269–341.

Hays, T. R. 1984. Predynastic development in Upper Egypt. In L. Krzyzaniak and M. Kobusiewicz (eds.), *Origin and Early Development of Food-producing Cultures in North-Eastern Africa*: 211-219. Poznan, Poznan Polish Academy of Sciences and Poznan Archaeological Museum.

Hays, T. R. and Hassan, F. A. 1974. Mineralogical Analysis of 'Sudanese Neolithic' Ceramics. *Archaeometry* 16 (1): 71-79.

Herbst, G. 2008. UCBS 00:01: preliminary assessments of a Late Neolithic Settlement at the Third Cataract. In W. Godlewski and A. Lajatar (eds.), *Between the Cataracts* (Part 2/1), *Proceedings of the 11th International Conference of the Society for Nubian Studies, Warsaw 27 August-2 September 2006*: 271–279. Warsaw, Warsaw University Press.

Hesse, A. 1996. Le néolithique de l'île de Saï, Soudan, 5000 ans d'histoire. *Dossiers d'Archéologique, Hors-Séries* 6: 36–37.

Hesse, A. and Chagny, B.-N. 1994. Relevé planimétrique de la surface du sol par photographie portée par un cerf-volant. *Revue d'archéométrie* 18: 5-11.

Hildebrand, E. A. 2006–2007. The significance of Sai Island for early plant food production in Sudan. In B. Gratien (ed.), *Mélanges offerts à Francis Geus. Cahiers de Recherches de l'Institut de Papyrologie et d'Egyptologie de Lille* 26: 173–181.

Hildebrand, E. A. and Schilling, T. M. 2016. Storage amidst early agriculture along the Nile: Perspectives from Sai Island, Sudan. *Quaternary International* 412, Part B: 81–95.

Hinkel, F. 1977. *The Archeological Map of the Sudan: A Guide to its Use and Explanation of its Principles*. Berlin, Akademie Verlag.

Hodder, I. 2006. *The leopard's tale*. London, Thames & Hudson.

Honneger, M. 1995. Kerma: note sur la reprise des fouilles de l'agglomération pré-Kerma. In C. Bonnet and collab. (eds.), Les fouilles archéologiques de Kerma (Soudan), *Genava* XLIII: 58-59.

Honegger, M. 1997. Kerma: l'agglomération pré-Kerma. *Genava* XLV: 113-118.

Honegger, M. 1999. Kerma: les occupations néolithiques et Pré-Kerma de la nécropole orientale. *Genava* XLVII: 77–82.

Honegger, M. 2003. Exploitation du territoire et habitat dans les sociétés pastorales du Soudan: l'exemple de Kerma entre les 5e et 3e millénaires av. J.-C. In M. Besse, L.-I. Stahl-Gretsch and P. Curdy (eds.), *Constellation: hommage à Alain Gallay. Cahiers d'archéologie romande* 95: 341–352. Lausanne.

Honegger, M. 2004a. Settlements and cemeteries of the Mesolithic and Early Neolithic at el-Barga (Kerma region). *Sudan & Nubia* 8: 27–32.

Honegger, M. 2004b. The Pre-Kerma: a cultural group from Upper Nubia prior to the Kerma civilisation. *Sudan & Nubia* 8: 38–46.

Honegger, M. 2004c. The Pre-Kerma Settlement: New Elements Throw Light on the Rise of the First Nubian Kingdom. In T. Kendall (ed.), *Nubian Studies: Proceedings of the Ninth Conference of the International Society for Nubian Studies, Boston 21-26 August 1998*: 83-94. Boston, Museum of Fine Arts and Northeastern University.

Honegger, M. 2004d. The Pre-Kerma period. In D. A. Welsby and J. R Anderson (eds.), *Sudan Ancient Treasures: An Exhibition Of Recent Discoveries From The Sudan National Museum*: 61-69. London, British Museum Press.

Honegger, M. 2005. El-Barga: un site clé pour la compréhension du Mésolithique et du début du Néolithique en Nubie. Revue de paléobiologie, vol. spécial 10 (Hommage à Louis Chaix): 95–104.

Honegger, M. 2006. Habitats préhistoriques en Nubie entre le 8e et le 3e millénaire av. J.-C.: l'exemple de la région de Kerma. In I. Caneva and A. Roccati (eds.), *Acta Nubica: Proceedings of the X International Conference of Nubian Studies, Rome 9-14 September 2002*: 3–13. Roma, Istituto Poligrafico e Zecca dello Stato.

Honegger, M. 2010. La Nubie et le Soudan: un bilan des 20 dernières années de recherche sur la Pré- et Protohistoire. *ArchéoNil* 20: 76–86.

Honegger, M., 2014. Recent advances in our understanding of prehistory in Northern Sudan. In J. R. Anderson and D. A. Welsby (eds.), *The Fourth Cataract and beyond. Proceedings of the 12th International Conference for Nubian Studies*: 19-30. Leuven, British Museum Publications on Egypt and Sudan 1.

Honegger, M. and Williams, M. 2015. Human occupations and environmental changes in the Nile valley during the Holocene: The case of Kerma in Upper Nubia (northern Sudan). *Quaternary Science Reviews* 130: 141-154.

Huysecom, E., Rasse, M., Lespez, L., Neumann, K., Fahmy, A., Ballouche, A., Ozainne, S., Maggetti, M., Tribolo, C. and Soriano, S. 2009. The emergence of pottery in Africa during the tenth millennium cal BC: new evidence from Ounjougou (Mali). *Antiquity* 83: 905-917.

Jesse, F. 2002. Wavy Line Ceramics: Evidence form Northeastern Africa. In K. Nelson and Associates (eds.), *Holocene Settlement of the Egyptian Sahara. Volume 2: The Pottery of Nabta Playa*: 79–96. New York, Kluwer Academic/Plenum Publishers.

Jesse, F. 2004. No link between the central Sahara and the Nile Valley? (Dotted) Wavy Line ceramics in the Wadi Howar, Sudan. In T. Kendall (ed.), *Nubian Studies: Proceedings of the Ninth Conference of the International Society for Nubian Studies, Boston 21-26 August 1998*: 296-308. Boston, Museum of Fine Arts and Northeastern University.

Jesse, F. 2006. A permanent link? The Wadi Howar region and the Nile. In I. Caneva and A. Roccati (eds.), *Acta Nubica: Proceedings of the X International Conference of Nubian Studies, Rome 9-14 September 2002*: 73-80. Roma, Istituto Poligrafico e Zecca dello Stato.

Jesse, F. 2008. Time of experimentation? - The 4th and 3rd millennia BC in Lower Wadi Howar, Northwestern Sudan. In W. Godlewski and A. Lajatar (eds.), *Between the Cataracts* (Part 2/1), *Proceedings of the 11th International Conference of the Society for Nubian Studies, Warsaw 27 August -2 September 2006*: 49-74. Warsaw, Warsaw University Press.

Keding, B. 2006. Pottery of the Wadi Howar - traditions, transformations and their implications. In K. Kroeper, M. Chlodnicki and M. Kobusiewicz (eds.), *Archaeology of Early Northeastern Africa*: 235-259. *In Memory of Lech Krzyzaniak*. Poznan, Poznan Archaeological Museum.

Khabir, A. M. 1991. A qualitative change in the texture of temper of Neolithic ceramics from the Central Nile Valley. *Sahara* 4: 145–48.

Klein, M., Jesse, F., Kasper, H. U. and Gölden, A. 2004. Chemical characterization of ancient pottery from Sudan by X-Ray Fluorescence Spectrometry (XRF), Electron Microprobe Analyses (EMPA) and Inductively Coupled Plasma Mass Spectrometry (ICP-MS). *Archaeometry* 46 (3): 339-356.

Klemm, D., Klemm R. and Murr, A. 2001. Gold of the Pharaohs - 6000 years of gold mining in Egypt and Nubia. *African Earth Sciences* 33: 643-659.

Kretz, R. 1983. Symbols for rock-forming minerals. *American Mineralogist* 68: 277–279.

Kröpelin, S. 2007. The Wadi Howar. In O. Bubenzer, A. Bolten and F. Darius (eds.), *Atlas of Cultural and Environmental Change in Arid Africa, Africa Praehistorica* 21: 38-39. Köln, Heinrich-Barth-Institut.

Kuper, R. 2002. Routes and roots in Egypt's western desert: the early Holocene resettlement of the eastern Sahara. In R. Friedman (ed.), *Egypt and Nubia. Gifts of the Desert*. London, British Museum Press: 1-12.

Kuper, R. and Kröpelin, S. 2006. Climate-controlled Holocene occupation in the Sahara: motor of Africa's evolution. *Science* 313: 803-807.

Lange, M. 2003. A-Group settlements sites from the Laqiya-Region. In L. Krzyzaniak, K. Kroeper and M. Kobusiewicz (eds.) *Cultural Markers in the Later Prehistory of Northeastern Africa and Recent Research. Studies in African Archaeology* 8: 105-127. Poznan, Poznan Archaeological Museum.

Lange, M. 2004. Wadi Shaw 82/52 - a Peridynastic Settlement Site in the Western Desert and its Relation to the Nile Valley. In T. Kendall (ed.), *Nubian Studies: Proceedings of the Ninth Conference of the International Society for Nubian Studies, Boston 21-26 August 1998*: 315-324. Boston, Museum of Fine Arts and Northeastern University.

Lange, M. 2006. The archaeology of the Laqiya Region (NW-Sudan): ceramics, chronology and cultures. In I. Caneva and A. Roccati (eds.) *Acta Nubica. Proceedings of the X International Conference on Nubian Studies, Rome 9-14 September 2002*: 107-115. Roma, Istituto Poligrafico e Zecca dello Stato.

Lange, M. 2006-2007. Development of pottery production in the Laqiya-Region, Eastern Sahara. In B. Gratien (ed.), *Mélanges offerts à Francis Geus. Cahiers de Recherches de l'Institut de Papyrologie et d'Egyptologie de Lille* 26: 243-381.

Lange, M, Nordström, H.-Å 2006. Abkan connections - The relationship between the Abkan cultures in the Nile valley and Early Nubian sites from the Laqiya Region (Eastern Sahara, Northwest-Sudan). In K. Kroeper, M. Chlodnicki and M. Kobusiewicz (eds.), *Archaeology of Early Northeastern Africa. In Memory of Lech Krzyzaniak*: 297-312. Poznan, Poznan Archaeological Museum.

Laviano, R. 2002. Diffrazione di raggi X. *Metodologie Analitiche Mineralogiche e Petrografiche. Dispense per il Corso di Laurea triennale in Scienza e Tecnologia per la Diagnostica e la Conservazione dei Beni culturali*. Dipartimento Geomineralogico, Università degli Studi di Bari: 73-82.

Lechtman, H. 1977. Style in technology: some early thoughts. In H. Lechtman and T. S. Merrill (eds.), *Material culture: style, organization, and dynamics of technology*: 3–20. St Paul, West Publishing Company.

Leoni, L. and Saitta, M. 1976. X-ray fluorescence analysis of 29 trace elements in rock and mineral standards. *Rendiconti della Società Italiana di Mineralogia e Petrologia* 32 (2): 497-510.

Leroi-Gourhan, A. 1964. *Le geste et la parole I: techniques et langage*. Paris, A. Michel.

Leroi-Gourhan, A. 1965. *Le geste et la parole II: la mémoire et les rythmes*. Paris, A. Michel.

Linseele, V. 2010. Did specialized pastoralism develop differently in Africa than in the Near East? An example from the west African Sahel. *Journal of World Prehistory* 23: 43–77.

Linseele, V., Van Neer, W., Thys, S., Phillipps, R., Cappers, R., Wendrich, W., Holdaway, S., 2014. New archaeozoological data from the Fayum 'Neolithic' with a critical assessment of the evidence for early stock keeping in Egypt. *PLoS One* 9 (10): e108517.

Livingstone Smith, A. 2000. Processing clay for pottery in Northern Cameroon: Social and Technical requirements. *Archaeometry* 42 (1): 21–42.

Livingstone Smith, A. 2001. Pottery manufacturing processes: reconstruction and interpretation. In E. A. A. Garcea (ed.), *Uan Tabu in the Settlement History of the Libyan Sahara*: 111-150. Firenze, All'Insegna del Giglio.

London, G. 1981. Dung tempered clay. *Journal of Field Archaeology* 8: 189-195.

MacDonald, K.C., Manning, K. 2010. Cord-wrapped roulette/Roulette de cordelette enroulée. In A. Haour, K. Manning, N. Arazi, O. Gosselain, S. Guèye, D. Keita, A. Livingstone Smith, K. MacDonald, A. Mayor, S. McIntosh and R. Vernet (eds.), *African Pottery Roulettes Past and Present: Technique, Identification and Distribution*: 144-156. Oxford, Oxbow Books.

Macklin, M. G., Woodward, J. C., Welsby, D. A., Duller, G. A. T., Williams, F. M. and Williams, M. A.J. 2013. Reach-scale river dynamics moderate the impact of rapid Holocene climate change on floodwater farming in the desert Nile. *Geology* 41: 695-698.

Macklin, M. G., Woodward, J. C., Toonen, W. H. J., Williams, M. A. J., Flaux, C., Marriner, N., Nicoll, K., Vertsaeten, G., Spencer, N. and Welsby, D. A. 2015. A new model of river dynamics, hydroclimatic change and human settlement in the Nile Valley derived from meta-analysis of the Holocene fluvial archive. *Quaternary Science Reviews* 130: 109-123.

Madella, M., García-Granero, J. J., Out, W. A., Ryan, P. and Usai, D. 2014. Microbotanical evidence of domestic cereals in Africa 7000 years ago. *PLoS One* 9: e110177.

Maggetti, M. 1982. Phase Analysis and its Significance for Technology and Origin. In J. S. Olin and A. D. Franklin (eds.), *Archaeological Ceramics*: 121-133. Washington, Smithsonian Institution Press.

Maggetti, M. 1990. Il contributo delle analisi chimiche alla conoscenza delle ceramiche antiche. In T. Mannoni and A. Molinari, A. (eds.), *Scienze in Archeologia*: 65-88. Firenze, All'Insegna del Giglio.

Maley, J. 1970. Introduction à la géologie des environs de la Deuxième Cataracte du Nil, au Soudan. In J. Vercoutter (ed.), *Mirgissa I, Mission Archéologique Française au Soudan. Direction générale des relations culturelles, scientifiques et techniques*: 122-157. Paris, Direction générale des relations culturelles, scientifiques et techniques.

Maley, J. 1991. The African rain forest vegetation and palaeoenvironments during the late Quaternary. *Climate Change* 19: 79–98.

Manning, K. and Timpson, A. 2014. The demographic response to Holocene climate change in the Sahara. *Quaternary Science Reviews* 101: 28–35.

Manzo, A., Coppa, A., Alemseged Beldados Aleho and Zoppi, V. 2010. *Italian Archaeological Expedition to the Sudan of the University of Naples 'L'Orientale', 2010 Field Season*, Project Report. -. (Unpublished).

Marks, A. E. and Ferring, R. 1971. The Karat Group. In J. L. Shiner, A. Marks, V. Chmielewski, J. de Heinzelin and T. R. Hays (eds.), *The Prehistory and Geology of Northern Sudan. Report to the National Science Foundation*: 187–277. A Report for the National Science Foundation.

Matthew, A. J., Woods, A. J. and Oliver, C. 1991. Spots before the eyes: New comparison charts for visual per-centage estimation in archaeological material. In A. Middleton and I. Freestone (eds.), *Recent Developments in Ceramic Petrology*: 211–263. London, British Museum Occasional Paper no. 81.

McGee, D., DeMenocal, P. B., Winckler, G., Stuut, J. B. W. and Bradtmiller, L. I. 2013. The magnitude, timing and abruptness of changes in North African dust deposition over the last 20,000yr. *Earth and Planetary Science Letters* 371–372: 163–176.

Mills, A. J. 1965. The Reconnaissance Survey from Gamai to Dal. A Preliminary Report for 1963-64 Season. *Kush* XIII: 1-12.

Mills, A. J. 1968. The Archaeological Survey from Gamai to Dal. Report on the 1965-66 Season. *Kush* XV: 200-210.

Mills, A. J. and Nordström H.-Å. 1966. The Archaeological Survey from Gamai to Dal. Preliminary Report on the Season 1964-65. *Kush* XIV: 1-15.

Mohammed-Ali, A. S. and Khabir, A.-R.M. 2003. The Wavy Line and the Dotted Wavy Line Pottery in the Prehistory of the Central Nile and the Sahara-Sahel Belt. *African Archaeological Review* 20: 25-58.

Moore, A. M. T. 1995. The inception of potting in western Asia and its impact on economy and society. In W. K. Barnett and J. W. Hoopes (eds.), *The emergence of pottery. Technology and innovation in ancient societies*: 39-54. Washington (DC), Smithsonian Institution Press.

Myers, O. H. 1958. Abka re-excavated. *Kush* VI: 131-141.

Myers, O. H. 1960. Abka again. *Kush* VIII: 174-181.

Nelson, K. 2002a. Introduction. In K. Nelson and Associates (eds.), *Holocene Settlement of the Egyptian Sahara. Volume 2: The Pottery of Nabta Playa*: 1–8. New York, Kluwer Academic/Plenum Publishers.

Nelson, K. 2002b. Ceramic Types of the Nabta-Kiseiba Area. In K. Nelson, K and Associates (eds.), *Holocene Settlement of the Egyptian Sahara. Volume 2: The Pottery of Nabta Playa*: 9-19. New York, Kluwer Academic/Plenum Publishers.

Nelson, K. 2002c. Ceramic Assemblages of the Nabta-Kiseiba Area. In K. Nelson and Associates (eds.), *Holocene Settlement of the Egyptian Sahara. Volume 2: The Pottery of Nabta Playa*: 21-50. New York, Kluwer Academic/Plenum Publishers.

Nelson, K. 2002d. Appendix I and II. In K. Nelson and Associates (eds.), *Holocene Settlement of the Egyptian Sahara. Volume 2: The Pottery of Nabta Playa*: 102-108. New York, Kluwer Academic/Plenum Publishers.

Neumann, K. 1989. Holocene vegetation of Eastern Sahara: charcoal from prehistoric sites. *African Archaeological Review* 7: 97-116.

Nicoll, K. 2004. Recent environmental change and prehistoric human activity in Egypt and northern Sudan. *Quaternary Science Reviews* 23: 561-580.

Nieuwenhuyse, O. P., Akkermans, P. M. M. G. and der Plicht, J. 2010. Not so coarse, nor always plain – the earliest pottery of Syria, *Antiquity* 84: 71-85.

Nordström, H.-A. 1962. Archaeological Survey on the West Bank of the Nile. Excavations and Survey in Faras, Argin and Gezira Dabarosa. *Kush* X: 34-61.

Nordström, H. - Å 1972. *Neolithic and A-Group sites*. Uppsala, Scandinavian University.

Nordström, H. - Å. 2006. The discovery of the Neolithic in Nubia. *ArchéoNil* 16: 31-39.

Orton, C., Tyers, P. and Vince, A. 1993. *Pottery in Archaeology*. Cambridge, Cambridge University Press.

Out, W. A., Ryan, P., García-Granero, J. J., Barastegui, J., Maritan, L., Madella, M. and Usai, D. 2016. Plant exploitation in Neolithic Sudan: a review in the light of new data from the cemeteries R12 and Ghaba. *Quaternary International* 412, Part B: 36–53.

Pachur, H. J., Kröpelin, S., Hoelzmann, P., Goschin, M. and Altmann, N. 1990. Late Quaternary fluvio-lacustrine environments of western Nubia. *Research in Egypt and Sudan* 120: 203-260.

Pachur, H. J. and Hoelzmann, P. 2000. Late Quaternary palaeoecology and palaeoclimates of the eastern Sahara. *Journal of African Earth Sciences* 30: 929-939.

Peacock, D. P. S. 1977. Ceramics in Roman and Medieval Archaeology. In D. P. S. Peacock (ed.), *Pottery and Early Commerce*: 21-34. London, Academic Press.

Peters, J., 1986. A revision of the faunal remains from two Central Sudanese sites: Khartoum Hospital and Esh Shaheinab. *Archaeozoologia, Mélanges publiés à l'occasion du 5° Congrès international d'archéozoologie, Bordeaux, Août 1986*. Grenoble, La Pensée Sauvage Editions: 11–33.

Prasad, G., Lejal-Nicol, A. and Vaudois-Mieja, N. 1986. A Tertiary age for Upper Nubia Sandstone Formation, central Sudan. *AAPG bulletin* 70: 138-142.

Quinn, P. S. 2013. *Ceramic Petrography: The Interpretation of Archaeological Pottery & Related Artefacts in Thin Section*. Oxford, Archaeopress.

Redman, C. L. 1978. *The rise of civilization: from early farmers to urban society in the ancient Near East*. San Francisco (CA), Freeman and Co.

Reimer, P. J, Bard, E., Bayliss, A., Beck, J. W., Blackwell, P. G., Bronk Ramsey, C., Buck, C. E., Cheng, H., Edwards, R. L., Friedrich, M., Grootes, P. M., Guilderson, T. P., Haflidason, H., Hajdas, I., Hatté, C., Heaton, T. J., Hoffmann, D. L., Hogg, A. G., Hughen, K. A., Kaiser, K. F., Kromer, B., Manning, S. W., Niu, M., Reimer, R. W., Richards, D. A., Scott, E. M., Southon, J. R., Staff, R. A., Turney, C. S. M. and van der Plicht, J. 2013. IntCal13 and Marine13 radiocarbon age calibration curves 0–50,000 years cal BP. *Radiocarbon* 55 (4): 1869–1887.

Reinold, J. 2000. *Archéologie au Soudan. Les civilisations de Nubie*. Paris, Editions Errance.

Reinold, J. 2001. Kadruka and the Neolithic in the Northern Dongola Reach. *Sudan & Nubia* 5: 2-10.

Reinold, J. 2004. Kadruka. In D. A. Welsby and J. R. Anderson (eds.), *Sudan Ancient Treasures: An Exhibition Of Recent Discoveries From The Sudan National Museum*: 42–48. London, British Museum Press.

Reinold, J. 2007. *La nécropole néolithique d'El-Kadada au Soudan Central. Les cimetières A et B du kom principal*. Paris, Culture France, Editions Recherche sur les civilisations.

Rice, P. M. 1984. Change and conservatism in Pottery-Producing Systems. In S. E. van der Leeuw and A. C. Pritchard (eds.), *The Many Dimensions of Pottery*: 231–293. University of Amsterdam, Amsterdam.

Rice, P. M. 1987. *Pottery analysis. A sourcebook*. Chicago, University of Chicago Press.

Riemer, H. 2007. When hunters started herding: Pastro-foragers and the complexity of Holocene economic change in the Western Desert of Egypt. In H.-P. Wotzka, M. Bollig and R. Vogelsang (eds.), *Proceedings of an International ACACIA Conference*: 105–144. Cologne, Heinrich-Barth-Institut.

Ritchie, J. C. and Haynes Jr., C.V. 1987. Holocene vegetation zonation in the eastern Sahara. *Nature* 330: 645–647.

Roset, J.-P. 1982. Tagalagal: un site à céramique au Xe millénaire avant nos jours dans l'Aïr (Niger). *Comptes Rendus de l'Académie des Inscriptions et Belles Lettres* 123 (3): 565-570.

Roset, J.-P. 1987. Néolithisation, Néolithique et post-Néolithique au Niger nord-oriental. *Bulletin de l'Association française d'étude sur le Quaternaire* 4: 203-214.

Roux, V. 2008. Evolutionary trajectories of technological traits and cultural transmission: A qualitative approach to the emergence and disappearance of the ceramic wheel-fashioning technique in the southern Levant during the fifth to the third millennia BC. In M. T. Stark, B. Bowser and L. Horne (eds.), *Cultural transmission and material culture. Breaking down boundaries*: 82–104. Tucson, Arizona University Press.

Roux, V. and Courty M. A. 2013. Introduction to discontinuities and continuities: theories, methods and proxies for an historical and sociological approach to evolution of past societies. *Journal of Archaeological Method and Theory* 20: 187–193.

Sackett, J. R. 1977. The Meaning of Style in Archaeology: A General Model. *American Antiquity* 42: 369-381.

Sadig, A. M. 2010. *The Neolithic of the Middle Nile Region. An Archaeology of Central Sudan and Nubia*. Kampala, Fountain Publishers.

Salvatori, S. 2012. Disclosing Archaeological Complexity of the Khartoum Mesolithic: New Data at the Site and Regional Level. *African Archaeological Review* 29 (4): 399-472.

Salvatori, S. and Usai, D. 2006-2007. The Sudanese Neolithic revised. *Cahiers de Recherches de l'Institut de Papyrologie et d' Egyptologie de Lille* 26: 323-333.

Salvatori, S., Usai, D., Abdelrahman, M. F., Di Matteo A., Iacumin, P., Linseele V. and Magzoub, M. 2014. Archaeology at el-Khiday: New Insight on the Prehistory and History of Central Sudan. In R. Anderson and D. A. Welsby (eds.), *The Fourth Cataract and beyond. Proceedings of the 12th International Conference for Nubian Studies*: 243-258. Leuven, British Museum Publications on Egypt and Sudan 1.

Santacreu D. J. A. 2014. *Materiality, Techniques and Society in Pottery Production. The Technological Study of Archaeological Ceramics through Paste Analysis*. Warsaw/Berlin, De Gruyter Open.

Schiffer, M. B. 1988. The effects of surface treatment on permeability and evaporative cooling effectiveness of pottery. In R. Farquhar, R. Hancock and L. Pavlish (eds.), *Proceedings of the 26th international archaeometry symposium*: 23–29. Toronto, University of Toronto press.

Schiffer, M. B. and Skibo, M. 1987. Theory and Experiment in the Study of Technological Change. *Current Anthropology* 28 (5): 595–622.

Schiffer, M. B., Skibo, M., Boelke, T. C., Neupert, M. A. and Aronson, M. 1994. New Perspectives on Experimental Archaeology: Surface Treatments and Thermal Response of the Clay Cooking Pot. *American Antiquity* 59: 197–217.

Schild, R and Wendorf, F. 2001. Combined Prehistoric Expedition's Radiocarbon Dates Associated with Neolithic Occupations in the Southern Western Desert of Egypt. In F. Wendorf, R. Schild, and Associates (eds.) *Holocene settlement of the Egyptian Sahara: The Archaeology of Nabta Playa*: 51–56. New York, Kluwer Academic/Plenum Publishers.

Schild, R., Chmielewska, M. and Wieckowska, H. 1968. The Arkinian and Shamarkian industries. In F. Wendorf (ed.), *The Prehistory of Nubia*: 651-767. Dallas, Fort Burgwin Research Center and Southern Methodist University Press.

Sereno, P. C., Garcea, E. A. A, Jousse, H., Stojanowski, C. M., Saliège, J. F., Maga, A., Ide, O. A., Knudson, K. J., Mercuri, A. M., Stafford Jr, T. W., Kaye, T .G, Giraudi, C., N'siala, I. M., Cocca, E., Moots, H. M., Dutheil, D. B. and Stivers, J. P. 2008. Lakeside cemeteries in the Sahara: 5000 years of Holocene population and environmental change. *PLoS ONE* 3 (8): e2995.

Shang, C. K., Satir, M., Morteani, G. and Taubald, H. 2010. Zircon and titanite age evidence for coeval granitization and migmatization of the early Middle and early Late Proterozoic Saharan Metacraton; example from the central North Sudan basement. *Journal of African Earth Sciences* 57: 492–524.

Shiner, J. L. 1968a. The Khartoum Variant industry. In F. Wendorf (ed.), *The Prehistory of Nubia, vol. 2*: 768-790. Dallas, Fort Burgwin Research Center and Southern Methodist University Press.

Shiner, J. L. 1968b. The Cataract Tradition. In F. Wendorf (ed.), *The Prehistory of Nubia, vol. 2*: 535–629. Dallas, Fort Burgwin Research Center and Southern Methodist University Press.

Sillar, B. and Tite, M. S. 2000. The Challenge of 'Technological Choices' for Materials Science Approaches in Archaeology. *Archaeometry* 42 (2): 2-20.

Sinopoli, C. M. 1991. *Approaches to Archaeological Ceramics*. New York, Plenum Press.

Skibo, J. M. 2013. *Understanding Pottery Function*. Manuals in Archaeological Method, Theory and Technique. New York, Springer-Verlag.

Skibo, J. M., Schiffer, M. B. and Reid, K. C. 1989. Organic-tempered pottery: an experimental study. *American Antiquity* 54 (1): 122-146.

Smith, S. T. and Herbst, G. 2008. Neolithic through Kerma settlement at Ginefab. In B. Gratien, B. (ed.), *Actes de la 4ᵉ Conférence Internationale sur l'Archéologie de la 4ᵉ Cataracte du Nil. Supplément Cahiers de Recherches de l' Institut de Papyrologie et d'Egyptologie de Lille 7*: 203-216. Lille.

Soper, R. 1985. Roulette decoration on African pottery: technical considerations, dating and distributions. *African Archaeological Review* 3: 29-51.

Stark, M. T. (ed.). 1998. *The Archaeology of Social Boundaries*. Washington, DC, Smithsonian Institution Press.

Stark, M. T., Bishop, R. L., and Miksa, E. 2000. Ceramic technology and social boundaries: Cultural practices in Kalinga clay selection and use. *Journal of Archaeological Method and Theory* 7: 295-331.

Thissen, L. C. 2007. Die Anfänge der Keramikproduktion in der Türkei – ein Überblick. In Badisches Landesmuseum (ed.), *Vor 12000 Jahren in Anatolien: die ältesten Monumente der Menschheit*: 218-29. Karlsruhe, Ausstellungskatalog Badisches Landesmuseums Karlsruhe.

Tierney, J. E. and deMenocal, P. B. 2013. Abrupt shifts in Horn of Africa hydroclimate since the Last Glacial Maximum. *Science* 342: 843–846.

Tite, M. S. 2008. Ceramic Production, Provenance and Use - A Review. *Archaeometry* 50 (2): 216-231.

Tsakos, A. and Hafsaas-Tsakos, H. 2014. A note on the Medieval period of Sai Island. In J. R. Anderson and D. A. Welsby (eds.), *The Fourth Cataract and beyond. Proceedings of the 12th International Conference for Nubian Studies*: 985-987. Leuven, British Museum Publications on Egypt and Sudan 1.

Usai, D. 1998. Prehistoric evidence from the Letti Basin area. *Archéologie du Nil Moyen* 8: 145–156.

Usai, D. 2004. Early Khartoum and Related Groups. In T. Kendall (ed.), *Nubian Studies: Proceedings of the Ninth Conference of the International Society for Nubian Studies, Boston 21-26 August 1998*: 419-435. Boston, Museum of Fine Arts and Northeastern University.

Usai, D. 2005. Early Holocene seasonal movements between the desert and the Nile Valley: Details from the lithic industry of some Khartoum Variant and some Nabta/Kiseiba sites. *Journal of African Archaeology* 3 (1): 103–115.

Usai, D. 2014. Recent advances in understanding the prehistory of Central Sudan. In J. R. Anderson and D. A. Welsby (eds.), *The Fourth Cataract and beyond. Proceedings of the 12th International Conference for Nubian Studies*: 31-44. Leuven, British Museum Publications on Egypt and Sudan 1.

Van der Leeuw S. E. 1984. Dust to dust: a transformational view of the ceramic cycle. In Van der Leeuw S. E. and Pritchard A. C. (eds.), *The Many dimensions of pottery: ceramics in archaeology and anthropology*: 707-792. Amsterdam, University of Amsterdam.

Van Peer, P., Fullagar, R., Stokes, S., Bailey, R. M, Moeyersons, J., Steenhoudt, F., Geerts, A., Vanderbeken, T., De Dapper, M. and Geus, F. 2003. The Early to Middle Stone Age transition and the emergence of modern human behaviour at site 8-B-11, Sai Island, Sudan. *Journal of Human Evolution* 45 (2): 187-193.

Van Peer, P., Herman, C. F. 2006. L'occupation paléolithique de l'île de Saï: résultats de trois campagnes de prospection 1996-1998. *ArchèoNil* 16: 41-60.

Velde, B. and Druc, I. C. 1999. *Archaeological Ceramic Material. Origin and Utilization*. Berlin Heidelberg, Springer-Verlag.

Vercoutter, J. 1958. Excavations at Sai 1955-57. A Preliminary Report. *Kush* VI: 144-169.

Vercoutter, J. 1972. Le Soudan nilotique: l'île de Saï. *Archéologia* 50: 62-70.

Vercoutter, J. 1985. Préface. L'archéologie de l'île de Saï. In B. Gratien (ed.), *Saï I - La nécropole Kerma*: 9-17. Paris, Éditions du CNRS.

Vermeersch, P. M. 2002. The Egyptian Nile Valley during the Early Holocene. In Jennerstrasse 8 (ed.), *Tides of the Desert. Gezeiten der Wüste, Africa Praehistorica* 14: 27-40. Köln, Heinrich Barth Institut.

Verwers, G. J. 1962. Archaeological Survey of the West Bank of the Nile. The Survey from Faras to Gezira Dabarosa. *Kush* X: 19–33.

Vila, A. 1975. *La prospection archéologique de la Vallée du Nil au sud de la cataracte de Dal: Nubie soudanaise 2. Les districts de Dal (rive gauche) et de Sarkomatto (rive droit)*. Paris, Centre Nationale de la Recherche Scientifique.

Vila, A. 1976a. *La prospection archéologique de la Vallée du Nil au sud de la cataracte de Dal: Nubie soudanaise 3. District de Ferka (Est et Ouest)*. Paris, Centre Nationale de la Recherche Scientifique.

Vila, A. 1976b. *La prospection archéologique de la Vallée du Nil au sud de la cataracte de Dal: Nubie soudanaise 4. District de Mograkka (Est et Ouest). District de Kosha (Est et Ouest)*. Paris, Centre Nationale de la Recherche Scientifique.

Vila, A. 1977a. *La prospection archéologique de la Vallée du Nil au sud de la cataracte de Dal: Nubie soudanaise 5. Le district de Ginis, Est et Ouest*. Paris, Centre Nationale de la Recherche Scientifique.

Vila, A. 1977b. *La prospection archéologique de la Vallée du Nil au sud de la cataracte de Dal: Nubie soudanaise 6. Le district d'Attab, Est et Ouest*. Paris, Centre Nationale de la Recherche Scientifique.

Vila, A. 1977c. *La prospection archéologique de la Vallée du Nil au sud de la cataracte de Dal: Nubie soudanaise 7. Le district d'Amara Ouest*. Paris, Centre Nationale de la Recherche Scientifique.

Vila, A. 1977d. *La prospection archéologique de la Vallée du Nil au sud de la cataracte de Dal: Nubie soudanaise 8. Le district d'Amara Est*. Paris, Centre Nationale de la Recherche Scientifique.

Vila, A. 1978a. *La prospection archéologique de la Vallée du Nil au sud de la cataracte de Dal: Nubie soudanaise 9. L'île d'Arnyatta. Le district d'Abri (Est et Ouest). Le district de Tabaj (Est et Ouest)*. Paris, Centre Nationale de la Recherche Scientifique.

Vila, A. 1978b. *La prospection archéologique de la Vallée du Nil au sud de la cataracte de Dal: Nubie soudanaise 10. Le district de Koyekka (rive droite). Les districts de Morka et de Hamid (rive gauche). L'île de Nilwatti*. Paris, Centre Nationale de la Recherche Scientifique.

Vila, A. 1979. *La prospection archéologique de la Vallée du Nil au sud de la cataracte de Dal: Nubie soudanaise 11. Récapitulations et conclusions, appendices*. Paris, Centre Nationale de la Recherche Scientifique.

Weiner, S. 2010. *Microarchaeology: Beyond the Visible Archaeological Record*. Cambridge, Cambridge University Press.

Welsby, D. A. (ed.) 2001. *Life on the Desert Edge: Seven Thousand Years of Human Settlement in the Northern Dongola Reach, Sudan*. London, SARS Publications.

Welsby, D. A., Macklin, M. G. and Woodward, J. C. 2002. Human responses to Holocene environmental changes in the Northern Dongola Reach of the Nile, Sudan. In R. Friedman (ed.), *Egypt and Nubia: Gifts of the Desert*. London, British Museum Press: 28-38.

Wendorf, F. (ed.) 1968. *The Prehistory of Nubia*. Dallas, Fort Burgwin Research Center and Southern Methodist University Press.

Wendorf, F., Schild, R. and Associates (eds.) 2001. *Holocene Settlement of the Egyptian Sahara. Volume 1: The Archaeology of Nabta Playa*. New York, Kluwer Academic/Plenum Publishers.

Wendorf, F., Schild, R. and Haas, H. 1979. A New Radiocarbon Chronology for Prehistoric Sites in Nubia. *Journal of Field Archaeology* 6 (2): 219-223.

Whiteman, A. J. 1971. *The Geology of the Sudan Republic*. Oxford, Clarendon Press.

Wickens, G. E. 1975. Changes in the Climate and Vegetation of the Sudan since 20,000 BP. *Boissiena* 24: 43-65.

Williams, B. B. 1989. *Excavations between Abu Simbel and the Sudanese Frontier. Neolithic, A-Group and Post A-Group Remains from Cemeteries W, V, S, Q, T and a Cave East of Cemetery K. Oriental Institute Nubian Expedition*, vol. IV. Chicago, The University of Chicago Press.

Williams, M. A. J. and Adamson, D. A. 1980. Late Quaternary depositional history of the Blue and White Nile Rivers in central Sudan. In M. A. J. Williams and H. Faure (eds.), *The Sahara and the Nile*: 281-362. Rotterdam, A. A. Balkema.

Williams, M. A. J., Williams, F. M., Duller, G. A. T., Munro, R. N., El Tom, O. A. M., Barrows, T. T., Macklin, M., Woodward, J. C., Talbot, M. R., Haberlah, D. and Fluin, J. 2010. Late quaternary floods and droughts in the Nile Valley, Sudan: new evidence from optically stimulated luminescence and AMS radiocarbon dating. *Quaternary Science Reviews* 29: 1116-1137.

Wobst, H. M. 1977. Stylistic Behavior and Information Exchange. In C. E. Cleland (ed.), *Papers for the Director: Research Essays in Honor of James B. Griffin*: 317-342. Ann Arbor, University of Michigan, Museum of Anthropology, Anthropological Papers 61.

Woodburn, J. 1982. Egalitarian Societies. *Man, New Series*, Vol. 17 (3): 431–451.

Woodburn, J. 1988. African Hunter-gatherer Social Organization: is it Best Understood as a Product of Encapsulation? In T. Ingold, D. Riches and J. Woodburn (eds.), *Hunters and Gatherers* Volume 1: *History, Evolution and Social Change*: 31-64. Oxford, Berg Publishers.

Woodward, J. C., Macklin, M. G. and Welsby, D. A. 2001. The Holocene fluvial sedimentary record and alluvial geoarchaeology in the Nile Valley of Northern Sudan. In D. M. Maddy, M. G. Macklin and J. C. Woodward (eds.), *River Basin Sediment Systems: Archives of Environmental Change*. Abingdon, A. A. Balkema: 327-356.

Woodward J. C., Macklin, M., Fielding, L., Millar, I., Spencer, Welsby, D. A. and Williams, M. 2015. Shifting sediment sources in the world's longest river: A strontium isotope record for the Holocene Nile. *Quaternary Science Reviews* 130: 124–140.

Zedeňo, M. N. 2002. Neolithic ceramic production in the Eastern Sahara of Egypt. In K. Nelson, K. and Associates (eds.), *Holocene Settlement of the Egyptian Sahara. Volume 2: The Pottery of Nabta Playa*: 51-64. New York, Kluwer Academic/Plenum Publishers.